T0214648

Smart Cement

Smart Cement

Development, Testing, Modeling and Real-Time Monitoring

Cumaraswamy Vipulanandan

CRC Press
Taylor & Francis Group
Boca Raton London New York

CRC Press is an imprint of the
Taylor & Francis Group, an **informa** business

First edition published 2022
by CRC Press
6000 Broken Sound Parkway NW, Suite 300, Boca Raton, FL 33487-2742

and by CRC Press
2 Park Square, Milton Park, Abingdon, Oxon OX14 4RN

© 2022 Taylor & Francis Group, LLC

CRC Press is an imprint of Taylor & Francis Group, LLC

Library of Congress Cataloging-in-Publication Data
Names: Vipulanandan, Cumaraswamy, 1956– author.
Title: Smart cement : development, testing, modeling and real-time monitoring / Cumaraswamy Vipulanandan.
Description: First edition. | Boca Raton, FL : CRC Press, 2021. | Includes bibliographical references and index.
Identifiers: LCCN 2021004199 (print) | LCCN 2021004200 (ebook) | ISBN 9780367278373 (hbk) | ISBN 9781032039695 (pbk) | ISBN 9780429298172 (ebk)
Subjects: LCSH: Cement. | Concrete–Additives. | Concrete–Testing. | Smart materials.
Classification: LCC TA434 .V57 2021 (print) | LCC TA434 (ebook) | DDC 624.1/833–dc23
LC record available at https://lccn.loc.gov/2021004199
LC ebook record available at https://lccn.loc.gov/2021004200

ISBN: 978-0-367-27837-3 (hbk)
ISBN: 978-1-032-03969-5 (pbk)
ISBN: 978-0-429-29817-2 (ebk)

DOI: 10.1201/9780429298172

Typeset in Sabon
by Newgen Publishing UK

Contents

Figures and Tables

FIGURES

TABLES

Preface

Cement is an inorganic binding material and is the largest volume of manufactured material in the world today. It is used in the construction and maintenance of facilities and infrastructures both onshore and offshore. Also, cement has the longest history of development and applications since it has been in use for more than 5,000 years. During the initial time period, natural cement was used by Egyptians, Romans, and Greeks for construction. Over the past hundred years, manufactured cement has been used in the construction of buildings, bridges, highways, shallow and deep foundations, pipelines, tunnels, storage facilities, and also oil, gas, and water wells. Cement is also used to close out abandoned wells and also in the sequestration of CO_2 gas in various geological formations. During the past several decades, many cement-based testing standards have been developed. Enhancing the mechanical properties of cementitious materials with various inorganic and organic additives has been the focus of recent research. However, major failures of structures, highways, and oil wells due to problems with cement have occurred. Hence, to maintain and extend the service life of cement-based infrastructures, it is important to monitor the changing conditions in cementitious materials in real time.

In this book, information related to recently developed innovative technologies on highly sensing chemo-thermo-piezoresistive smart cement with real-time monitoring development, testing, characterisation, and modeling has been included. Also, the study was focused on many types of cements, including Portland cements and oil well cements. New theories have been developed to characterise materials. Based on the electrical impedance-frequency response, smart cement was characterised as a resistive material. Hence, the electrical resistivity of smart cement was selected as the sensing property since this property can be easily continuously monitored in the field from the time cement slurry is prepared through the entire service life of the relevant infrastructures. Based on experimental results, theory has been developed to make this cement piezoresistive without changing its basic resistivity. A new Vipulanandan rheological model has been developed to characterise shear-thinning cement slurries with various additives, and predictions of the experimental data are compared to the currently used models. The new Vipulanandan rheological model has a limit to the maximum shear stress tolerance for shear thinning cement slurry, whereas no other rheological models have this limit.

Initial electrical resistivity, a bulk material property, can be used as the quality control in the mixing of cement and concrete with various additives. Also, changes in electrical resistivity with various new parameters have been used to monitor the curing of various types of cements and concrete. A new Vipulanandan p-q curing model has been employed to quantify continuous changes in the resistivity of curing cement and concrete. Moreover, a new resistivity-based index has been developed to predict the strength of cement and concrete.

Chapter 5 is focused on the development and characterisation of piezoresistive smart cement. In this chapter, compressive, tensile, and bending loading behavior properties are compared to those of unmodified cement. Smart cement was developed by adding less than 0.1% of conductive filler (no nanoparticles) to the cement. The addition of less than 0.1% conductive or semi-conductive fibers (dispersed carbon fibers, basaltic fibers, or mixtures) to the cement substantially enhanced the tensile and compressive piezoresistivity behaviors (stress-piezoresistive strain relationships) of the cement to make it highly sensing. Compared to the cement compressive failure strain of 0.2%, change in the long-term piezoresistive strain was over 500 times (50,000%) higher, making smart cement very sensing. Also, a new theory for producing piezoresistive cement has been developed and verified through experimental results. Both under tensile and compressive loading the resistivity increased due to deviatoric stress (shear stress), another major shift in the technology and understanding of the material's piezoresistive behavior. An increase in the resistivity with the loading also makes the material better and more stable. Moreover, stress-strain and stress-piezoresistive strain relationships were modeled using Vipulanandan p-q models.

The effects of various additives (fly ash, meta kaolin, sodium silicate), water-to-cement ratios, curing conditions, curing time, and temperatures on the piezoresistive behavior of chemo-thermo-piezoresisitive smart cement have been tested and quantified. The effects of montmorillonite clay, drilling mud, and CO_2 contaminations on smart cement behavior were also investigated. As well, the effects of adding various types of nanoparticles, including iron oxide, sodium aluminate, and silicon dioxide on the chemo-thermo piezoresistive behavior of smart cement were investigated and quantified. The effectiveness of various additives to reduce fluid loss from cement slurries was investigated for the short term and long term and modeled using the Vipulanandan fluid loss model and compared to the American Petroleum Institute (API) model.

The sensitivity of smart cement in detecting and quantifying gas leaks was investigated. Also, the new Vipulanandan fluid flow model (generalised Darcy's Law) was verified through gas leak in smart cement slurry and solidified smart cement. During gas leak in cement slurry, electrical resistivity increased but in solidified smart cement it decreased, representing changes in stresses within smart cement. Resistivity has been related to gas leak rates in cement slurry and solidified cement as well.

Also the sensitivity of smart cement in detecting dynamic impact loading and cyclic loading has been experimentally verified at different frequencies and quantified. Smart cement can also detect pre-earthquake signals.

In order to demonstrate the performance of smart cement, laboratory and field well model tests were performed using a new instrumentation method outside the casing that was developed during this study to monitor electrical resistance changes in drilling muds and cement using the two-probe concept. When installing the wells, resistivity was monitored to determine the flow of drilling mud, spacer fluid, and cement. Also the piezoresistive response of the cement sheath was verified by applying pressure in the casing. Change in the resistances of the hardened smart cement with depth around the model wells have been continuously monitored for years. Using artificial intelligence (AI) models, the changes in cement resistivity regarding depth and time were modeled and compared it to the Vipulanandan Curing Model. Also pressure in the casing was predicted from changes in the vertical resistivity of smart cement, proving it to be a three-dimensional (3D) sensor in the field.

Applications of smart cement in grouts, foam cement, and concrete have been tested and verified. Also, the potential applications of smart cement grouts in repairing damaged cements have been investigated. Effects of foam contents on the behavior of smart cement have been tested and quantified as well. When smart cement was used as the binder in concrete, it made the concrete highly sensing. Changes in compressive pulse velocity in the concrete were compared to changes in the resistivity of the curing concrete, and the resistivity changes were much higher than the changes in the pulse velocity. The piezoresistive behavior of the concrete with varying aggregate contents and smart cement binder have been tested and modeled. With smart cement binder in concrete, the stress distribution between the smart cement binder and the aggregates in the concrete have been quantified. In some concretes the aggregates were stressed much higher than the strength of the concrete and showed the need for using stronger aggregates in the concrete.

Also a new Vipulanandan failure model for cement and concrete has been developed and verified with the available experimental data.

Acknowledgments

This book highlights the integration of knowledge, experiences, and new technologies to make chemo-thermo-piezoresistive smart cement highly sensing without any buried sensors and integrated with a real-time monitoring system. Smart cement will be similar to human skin, highly sensitive to many changes, and real-time monitoring will be similar to the human brain.

In order to build and maintain safe and resilient infrastructure systems around the world, it is important to monitor changes to maintain these systems. Also, the integrity of infrastructures is impacted by natural and human-made disasters. Hence, making cement be a 3D sensor is a paradigm shift in technology and can have a major impact in the very near future around the world.

The funding provided to support continuous research at the Center for Innovative Grouting Materials and Technology (CIGMAT) and the Texas Hurricane Center for Innovative Technology (THC-IT) by the US Department of Energy (US DOE), coordinated by the Research Partnership for Secure Energy for America (RPSEA), the National Science Foundation (NSF), the US Environmental Protection Agency (US EPA), the Texas Department of Transportation (TxDOT), the Texas Hazardous Waste Research Center, the City of Houston, and many industries (materials, grouts, construction, pipe, and oil) are very much appreciated.

The author would first like to thank all his graduate students for performing the tests at the Center for Innovative Grouting Materials and Technology (CIGMAT) laboratory and the Texas Hurricane Center for Innovative Technologies (THC-IT) at the University of Houston (Texas), and also monitoring the laboratory models, field well, pipelines, and piles supporting bridges for a number of years. Thousands of tests have been performed in the laboratory and also in the field to verify the several new analytical Vipulanandan models with experimental results in the past four decades. The graduate students are appreciated with their publications, which are included as references in this *Smart Cement* book.

Special thanks are extended to my graduate adviser, Professor Raymond J. Krizek from Northwestern University, and my mentor, Professor Michael W. O'Neill from the University of Houston. Also for guidance and support from Professor Leon Keer from Northwestern University; Professor Osman Ghazzaly from the University of Houston; David Magill, past president of Avanti International; Rafael Ortega, president of Aurora Technical Services; Kenneth Tand, principal engineer of Kenneth Tand and Associates; and Dr. Daniel Wong, chief executive officer of Tolunay-Wong Engineers. Also, thanks are extended to Sensytec Company, Ody De La Paz, and Sai Anudeep Reddy Maddi for commercialising the smart cement technology.

Thanks are extended as well to the University of Houston (Texas) for its continued support in providing the laboratory and field testing facilities.

Also thanking my family, Siblings and their families, parents, aunts, uncles and cousins for their continuous support and encouragement.

Cumaraswamy Vipulanandan (Vipu) is a professor of civil and environmental engineering with an endowed professorship at the University of Houston (Texas), where he is director of the Center for Innovative Grouting Materials and Technology and of the Texas Hurricane Center for Innovative Technology, and former department chairman. Recently, he received two patents for "Smart Cement" and "Detection and Quantification of Corrosion on Materials." He has received several national, state, and local awards for his research and teaching. He has served as thesis adviser to 35 PhDs and 90 Master's students.

Author

Cumaraswamy Vipulanandan ("Vipu") is a professor of civil and environmental engineering, holding an endowed professorship at the University of Houston (Texas), where he is also director of the Center for Innovative Grouting Materials and Technology and of the Texas Hurricane Center for Innovative Technology.

He was department chairman from 2001 to 2009, during which time he was elected to the Executive Committee of the US Civil Engineering Department Head Council for four years. He earned his MS and PhD in civil engineering from Northwestern University, Evanston, Illinois, USA, and his Bachelor's degree in civil engineering from the University of Moratuwa, Sri Lanka, after graduating from Royal College Colombo in Sri Lanka. His current research interests are in smart cements, smart grouts, smart orthopedic casting materials, polymer concrete, coatings and liners, polymer treatment of expansive clays, integrity of pipelines and pipe joints, grouting, deep foundations, corrosion detection and quantification, treatment of contaminated soils, recycling of waste water using multichamber microbial fuel, integrated with solar energy, real-time monitoring of materials, ground faulting conditions and flooding, annual prediction of hurricanes in the Gulf of Mexico, and disaster management with rapid recovery. He has been the principal investigator or co-principal investigator for more than 80 funded projects since 1984, amounting to more than $13 million, and his work has resulted in more than 300 refereed papers and over 200 conference presentations. He has served as thesis adviser to 35 PhD and 90 Master's students. Recently, he received two US patents for "Smart Cement" and "Detection and Quantification of Corrosion on Materials," and he was awarded the Most

Valuable Professional in 2011 by the Underground Construction Technology Association. He was the keynote speaker at the International Workshop on Deep Foundations-2007 in Japan, and in 2015 he received the Best Paper Presentation award at the Second Global Well Cementing Conference. He is a registered professional engineer in the state of Texas. He was the editor of seven American Society of Civil Engineers and one American Concrete Institute special publications on advances in deep foundations, grouting, soft soils, vegetation and expansive clays, recycled materials, pipelines, and fracture of concrete materials, and is the chief editor of the journal *Advances in Civil Engineering*.

Chapter 1

Introduction

1.1 HISTORY

Cement is an inorganic binding material and has evolved over 5,000 years from natural materials to industrial production with changes in chemical compositions and particle size distributions. Natural cement made from limestone containing clay minerals continued to be used until the nineteenth century. When water is added to cement, the cement will react with the water and bind with many types of inorganic materials to form durable composites for various types of applications. Historically, cement is the most valuable material developed by humans to enhance growth and development around the world. Initially, cement was used in the building of Egyptian pyramids. Also, around 300 BC, the Romans used cement with various types of admixtures to construct many types of buildings. The main focus over centuries has been on developing stronger and more durable cements. In 1824, Portland cement was invented by Joseph Aspdin from Leeds, England. In the 1850s, Louis Vicat from France developed the Vicat needle method to determine the setting time of cement, and this method is still being used. In America, the production of natural cements reached its peak in the 1890s, only to be overtaken by the Portland cement production. Currently, cement is used in multiple applications in both onshore and offshore construction. Now there are standards set for various types of Portland cements and oil-well cements to ensure quality of production and also help with the multiple applications for cement. Cement is the largest manufactured material around the world, and in recent years over 4 trillion mega grams have been manufactured annually.

1.2 CEMENTS

Cement is manufactured by combining clay or shale (aluminum silicate) with limestone (calcium based) and processed at around 1,450°C or higher temperatures to produce calcium silicate clinkers. At present, cements are broadly characterised based on the application, such as Portland cements and oil-well cements.

1.2.1 Portland Cement

The first American Society for Testing and Materials (ASTM) Portland Cement standard was developed in 1940. Now, the ASTM C150/C150M-19 covers ten types of Portland cements based on the applications and compositions, as summarised in Table 1.1. *Type I* is general-purpose cement used for construction purposes and, together with *Type II*, accounts for about 92% of United States-produced cement. *Type III* only accounts for about 3.5% of cement production, while *Type IV* is only available on special request and *Type V* is difficult to obtain because it is less than 0.5% of production. Table 1.1 summarises the ASTM Portland Cement classifications. and general uses are summarised in the remarks column.

1.2.2 Oil-Well Cement (OWC)

When used in oil wells, cement has multiple functions, which include structural integrity, as a protective seal to the casing, prevention of blowouts, and promotion of zonal isolation. Portland cement was first used in oil well construction in 1906. In 1948, the American Petroleum Institute (API) developed the first code for testing cement. The standards of the API suggest adopting the chemical requirements determined by the ASTM procedures and physical requirements determined in accordance with procedures outlined in API RP 10B and ASTM. Based on the API classification, currently there are six classes of cements (A, B, C, D, G, and H), which can be used for oil-well cementing based on the depths and downhole pressures and temperatures.

Cement slurry flowing ability (rheology) and stability are two of the major requirements of oil-well cementing. Oil-well cements (OWCs) are usually made from Portland cement clinkers or from blended hydraulic cements. OWCs are classified into grades by the API based upon their $Ca_3Al_nO_p$ (tricalcium aluminate – C_3A) content as ordinary (O), moderate sulphate resistant (MSR), and high sulphate resistant (HSR). Each class is applicable for a certain range of well depth, temperature, pressure, and sulphate environments, as summarised in Table 1.2.

The API, on the other hand, has specified requirements (API Specification 10A) for oil-well cements, which classify them into six classes and three grades (O, MSR, and HSR). The petroleum industry in the United States primarily uses cement classes A, B, C, G, and H, or conventional Portland cement mixed with suitable additives for well cementing. The API cement classifications and their uses with remarks are summarised in Table 1.2.

API specifications represent a realistic method of classifying Portland cement for use in oil wells by specifying the required properties.

Table 1.1 ASTM C150 standard on Portland cement classifications and remarks

Type	Compositions	Remarks
I	Maximum limits on Magnesium oxide (MgO), tricalcium aluminate, loss on ignition and insoluble residues.	General purpose, when there are no special properties. Compressive strength after 28 days is 28 MPa. Widely used regular cement for general applications. Similar to API class A.
IA	Air-entraining cement. Otherwise similar to Type I cement.	General purpose applications with air entraining agent. Compressive strength after 28 days is 22 MPa.
II	Maximum limits on aluminum oxide (Al_2O_3), ferric oxide (Fe_2O_3), magnesium oxide (MgO), tricalcium aluminate, loss on ignition and insoluble residues.	General purpose applications with moderate sulfate resistance. Compressive strength after 28 days is 28 MPa. Similar to API class B.
IIA	Air-entraining cement. Otherwise similar to Type II cement.	General purpose applications similar to Type II cement with air entraining agent. Compressive strength after 28 days is 22 MPa.
II (MH)	Moderate heat of hydration (MH) and moderate sulfate resistance. Otherwise similar to Type II cement.	General purpose applications similar to Type II cement with moderate heat of hydration and moderate sulfate resistance.
II (MH)A	Air-entraining cement. Otherwise similar to Type II (MH) cement.	General purpose applications similar to Type II (MH) cement with air entraining agent. Compressive strength after 28 days is 22 MPa.
III	Maximum limits on magnesium oxide (MgO), tricalcium aluminate, loss on ignition and insoluble residues.	When a high-early strength is required. Compressive strength after 1 day and 3 days is 12 MPa and 24 MPa respectively. Similar to API class C.
IIIA	Air-entraining cement. Otherwise similar to Type III cement.	When a high-early strength is required. Compressive strength after 1 day and 3 days is 10 MPa and 19 MPa respectively.
IV	Maximum limits on ferric oxide (Fe_2O_3), magnesium oxide (MgO), tricalcium aluminate, loss on ignition and insoluble residues.	When a low heat of hydration is desired (in massive structures). Compressive strength after 7 days and 28 days is 7.0 MPa and 17.0 MPa respectively.
V	Maximum limits on magnesium oxide (MgO), tricalcium aluminate, loss on ignition and insoluble residues.	When high sulfate resistance is required. Compressive strength after 3 days and 28 days is 8.0 MPa and 21.0 MPa respectively.

Table 1.2 API cement classifications and remarks

API class	Compositions	Operating temperatures °C (°F)	Remarks
A	Available in O grade only. Maximum limits on magnesium oxide (MgO), sulfur trioxide (SO₃), loss on ignition and insoluble residues. Materials in the amounts meet the requirements of ASTM C465.	25–77°C (80–170°F)	Good for 0–1,800 m (0–6,000 ft) depths. Used when special properties are not required. Similar to ASTM C150 *Type I*. Recommended w/c ratio is 0.46. The compressive strength after 8 hours and 24 hours is 1.7 MPa and 12.4 MPa respectively.
B	Available in MSR and HSR grades. Maximum limits on magnesium oxide (MgO), sulfur trioxide (SO₃), tricalcium aluminate, loss on ignition and insoluble residues. Materials in the amounts meet the requirements of ASTM C465.	25–77°C (80–170°F)	Good for 0–1,800 m (0–6,000 ft) depths. Used for moderate to high sulphate resistance. Similar to ASTM C150 *Type II*. Recommended w/c ratio is 0.46. The compressive strength after 8 hours and 24 hours is 1.4 MPa and 10.3 MPa respectively.
C	Available in O, MSR, and HSR grades. Maximum limits on magnesium oxide (MgO), sulfur trioxide (SO₃), tricalcium aluminate, loss on ignition and insoluble residues. Materials in the amounts meet the requirements of ASTM C465.	25–77°C (80–170°F)	Good for 0–1,800 m (0–6,000 ft) depths. Used for moderate to high sulphate resistance and when high early strength is required. Similar to ASTM C150 *Type III*. Recommended w/c ratio is 0.56. The compressive strength after 8 hours and 24 hours is 2.1 MPa and 13.8 MPa respectively.
D	Available in MSR and HSR grades. Maximum limits on magnesium oxide (MgO), sulfur trioxide (SO₃), tricalcium aluminate, loss on ignition and insoluble residues. Materials in the amounts meet the requirements of ASTM C465.	77–110°C (170–230°F)	Good for 1,800–3,000 m (6,000–10,000 ft) depths. Used for moderate to high sulphate resistance and moderately high temperatures and pressures. Recommended w/c ratio is 0.38. The compressive strength after 8 hours and 24 hours depends on the curing temperature and pressure.
G	Available in MSR and HSR grades. Maximum limits on magnesium oxide (MgO), sulfur trioxide (SO₃), tricalcium aluminate, loss on ignition and insoluble residues. Chemical additives required for chromium (VI) reduction.	27–93°C (80–200°F)	Good for 0–2,400 m (0–8,000 ft) depths. Used for moderate to high sulphate resistance. Has improved slurry acceleration and retardation. Recommended w/c ratio is 0.44. The compressive strength after 8 hours depends on the curing temperature.

(continued)

Table 1.2 Cont.

API class	Compositions	Operating temperatures °C (°F)	Remarks
H	Available in MSR and HSR grades. Maximum limits on magnesium oxide (MgO), sulfur trioxide (SO₃), tricalcium aluminate, loss on ignition and insoluble residues. Chemical additives required for chromium (VI) reduction.	27–93°C (80–200°F)	Good for 0–2,400 m (0–8,000 ft) depths. Used for moderate to high sulphate resistance. Has improved slurry acceleration and retardation. Recommended w/c ratio is 0.38. The compressive strength after 8 hours depends on the curing temperature.

1.3   CEMENT CHEMISTRY AND HYDRATION

Cement reaction with water and other additives is very much influenced by not only the shape and size distribution but also the chemical composition of the cement particles. Average particle size (d_{50}) of cement is about 40 μm. Specific gravities of cements are in the range of 3.05 to 3.20, and their Blaine-specific surface areas are in the range of 200 m^2/kg to 550 m^2/kg (Bogue 1955; Taylor 1997). Based on the image analyses of this current study, the scanning electron micrograph (SEM) image of a typical cement powder with a magnification of 35,000 is shown in Figure 1.1.

It is also important to determine the major chemical composition of typical cement. In this current study, an X-ray diffraction (XRD) analysis was performed in order to determine the representative chemical composition of oil-well cement at 25°C. The XRD pattern of the cement powder was obtained by using a Siemens D5000 powder x-ray diffraction device. An XRD analysis was performed on cement passing sieve No. 200 (75 μm). The powder (≈2 g) was placed in an acrylic sample holder about 3 mm deep. The sample was analyzed by using parallel beam optics with CuKα radiation at 40 kV and 30 mA. The sample was scanned for reflections (2θ) from 0° to 90° in steps of 0.02° and 2 sec count time per step.

The major constituents of cement are identified in Figure 1.2. The four principal compounds that constitute cement (Portland cement and oil-well cement) are:

- Tricalcium silicate (abbreviated as C_3S) Ca_3SiO_5,
- Dicalcium silicate (abbreviated as C_2S) Ca_2SiO_4,
- Tricalcium aluminate (abbreviated as C_3A) $Ca_3Al_2O_4$, and
- Tetracalcium aluminoferrite (abbreviated as C_4AF) $Ca_4Al_nFe_{2-n}O_7$.

Figure 1.1 SEM micrograph of cement particles (magnification 35,000x).

Table 1.3 Composition of cements in percentage

Type of cement	C_3S	C_2S	C_3A	C_4AF	Gypsum	Surface area (m^2/kg)
Class H	52–64	16–24	0.6–3	11–15	1.8	220–300
Class G	56–58	18–19	1–2	11–15	—	270–350
Portland Cement	50	25	12	8	3.5	320–380

When the cement is mixed with water, the binding phases in the cement (C_3S, C_2S, C_3A, and C_4AF) react in different ways based on the mixing methods and environmental conditions. During hydration of the cement, chemical reactions between cement and water take place, which leads to continuous cement slurry thickening and hardening. In the cement, more than 70% of the total material is of silicate phase, as summarised in Table 1.3. Calcium silicate hydrate (abbreviated as CSH) and calcium hydroxide are produced when both tricalcium silicate (Ca_3SiO_5) and dicalcium silicate (Ca_2SiO_4) react with water, as follows (Bogue 1955; Ramachandran 1984; Taylor 1997):

$$2\ Ca_3SiO_5 + 7\ H_2O \rightarrow 3\ CaO \cdot 2\ SiO_2 \cdot 4\ H_2O + 3\ Ca(OH)_2, \qquad (1.1)$$

$$2\ Ca_2SiO_4 + 4\ H_2O \rightarrow 3\ CaO \cdot 2\ SiO_2 \cdot 4\ H_2O + Ca(OH)_2. \qquad (1.2)$$

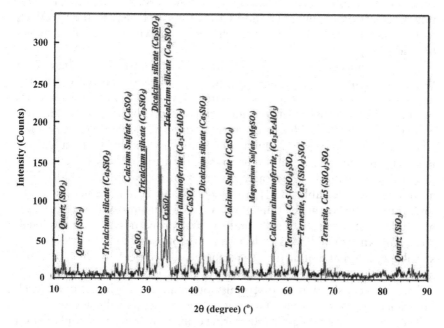

Figure 1.2 Typical X-ray diffraction (XRD) pattern for cement.

The resulting product, 'calcium silicate hydrate' gel, is the principal binder of hardened cement. Also calcium silicate hydrate comprises almost 70% of fully hydrated Portland cement. At short hydration times, $Ca_3Al_2O_4$ (aluminate phases) are the most reactive. Aluminate phases have a significant influence on the rheology of cement slurry and the early strength development of set cement. But their presence is relatively small compared to the silicates. Tricalcium aluminates ($Ca_3Al_2O_4$) hydration reaction is as follows:

$$2Ca_3Al_2O_4 + 27\,H_2O \rightarrow Ca_2 \cdot Al_2O_3 \cdot 8H_2O + Ca_4 \cdot Al_2O_3 \cdot 19H_2O. \quad (1.3)$$

The calcium aluminate hydrates (C_2AH_8 and C_4AH_{19}) will be converted into the more stable form $Ca_3 \cdot Al_2O_3 \cdot 6H_2O$ from its metastable form by the following reaction:

$$Ca_2 \cdot Al_2O_3 \cdot 8H_2O + Ca_4 \cdot Al_2O_3 \cdot 19H_2O \rightarrow 2\ Ca_3 \cdot Al_2O_3 \cdot 6H_2O + 15H_2O. \quad (1.4)$$

$Ca_3Al_2O_4$ hydration is controlled by the addition of 3% to 5% gypsum to the cement clinker before grinding. Upon contact with water, part of the gypsum dissolves. The calcium and sulfate ions released into the solution react with the aluminate and hydroxyl ions released by the $Ca_3Al_2O_4$ to form a calcium trisulfoaluminate hydrate, known as mineral ettringite (Bogue 1955; Taylor 1997).

Oil-well cements such as class G and class H, considered to be two of the most popular cements, are used in deep oil- and gas-well cementing applications and also in closing abandoned wells. These cements are produced by pulverising clinker consisting essentially of calcium silicates ($Ca_nSi_mO_p$) with the addition of calcium sulphate ($CaSO_4$) (John 1992). Class H cement is produced by a similar process, except that the clinker and gypsum are ground relatively coarser than for a Class G cement, to give a cement with a surface area generally in the range 220–300 m^2/kg, as summarised in Table 1.3 (John 1992; Zhang 2010). Cementing is an important operation at the time of oil well construction. When admixtures are added with cement, tensile and flexural properties will be modified. Also, admixtures will have an effect on the rheological, corrosion resistance, shrinkage, thermal conductivity, specific heat, electrical conductivity, and absorbing (heat and energy) properties of oil-well cement (Bao-guo 2008). Oil-well cement slurry is used several thousand meters below ground level, and hence determining cement setting time is always a challenge.

There is an urgent need for monitoring cement hydration in situ based on changes in the material properties. One of the currently used methods is by monitoring temperature changes, which is not a material property.

1.4 APPLICATIONS

In the past two centuries, cement and concrete have been widely used in many applications and these have been well documented (McCarter 1994, 1996; Vipulanandan et al.2005a, e, 2009a, 2012a, c; Mangadlao et al. 2015). Cement slurries and grouts, based on the water-to-cement (w/c) ratio, have been used in the construction of shallow and deep oil, gas, and water wells both onshore and offshore (Nelson 1990; Calvert et al. 1990; Liao et al. 2014; Vipulanandan et al. 2014a through 2019a). Also, cement slurries are used to bond the pipes to the formation in horizontal directional drilling. Cement slurries are used to bond steel casings and pipes to the various geological formations in the wellbore and also isolate the formations. In the well application, cement has to bond very well with highly varying natural geological formations with depth and to human-made steel casing and pipes, and also has to perform for many decades under varying loading conditions, temperatures, pressures, and seismic activities. Hence it is important to monitor the performance of the cement from the time of mixing to the entire service life in situ.

Concrete is a mix of natural aggregates with human-made cement, which plays a critical role in achieving the strength and other properties needed for applications. Cement is the most critical component in concrete since it is the binding agent for all the coarse and fine aggregates used in the concrete. Concrete bridges, piles, streets, dams, buildings, pipes, and homes have been built using both fresh concrete and precast concrete (Vipulanandan et al. 2001a, 2005b, c, d, e, 2009a, 2014e, f).

1.5 STANDARDS

There are many types of cements manufactured based on local and international standards for various applications (Table 1.1 and Table 1.2). Portland cement is used in infrastructure construction and maintenance and also in treating soils to improve performance (Vipulanandan et al. 1993, 1995a, b, 1996b; Wang et al. 1996; Vipulanandan 1998, 2001a, b, 2007a, b). Cement is also used in treating contaminated soils and waste materials (Wang et al. 1996; Kim et al. 2003 and 2006). The type of Portland cement used is selected based on the type of applications and also long-term performance in the environment it will be subjected to during its service life.

Cements such as class G and class H are used in both vertical and horizontal well-cementing applications. These cements are produced by pulverising clinker consisting essentially of calcium silicates with the addition of calcium sulphate (John 1992). Cementing is an important operation at the time of oil well construction to ensure safety and long-term stability during the entire service life. When admixtures are added with cement, tensile and flexural properties will be modified. Also, admixtures will have an effect on the rheology, corrosion resistance, drying shrinkage, thermal conductivity, specific heat, and electrical resistivity of oil-well cement (Bao-guo 2008; Vipulanandan et al. 2014a through 2017b; Amani et al. 2020). Oil-well cement slurry is used several thousand feet below ground level. Hence, determining cement setting time is always a challenge.

The quality of oil-well cement is important as it will directly impact on various operations during well installation. If the quality of cement is bad, it will add unnecessary cost to well installation operations. But the high quality of cement will provide a high-quality well, which in turn will have long-term durability. The setting time of oil-well cement is very important to control cementing operations (Zhang et al. 2010). It will be difficult to pump, if the cement sets too fast. Hence, it is important that this limit not be reached before the cement slurry fills the annulus. The w/c, the composition and particle size distribution of cement, the presence of additives (mineral and chemical), temperature, and pressure will affect the setting time of the cement (Bogue 1955; Taylor 1997; Vipulanandan et al. 2014a through 2020b). The cement will hydrate and become a rigid material to meet the various performance requirements.

The Vicat setting time test (ASTM C191) is used to determine the initial and final setting times for hydrating cementitious mixtures. It measures the change in the penetration depth of a plunger with a diameter of 1.13 ± 0.05 mm under a constant applied load of 300 g and the increasing structure formation with curing acts to reduce the extent of penetration into the specimen. The initial setting time and the final setting time are determined at a penetration depth of 25 mm and 0.5 mm respectively.

Also there are ASTM standards, API Recommended Practice 10B, and Center for Innovative Grouting Materials and Technology (CIGMAT) standards to test cements in compression, tension and bending, rheology and fluid loss, coatings, grouts, and polymer composites (CIGMAT CT1 & CT3 2006; CIGMAT PC1 & GR2 2002).

1.6 ADDITIVES

Many types of inorganic and organic additives are used to modify the behavior of cement during mixing, hydration, and also after solidification. Typical modifiers used in cement slurries can be categorised as follows:

1.6.1 Setting Time and Thickening Time Altering Additives

The setting time of cement can usually be increased by reducing the amount of tricalcium aluminate (C_3A). Accurate control of the thickening time, defined as the time after initial mixing at which the cement can no longer be pumped, is crucial in the oil-well cementing process and repairing operations using cement grouts. The thickening time is usually controlled by using retarders. The addition of carbohydrates such as sucrose can significantly extend the thickening time or even prevent setting completely. Other retarders used in cements include cellulose derivatives, organophosphates, and inorganic compounds. The thickening time of cement slurries increases with the addition of polyvinyl alcohol (PVA) latex. Ligno-sulfonates and hydroxyl carboxylic acids are retarders that have been used with cements. Salts of carbonates, aluminates, nitrates, sulphates, and thiosulphates, as well as alkaline bases such as sodium hydroxide (NaOH), potassium hydroxide (KOH), and ammonium hydroxide (NH_4OH) have been used to accelerate the setting time. In recent years, various types of nanoparticles have been used to modify the setting time of cement slurries.

1.6.2 Weighting Agents

Density-altering additives (weighting agents or extenders) are used in cement to achieve specific density requirements. Weighting agents add weight to the slurry to achieve higher density, while extenders are low-specific-gravity materials that are used to reduce the slurry density and to increase slurry yield. For instance, barite, sand, and micro-sand have relatively high specific gravity and are finely divided solid materials used to increase the density of cement slurry. Barite ($BaSO_4$) is the most commonly available weighting agent in oil/gas well cementing. Also, hematite, calcium carbonate, siderite, limonite, and manganese tetraoxide are other types of weighting agents used in cementing slurries. Moreover, bentonite is used to avoid segregation of heavy constituents in the cement slurry.

1.6.3 Enhance Mechanical Properties

The compressive strength and the shear bond strength of cement are modified with the addition of salts and metal oxides. An increase in flexural strength and in energy absorption before fracture was achieved by incorporating relatively small amounts of polymer latex together with short fibers. Various types of polymeric and glass fibers have been used to modify solidified cement properties. Also, in recent years various types of nanoparticles have been added to enhance the mechanical properties of cement (Vipulanandan et al. 2016a through 2018c).

1.6.4 Other Extenders

Bentonite helps in fluid loss control, but hardened cement becomes less resistant to corrosive fluids due to permeability issues. Adding fly ash (ASTM C618) can add value to the cement and also resist corrosive fluids. Based on the image analyses of the current study, the scanning electron

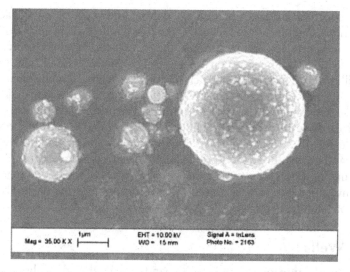

Figure 1.3 SEM micrograph of fly ash particles (magnification 35,000x).

microscopy (SEM) image of fly ash particles (spherical, 0.5 to 4 µm) is shown in Figure 1.3. Also, metakaolin (dehydrated clay) is used in cement (Vipulanandan et al. 1992, 2012c). Sodium silicates provide sufficient viscosity to allow the use of large quantities of mix water without excessive free water separation (Vipulanandan et al. 2016a). Silica fumes aid in obtaining low-density cement systems with a high rate of compressive strength development and improve fluid loss control. Foaming agents are used to reduce the modulus and permeability. Surfactants are used to improve the flowability and wetability of cement slurry.

Also, carboxymethylcellulose (CMC) polymer along with some other additives have been used with oil-well cement to improve the early compressive strength and the rheological properties (Vipulanandan et al. 2019a). Moreover, it reduced the permeability of oil-well cement. As reported by Choolaei (2012), when nanosilica was mixed with cement mortars, the rheological properties of cement slurry were improved, producing an increase in the compressive and flexural strengths of the cement mortars. Nanosilica also helped eliminate free water in the designed cement slurries. The setting time and the length of the dormant period were decreased with the addition of nanosilica to the cement. As well, in recent years, many types of nanoparticles have been added to cement to enhance its performance (Vipulanandan et al. 2014a through 2020b; Mohammed et al. 2018).

1.7 FAILURES

1.7.1 Infrastructures

Many concrete bridges, highways, dams, buildings, storage facilities, foundations, and pipes have failed over the past hundred years due to loadings, earthquakes, fires, and aging. In 1926, the St. Francis Dam unreinforced concrete gravity-arch dam was constructed in Los Angeles, California, to store 163 million cubic meter (l32,000 acre-ft.) of water with a dam height of about 62 m (205 ft.). On March 12, 1928, at midnight, there was a major dam failure in Los Angeles with over 400 deaths (Petrosk 2003). A bridge on the New York State thruway over Schoharie Creek collapsed in 1987 resulting in 10 deaths. In 2013, there was a building failure in Bangladesh with over a thousand deaths. In 2018, a bridge failure occurred in Italy with 37 deaths and many injuries. Failures can result in many types of losses and impact the economy, and hence there is a need for real-time monitoring of the changing conditions in infrastructures.

1.7.2 Wells

As the production of oil and gas expands on land and offshore around the world, there are many challenges in well construction beginning at the ground surface and the seafloor respectively. Recent case studies on well failures have clearly identified several issues that resulted in various types of delays in cementing operations. Also, preventing the loss of fluids to the formations and proper well cementing have become critical issues in well construction to ensure wellbore integrity because of varying down-hole conditions (Fuller et al. 2002; Gill et al. 2005; Ravi et al. 2004; Labibzadeh et al. 2010). Moreover, the environmental friendliness of cements is a critical issue that is becoming increasingly important (Durand et al. 1995; Thaemlitz et al. 1999; Dom et al. 2007). Lack of cement returns may compromise the casing support, and excess cement returns can cause problems with flow and control lines (Fuller et al. 2002; Gill et al. 2005; Ravi et al. 2004). Hence there is a need for monitoring the cementing operation in real time. At present there is no technology available to monitor the cementing operation in real time, from the time of placement through the entire service life of wells. Moreover, there is no reliable method to determine the length of the competent cement supporting the casing.

Production of oil and gas wells is placed into deep waters and hence there are unique challenges in monitoring the well construction beginning at the seafloor. Two separate studies have been performed on oil well blowouts offshore in the United States, with one done between the years 1971 and 1991, and the other done during the period 1992 to 2006, before the *Deepwater Horizon* blowout in the Gulf of Mexico in 2010. The two studies clearly identified cementing failures as the major cause for blowouts (Izod et al. 2007). Cementing failures increased significantly during the second period of study when 18 of the 39 blowouts were due to cementing problems (Izod et al. 2007). Also the *Deepwater Horizon* blowout in the Gulf of Mexico in 2010, which led to 11 fatalities, was due to cementing issues (Carter et al. 2014). With some of the reported failures and growing interest in environmental safety and economic concerns in the oil and gas industry, the integrity of the cement sheath is of major importance. In the past there was no technology available to monitor the cementing operation in real time from the time of placement through the entire service life of the wells. Also, there was no reliable method to determine the length of the competent cement supporting the casing. With fluctuating oil prices, it is even more important to develop technologies for safe operations and efficient oil and gas production.

In recent years, smart cement technology with a real-time monitoring system has been developed to monitor cement performance in various types of infrastructures, including oil wells.

1.8 SMART CEMENT

It is important to make cement highly sensing to monitor changes in the stresses, cracking, temperature, erosion, and also contamination during its service life. Recently, chemo-thermo-piezoresistive smart cement has been developed, and its performance has been tested and verified and results will be discussed in the following chapters in this book.

1.8.1 Piezoresistive Behavior

The change in electrical resistance in metal devices due to an applied mechanical load was first discovered in 1856 by Lord Kelvin. With single crystal silicon becoming the material of choice for the design of analog and digital circuits, the large piezoresistive effect in silicon and germanium was first discovered in 1954 (Smith 1954).

Usually the resistance change in metals is mostly due to the change in geometry resulting from applied mechanical stresses. However, even though the piezoresistive effect is small in those cases, it is often not negligible. Strain gauges are a good example of a piezoresistive material where with the application of strain to the attached material the electrical resistance will change in the strain gauges because how the metal strain gauges are configured. Also in strain gauges the resistance change will be positive under tensile stress or strain, and negative under compressive stress or strain. In the past few decades, various investigations have been performed to make polymers and cement composites be piezoresistive (Chung et al. 1995, 2000, 2001; Vipulanandan et al. 2002, 2005a through 2008). In the recently developed smart cement by Vipulanandan (U.S. Patent Number 10,481,143 (2019)), the resistivity change is positive under both tensile and compressive loading because the changes are dominated by the deviatoric (shear) stresses in the cement (Vipulanandan et al. 2014a through 2019b).

1.8.2 Thermo-resistive Behavior

In the sensing element, electrical resistance will change due to temperature change in the operating temperature ranges. In 1871, platinum was proposed by Sir William Siemens as the most suitable material (Siemens 1871). Also, nickel and copper have been developed as temperature sensors due to measurable changes in electrical resistance. The recent development of smart cement by Vipulanandan also can be used to sense temperature changes due to measurable changes in electrical resistivity (Vipulanandan et al. 2014b).

1.8.3 Chemo-Resistive Behavior

Chemo-resistive materials are a class of sensors that change in electrical resistance in response to changes in the surrounding chemical environment. These are materials such as metal oxide semiconductors, conductive polymers, and nanomaterials like graphene, carbon nanotubes, and nanoparticles. As far back as 1965, there are reports on semiconductor materials exhibiting electrical resistivity changes due to ambient gases and vapors. In 1985, Wohltjen and Snow developed a copper compound to detect ammonia vapor at room temperature and the resistivity decreased (Wohltjen et al. 1985). The recently developed smart cement by Vipulanandan can be used also to sense chemical additives and contaminations based on changes in electrical resistivity (Vipulanandan et al. 2014b, 2018k).

1.9 BEHAVIOR MODELS

It is important to have behavior models for cements and concretes in order to not only clearly understand the behavior but also to integrate it with artificial intelligence (AI) networks and 3D printing applications. The past models developed for cement hydration and cement behavior under various loading conditions are empirical and limited to the ranges of variables investigated in the relevant studies. In concrete, cement is the binder that develops the strength and other relevant properties for the concrete. But the behavior models do not quantify the role of cement in the concrete. Recently, a new Vipulanandan rheological model has been developed to better characterise the rheological behavior of smart cement slurry, drilling muds, spacer fluids, and other fluids with and without various additives including nanoparticles (Afolabi et al. 2019; Tchameni et al. 2019; Montes 2019; Mohammed 2018; Vipulanandan et al. 2014a). Also analytical models have been developed to characterise the curing, stress-strain, and piezoresistive behaviors of smart cement (Vipulanandan et al. 1990 through 2020b). The Vipulanandan fluid flow model (generalised Darcy's model) and fluid loss model have been developed and verified with experimental results. Also a new Vipulanandan failure model for cement and concrete has been developed and verified with experimental test results.

1.10 SUMMARY

Cement has evolved as the most popular construction material in the world and is used in both onshore and offshore construction and maintenance. There are many standards developed for manufacturing cement based on the potential applications. With a better understanding of chemical and physical cement behaviors and also advances in technologies, there is a need to make the cement a highly sensing bulk material so that health monitoring can be done in all its applications. In structural health monitoring, it is also important to identify a cement property that is highly sensitive and also can be easily monitored in the field. In recent years, chemo-thermo-piezoresistive smart cement has been developed with an integrated real-time monitoring system to be used in all the cement applications.

Chapter 2

Material Characterisation and Real-Time Monitoring

2.1 HISTORY

Real-time reporting of environmental conditions dates back to the invention of the telegraph in the 1830s. Until then, weather reports and forecasts could only be transmitted manually, usually involving a train journey from one location to another. Use of the telegraph enabled weather reports and warnings to be sent almost instantaneously to the same locations. With the advancement of technologies many real-time measurements are made using sensors. The first such sensor that was developed was the pH meter, a device that measures the quality of hydrogen ions in the solution being tested. It was produced in 1934. The principle behind a pH meter is similar to that of sensors that measure other chemicals. A pH meter measures the electrical potential between a chamber that is separated from the external environment by a membrane (originally the membrane was glass). This measurement requires comparison with a reference chamber, and pH meters now can incorporate the measurement and reference electrodes into the same probe. When cement is mixed with water, initially it is a slurry and a pH meter can be used to measure pH changes for a few hours, but when cement solidifies, a pH meter cannot be used.

Real-time monitoring is becoming popular with the advances in sensor technologies and wireless transmission of data. In the fields of medicine, chemical processing, weather forecast and security, monitoring is the core component of these various types of operations (Florinel-Gabriel 2012). It is not only collecting data at selected frequencies but also processing it to interpret the changing trends and storing the data that is all important. In the past, most monitoring was based on photo images for medical patients to land survey, but it has limited application due to accessibility under the foundations supporting buildings and bridges, and also oil and gas wells where cement plays a critical role.

In order to real-time monitor changes in cement, it is important to identify the parameters that can be monitored in real time from the time of mixing to the entire service life of the cement and cementitious composites including concrete, not only in the laboratory but also in the field. The cement will be a slurry initially and will hydrate and harden to meet the design requirements and will be in service for decades. It will be unique if one monitoring parameter can be identified to characterise and monitor the cement from the time of mixing to the entire service life (Carter et al. 2014; Vipulanandan et al. 2014 (b–d), 2015 (a–e)).

2.2 STANDARDS

There are several ASTM and API standards to mix cement paste based on the applications. The cement slurry property will be influenced by not only the cement composition and water-to-cement ratio (w/c) but also the method of mixing with various additives. Some of the mixing standards are as follows:

ASTM C305: Practice for Mechanical Mixing of Hydraulic Cement Pastes and Mortars of Plastic Consistency

ASTM C511: Specification for Mixing Rooms, Moist Cabinets, Moist Rooms, and Water Storage Tanks Used in the Testing of Hydraulic Cements and Concretes

ASTM C1738: Practice for High-Shear Mixing of Hydraulic Cement Pastes

API Specification 10A: Specifications for Cements and Materials for Well Cementing

For concrete mixing, there are other standards. Developing a material property parameter to monitor concrete mixing is also important to ensure quality. Currently used monitoring test such as the flow cone test (ASTM D6994) is not based only on the material property of the concrete but also the flow plate and the cone shape and size.

2.3 MATERIAL CHARACTERISATION

It is important to identify the type of testing material (example: metal, cement, concrete, plastic, wood, asphalt) so that the relevant material property can be measured and monitored in the field. Based on past experience and research, changes in electrical properties were selected to be the representative properties for cement and other materials so that it can be used for monitoring in multiple applications. Electrical properties of a material can be represented by the permittivity, resistivity, or a combination in the number of series or parallel electrical circuits.

2.3.1 Vipulanandan Impedance Model

2.3.1.1 Equivalent Circuits

Identification of the most appropriate equivalent circuit to represent the electrical properties of a material is essential to further understand its properties. In this study, an equivalent circuit to represent smart cement was required for better characterisation through analyses of the impedance spectroscopy (IS) data. There were many difficulties associated with choosing a correct equivalent circuit. It was necessary somehow to make a link between the different elements in the circuit and the different regions in the impedance data of the corresponding sample. Given the difficulties and uncertainties, researchers tend to use a pragmatic approach and adopt a circuit that they believe to be most appropriate based on their knowledge of the expected behavior of the material under study, and demonstrate that the results are consistent with the circuit used.

In this study, different possible equivalent circuits were analyzed to find an appropriate equivalent circuit to represent the smart cement with and without additives under investigation. Also this method can be used to characterise all types of fluids and solids.

2.3.1.2 CASE-1: General Bulk Material—Resistance and Capacitor

In the equivalent circuit for CASE-1, the contacts and the bulk material were connected in series, and also both the contacts and the bulk material were represented using a capacitor (permittivity) and a resistor (resistivity) connected in parallel, as shown in Figure 2.1.

In the equivalent circuit for CASE-1, R_b and C_b are the resistance and capacitance of the bulk material respectively. The R_c and C_c are the resistance and capacitance of the contacts respectively. Both contacts are represented with the same resistance (R_c) and capacitance (C_c) if they are identical, and if not, they will be different. Total impedance of the equivalent circuit for CASE-1 (Z_1) can be represented as follows (Vipulanandan et al. 2013a):

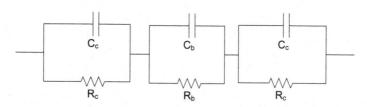

Figure 2.1 Equivalent electrical circuit for CASE-1.

$$z_1(\sigma) = \frac{R_b(\sigma)}{1 + \omega^2 R_b^2 C_b^2} + \frac{2R_c(\sigma)}{1 + \omega^2 R_c^2 C_c^2} - j\left\{\frac{2\omega R_c^2 C_c(\sigma)}{1 + \omega^2 R_c^2 C_c^2} + \frac{\omega R_b^2 C_b(\sigma)}{1 + \omega^2 R_b^2 C_b^2}\right\},$$

(2.1)

$$= R_1 + j\,X_1,$$

(2.2)

where ω is the angular frequency of the applied alternative current (AC) signal. Also, the term R_1 in Eqn. (2.2) represents the real part of the impedance (Z_{real} of Z_1) and X_1 represents the imaginary part of the impedance (Z_1). Additionally, σ is the independent variable representing the stress, temperature, contamination, and other variables based on the applications. When the frequency of the applied signal is very low, $\omega \rightarrow 0$, $Z_1 = R_1 = R_b + 2R_c$, and when it is very high, $\omega \rightarrow \infty$, $Z_1 = 0$ and also R_1 and X_1 (Eqn. 2.2) will be equal to zero.

2.3.1.3 CASE-2: Special Bulk Material—Resistance Only

In CASE-2, as a special case of CASE-1, the capacitance of the bulk material (C_b) was assumed to be negligible, as shown in Figure 2.2.

The total impedance of the equivalent circuit for CASE-2 (Z_2) is as follows:

$$z_2(\sigma) = R_b(\sigma) + \frac{2R_c(\sigma)}{1 + \omega^2 R_c^2 C_c^2} - j\frac{2\omega R_c^2 C_c(\sigma)}{1 + \omega^2 R_c^2 C_c^2}.$$

(2.3)

$$= R_2 + j\,X_2$$

(2.4)

The term R_2 in Eqn. (2.4) represents the real part of the impedance (Z_{real} of Z_2) and X_2 represents the imaginary part of the impedance (Z_2). When the frequency of the applied signal is very low, $\omega \rightarrow 0$, $Z_2 = R_2 = R_b + 2R_c$, and when it is very high, $\omega \rightarrow \infty$, $Z_2 = R_2 = R_b$ and X_2 will be equal to zero, as shown in Figure 2.3. In CASE-2, if the impedance is measured at very high frequency, it will measure the resistance (R_b) in the material and eliminate the effects of the contacts, and also it is frequency independent. This

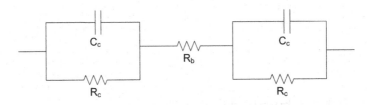

Figure 2.2 Equivalent electrical circuit for CASE-2.

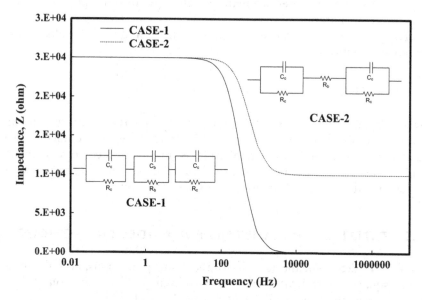

Figure 2.3 Comparison of typical responses of equivalent circuits for CASE-I and CASE-2.

becomes another unique advancement in measurement and also monitoring since the resistance is independent of the very high frequency of measurement, as shown in Figure 2.3.

It is important to identify the electrical material properties of the testing material (bulk material) using the impedance-frequency relationship as shown in Figure 2.3.

2.4 VIPULANANDAN CORROSION/CONTACT INDEX

The contact resistance (R_c) and contact capacitance (C_c) for the probe, which is in contact with the material, can be defined as:

$$R_c = \rho_c K = \rho \frac{L}{A} \tag{2.5}$$

$$C_c = \epsilon_c \frac{A}{L}, \tag{2.6}$$

where A = interface/contact area between the probe and the material; L = interface/contact length between the probe and the material;

ρ_c = resistivity of the interface material; and ϵ_c = absolute permittivity of the interface material.

The product of the equations given in (2.5) and (2.6) results in the following relationship:

$$R_c C_c = \rho_c \epsilon_c. \tag{2.7}$$

where ρ_c and ϵ_c in Eqns. (2.5) and (2.6) are interface/contact material properties, and hence $R_c C_c$ in Eqn. (2.7) at the point of contact is also material property and will be referred to as the electrical corrosion/contact index. This parameter can be used to characterise the contact at each point (Vipulanandan et al. 2018d, 2019b; U.S. Patent 2020).

2.5 STATISTICAL PARAMETERS FOR MODEL PREDICTIONS

In order to determine the accuracy of the model predictions, both the root-mean-square error (RMSE) and the coefficient of determination (R^2) in curve fitting are defined in Eqns. (2.8) and (2.9) as follows:

$$RMSE = \sqrt{\frac{\sum_{i=1}^{n}(y_i - x_i)^2}{N}} \tag{2.8}$$

$$R^2 = \left(\frac{\sum_i (x_i - \bar{x})(y_i - \bar{y})}{\sqrt{\sum_i (x_i - \bar{x})^2} * \sqrt{\sum_i (y_i - \bar{y})^2}}\right)^2 \tag{2.9}$$

where y_i = actual value; x_i =calculated value from the model; $\bar{y}$ =mean of the actual values; $\bar{x}$= mean of the calculated values; and N is the number of data points.

2.6 TESTING

2.6.1 Smart Cement

Smart Portland cements and smart oil-well cements with less than 0.1% of carbon fibers and/or basaltic fibers and w/c of 0.38 were used in this study. In order to further characterise the materials, smart cement samples with various types of additives were also prepared by using a table top blender mixer at a speed of 2,000 rpm for five minutes or the API mixing method.

Cement slurries with and without selected additives were tested to investigate the characteristics of the materials as CASE-1 or CASE-2.

2.6.1.1 Specimen Preparation

After mixing the smart cement with and without additives and aggregates (concrete), the specimens were prepared in plastic cylindrical molds with a diameter of 50 mm and a height of 100 mm. Two pairs of conductive flexible wires (monitoring probes) were placed about 50 mm vertically apart on two sides of the molds, as shown in Figure 2.4(a). After preparing the specimen, based on the weight and volume, the unit weight was determined for the cementitious materials investigated in this study. Electrical impedance was measured at various frequencies using an inductance (L), capacitance (C), and resistance(R) meter, referred to as an LCR meter, as shown in Figure 2.4.

2.6.1.2 Curing

The specimens were cured under room condition at 23°C and relative humidity (RH) of 50%.

2.6.2 Results

The LCR meter with alternative current (AC) and frequency from 20 Hz to 300 kHz was used to investigate the characteristics of cement without and with 50% fly ash, 10% metakaolin, 20% foam, and 75% aggregates. This was done to identify their electrical properties immediately after mixing and at different curing ages to characterise the materials as either CASE-1 or CASE-2. This is also important to identify the electrical material property that can be monitored.

2.6.2.1 Smart Cement Only

Studies were performed on the Portland cements and oil-well cements to investigate the variation of impedance with the frequency at different curing times since cement transforms from a liquid phase to a solid phase. At least 15 frequency measurements were made for each test to clearly identify the trend. The impedance-frequency responses with the curing time clearly indicated that the cement slurry (immediately after mixing—0 day) and solid (cured for one day and more) were represented by CASE-2, as shown in Figure 2.5. The contact and bulk resistances (R_c and R_b) and contact capacitance (C_c) were determined using the test data in Eqn. (2.3) combined with minimising the root-mean-square error (RMSE), and the results are summarised in Table 2.1. The test results also indicated that both the bulk resistance (R_b) and the contact resistance (R_c) increased with the curing time.

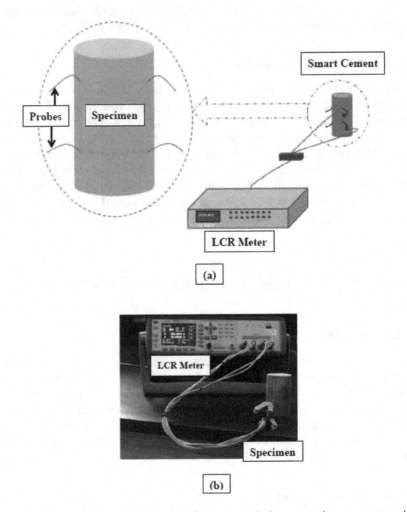

Figure 2.4 Experimental setup using the LCR meter with the two probes to measure the electrical impedance with frequency (a) schematic of the test configuration with the specimen, (b) photo of the LCR Meter connected to the specimen.

The bulk resistance of the cement (R_b) increased from 43.7 Ω to 1,390 Ω (Table 2.1) with the relative resistance change ($\Delta R_b/R_b$) of a 30.8-times increase (3,080%) in 28 days of curing. Hence, the relative resistance change of 3,080% is a very sensitive indicator of the changes within the cement due to hydration. The contact resistance (R_c) increased from 750 Ω to 1,790 Ω with the relative resistance change ($\Delta R_c/R_c$) of a 1.4-times increase (140%) in 28 days of curing. The contact capacitance (C_c) increased from 2.4 x 10^{-6} F to 4.3 x 10^{-6} F with the relative capacitance change ($\Delta C/C_c$) of a 0.8 time

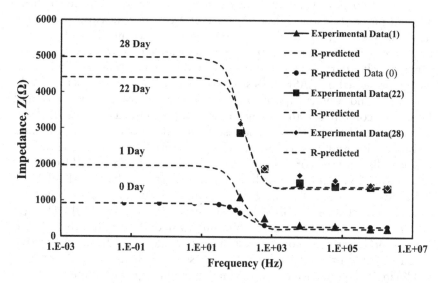

Figure 2.5 Impedance-frequency relationships for the smart cement with curing time.

Table 2.1 Impedance model parameters for different curing times for smart cement with additives

Curing Time	Materials		
	Smart Cement Only	Smart Cement with 10% Metakaolin	Smart Cement with 50% Fly ash
Immediately after mixing (0 Hour)	R_b = 43.7 Ω R_c = 750 Ω C_c = 2.4 x 10⁻⁶ F	R_b = 62.7 Ω R_c = 650 Ω C_c = 3x10⁻⁶ F	R_b = 65.2 Ω R_c = 400 Ω C_c = 1x10⁻⁶ F
28 days	R_b = 1,390 Ω R_c = 1,790 Ω C_c = 4.3x10⁻⁶ F	R_b = 528 Ω R_c = 1,158 Ω C_c = 7x10⁻⁷ F	R_b = 3,580 Ω R_c = 3,120 Ω C_c = 2.5x10⁻⁷ F

increase (80%) in 28 days of curing. Also, the new Vipulanandan contact index ($R_c C_c$), no size effect, increased from 1.8×10^{-3} ΩF to 7.7×10^{-3} ΩF, a 328% increase in 28 days of curing. Increase in the contact index is an indication of the potential corrosion of the monitoring metal wire within bulk cement material (Vipulanandan et al. 2018d, 2019b).

2.6.2.2 Smart Cement with 10% Metakaolin

Studies were performed on smart cement (using Portland cements and oil-well cements) with 10% metakaolin to investigate the variation of impedance with the frequency at different curing times since cement transforms from a liquid phase to a solid phase. At least 15 frequency measurements were made for each test to clearly identify the trend. The impedance-frequency responses with the curing time clearly indicated that the cement slurry (immediately after mixing—0 day) and solid (cured for one day and more) were represented by CASE-2, as shown in Figure 2.6. The contact and bulk resistances and contact capacitance were determined using the test data and Eqn. (2.3) combined with minimising the root-mean-square error (RMSE), as summarised in Table 2.1. The test results also indicated that the bulk resistance (R_b) and contact resistance (R_c) increased with the curing time.

The bulk resistance of cement with 10% metakaolin increased from 62.7 Ω (R_b) to 528 Ω (Table 2.1) with the relative resistance change ($\Delta R_b/R_b$) of a 7.4-times increase (740%) in 28 days of curing. Hence, the relative increase in resistance change of 740% is a very sensitive indicator of the changes within cement with 10% metakaolin due to hydration under the selected curing condition. The contact resistance (R_c) increased from 650 Ω (R_c) to 1,158 Ω with the relative increase in resistance change ($\Delta R_c/R_c$) of 0.78 (78%) in 28 days of curing. The contact capacitance (C_c) decreased from 3.0 x 10^{-6} F (C_{co}) to 7 x 10^{-7} F with the relative decrease in capacitance change ($\Delta C/C_c$) of 0.77 (77%) in 28 days of curing. Also, the new Vipulanandan contact index (R_cC_c), no size effect, decreased from 1.95 x 10^{-3} ΩF to 8.1 x 10^{-4} ΩF, a 58% decrease in 28 days of curing. Decrease in the contact index is a good indication of less potential for corrosion of the monitoring metal wire within bulk cement material.

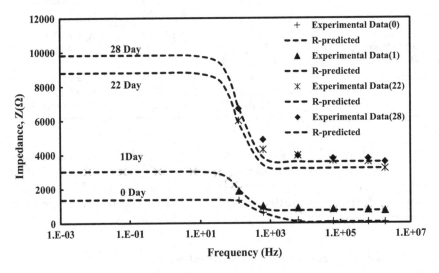

Figure 2.6 Impedance-frequency characterization of the smart cement modified with 10% metakaolin and curing time.

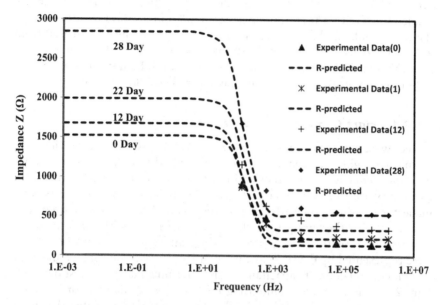

Figure 2.7 Impedance-frequency characterization of the smart cement modified by 50% fly ash and curing time.

2.6.2.3 Smart Cement with 50% Fly Ash

Studies were performed on Portland cements and oil-well cements to investigate the variation of impedance with frequency at different curing times since smart cement with 50% fly ash additive will transform from a liquid phase to a solid phase. At least 15 frequency and impedance measurements were made for each test to clearly identify the trend. The impedance-frequency responses with the curing time clearly indicated that cement slurry (immediately after mixing—0 day) and solid (cured for one day and more) were represented by CASE-2, as shown in Figure 2.7. The contact and bulk resistances and contact capacitance were determined using the test data and Eqn. (2.6) combined with minimising the root-mean-square error (RMSE), and the results are summarised in Table 2.1. The test results also indicated that both the bulk resistance (R_b) and the contact resistance (R_c) increased with the curing time.

The bulk resistance of the cement increased from 65.2 Ω (R_b) to 3,580 Ω (Table 2.1) with the relative resistance change ($\Delta R_b/R_b$) of a 53.9-times increase (5,390%) in 28 days of curing. Hence, the relative resistance change of 5,390% is a very sensitive indicator of the changes within cement with 50% fly ash additive due to hydration. The contact resistance (R_c) increased from 400 Ω (R_c) to 3,120 Ω with the relative resistance change ($\Delta R_c/R_c$) of a 6.8-times increase (680%) in 28 days of curing. The contact capacitance (C_c)

decreased from 1.0×10^{-6} F (C_c) to (2.5×10^{-7} F with the relative capacitance change ($\Delta C/C_c$) of a 0.75-times decrease (75%) in 28 days of curing. Also, the new Vipulanandan contact index (R_cC_c), no size effect, increased from 4.0×10^{-4} ΩF to 7.8×10^{-4} ΩF, a 95% increase in 28 days of curing. Increase in the Vipulanandan contact index is an indicator of higher corrosion potential in the monitoring wire (Vipulanandan et al. 2018d, 2019b).

2.6.2.4 Smart Cement with 20% Foam

Studies were performed on Portland cements and oil-well cements with 20% foam (by weight of cement slurry) to investigate the variation of impedance with the frequency at different curing times since cement transforms from a liquid phase to a solid phase. At least 15 frequency measurements were made for each test to clearly identify the trend. The impedance-frequency responses with the curing time clearly indicated that with 20% foam the cement slurry (within a day after mixing) and solid (28 days of curing) were represented by CASE-2, as shown in Figure 2.8. The contact and bulk resistances and contact capacitance were determined using the test data and Eqn. 2.6 combined with minimising the root-mean-square error (RMSE), and are summarised in Table 2.2. The test results also indicated that both the bulk resistance (R_b) and the contact resistance (R_c) increased with the curing time.

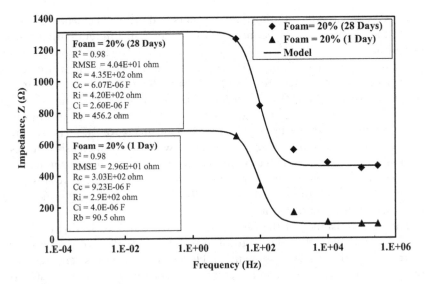

Figure 2.8 Impedance-frequency characterization of the smart cement with 20% foam (by weight).

Table 2.2 Impedance model parameters for 20% foam cement at different curing times

Curing Time	Smart Cement with 20% Foam
I day	R_b = 90.5 Ω R_c = 296 Ω C_c = 6.6 ×10^{-6} F R^2 = 0.98 RMSE = 29.6 Ω
28 days	R_b = 456.2 Ω R_c = 435 Ω C_c =4.3×10^{-6} F R^2 = 0.98 RMSE = 40.4 Ω

The bulk resistance of the cement increased from 90.5 Ω (R_b) to 456.2 Ω (Table 2.2) with the relative resistance change ($\Delta R_b/R_b$) of a 4-times increase (400%) in 28 days of curing. Hence the relative resistance change of 400% is a very sensitive indicator of the changes within the cement due to hydration. The contact resistance (R_c) increased from 296 Ω (R_c) to 435 Ω with the relative resistance change ($\Delta R_c/R_c$) of about a 0.47-times increase (47%) in 28 days of curing. The contact capacitance (C_c) increased from 6.6 × 10^{-6} F (C_{co}) to 4.3 × 10^{-6} F with the relative capacitance change ($\Delta C/C_c$) of a 0.35-times decrease (35%) in 28 days of curing. Also, the new Vipulanandan contact index ($R_c C_c$), no size effect, increased from 1.95 × 10^{-3} ΩF to 1.87 × 10^{-3} ΩF, a 4 328% decrease in 28 days of curing. A decrease in the Vipulanandan contact index is a good indication of less potential for corrosion of the monitoring metal wire within the bulk cement material.

2.6.2.5 Smart Cement with 75% Aggregates (Concrete)

Studies were performed on concrete with smart Portland cement with 75% aggregates to investigate the variation of impedance with the frequency at different curing times since concrete transforms from a liquid phase to a solid phase. At least 15 frequency measurements were made for each test to clearly identify the trend. The impedance-frequency responses with the curing time clearly indicated that the cement slurry (immediately after mixing) and solid (28 days of curing) were represented by CASE-2, as shown in Figure 2.9 and Figure 2.10 respectively. The contact and bulk resistances and contact capacitance were determined using the test data in Eqn. (2.6) combined with minimising the root-mean-square error (RMSE), and are summarised in Table 2.3. The test results also indicated that both the bulk resistance (R_b) and contact resistances (R_c) increased with the curing time.

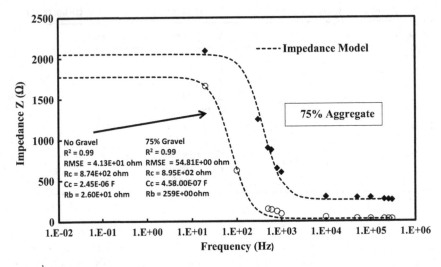

Figure 2.9 Impedance characterization of the smart cement concrete (75% gravel) immediately after mixing.

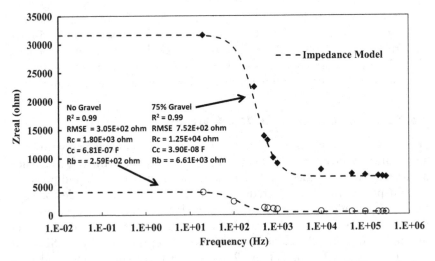

Figure 2.10 Impedance characterization of the smart cement concrete (75% gravel) after
28 days of curing.

Table 2.3 Impedance model parameters for smart cement concrete at different curing times

Curing Time	Concrete with Smart Cement Binder (75% Aggregate)
Immediately After Mixing	R_b = 259 Ω R_c = 895 Ω C_c = 4.6 $\times 10^{-7}$ F $R_c C_c$ = 4.1 $\times 10^{-4}$ Ω F R^2 = 0.99 RMSE = 58.1 Ω
28 days of Curing Under Room Condition (23°C and 50% Relative Humidity (RH))	R_b = 6,610 Ω R_c = 12,500 Ω C_c = 3.9 $\times 10^{-8}$ F $R_c C_c$ = 4.9 $\times 10^{-4}$ Ω F R^2 = 0.99 RMSE = 752 Ω

The bulk resistance of the smart cement concrete increased from 259 Ω (R_b) to 6,610 Ω (Table 2.3) with the relative resistance change ($\Delta R_b/R_b$) of a 24.5-times increase (2,450%) in 28 days of curing. Hence the relative resistance change of 2,450% is a very sensitive indicator of the changes within the concrete due to hydration of the smart cement binder. The contact resistance (R_c) increased from 895 Ω (R_c) to 12,500 Ω with the relative resistance change ($\Delta R_c/R_c$) of a 13-times increase (1,300%) in 28 days of curing. The contact capacitance (C_c) decreased from 4.6 $\times$ 10^{-7} F (C_c) to 3.9 $\times$ 10^{-8} F with the relative capacitance decrease ($\Delta C/C_c$) of a 0.92-times increase (92%) in 28 days of curing. Also, the new Vipulanandan contact index ($R_c C_c$), no size effect, increased from 4.1 $\times$ 10^{-4} ΩF to 4.9 $\times$ 10^{-4} ΩF, a 19.5% increase in 28 days of curing. An increase in the Vipulanandan contact index is an indicator of higher corrosion potential in the monitoring wire (Vipulanandan et al. 2018d, 2019b).

2.7 SUMMARY

A new method has been developed to characterise materials based on their electrical properties. Also, several tests have been performed on cement with and without additives to identify the important electrical properties, to characterise materials. Also in this study, using the new Vipulanandan corrosion/contact index, the corrosion potential for metal placed in cement and concrete, or in contact with cement and concrete, have been quantified. Based on the study, the following conclusions are advanced:

1 A new method has been developed to characterise materials based on their electrical properties since electrical property changes can be easily monitored in the field. Using the two-probe method with alternative current (AC), the impedance-frequency relationship can be experimentally developed to characterise the material. If bulk material properties are influenced by resistivity and permittivity (representing resistance and capacitance), the response was identified as CASE-1. If the bulk material property is represented by resistivity, it was identified as CASE-2. The Vipulanandan impedance model can be used to quantify the two-probe contacts and bulk material resistance and capacitance.

2 Smart cement slurry and solidified cement were identified as CASE-2 based on the impedance-frequency responses. This unified the characterisation of cement slurry and the solidified cement with the changes in electrical material property resistivity with time and other variables. Hence, for smart cement, electrical resistance will be the parameter to be monitored at a frequency of 300 kHz or higher based on the type of two-probe contacts used for monitoring.

3 Resistivity is a material property and also is a second-order tensor, and can be used to characterise cement behavior in three orthogonal directions (3D).

4 Smart cement with metakaoline, fly ash, foam, and aggregate additions were also represented by CASE-2. Hence, electrical resistivity was used to characterise the cementitious materials investigated in this study.

5 Also, using the Vipulanandan contact index, the changes in the two-probe contacts with cement curing with and without additives were quantified. In smart cement with and without fly ash and aggregates the contact index increased with the curing time, an indication for higher corrosion potential for the metal probes placed in the cement. Smart cement with 10% metakaolin and 20% foam showed that the Vipulanandan contact index decreased with time and hence there was less potential for corrosion of the probes placed in the cement for monitoring.

Chapter 3

Rheological Behavior of Cement with Additives

3.1 HISTORY

Rheological behavior represents the response of fluids to flow under various pressure gradients, temperature, and varying configurations of the flow through media. The word 'rheology' was coined by Eugene C. Bingham in 1920, and since then there have been number of rheological models developed to characterise the fluids. In practice, rheology is principally concerned with extending the continuum mechanics to characterise the flow of fluids. Viscosity is a material property that represents the fluid's resistance to flow and also its resistance to deformation under applied shear stress. Newtonian fluids can be characterised by a single coefficient of viscosity for a specific temperature. Although this viscosity will change with temperature, it does not change with the shear strain rate. Only a small group of fluids exhibit such constant viscosity. For most of fluids, the viscosity changes with the shear strain rate (the relative flow velocity), and these fluids are called non-Newtonian fluids. There are many textbooks totally dedicated to the rheology of various fluids.

The viscosity of cement slurry affects mixing and pumping processes. When mixing concrete and other composites, cement must be well mixed with water and aggregates to achieve the required properties with the curing time. The viscosity must be controlled to ensure pumpability of the slurry during the entire well installation period. In oil and gas wells, because of the increased temperature, the viscosity decreases due to thermal thinning. This can lead to unplanned flow characteristics affecting the pumping operations. Knowledge of the rheological properties of cement slurries is important to enhance the mixing with the aggregates in concrete and also to ensure displacement and safe circulating pressures in wells.

Much has been learned in recent years about the rheology of cement paste and how it relates to the microstructure. Such progress has been made possible in part by the use of specialised instruments for measuring dynamic rheological properties, developed to characterise viscoelastic materials.

This chapter reviews recent studies of cement paste in which rheological properties are used to quantify the effects of cement composition with water and various types of additives. Based on the additives and type of cement, cement slurry can be a shear-thinning or shear-thickening fluid. Not much has been learned about the rheology of concrete. Measuring the flow behavior of concrete presents interesting challenges, and concrete rheometers have been develop for this purpose, but they are not widely used. Concrete rheology is controlled by the rheology of its cement paste, although such links have not yet been directly established. Concrete rheology provides important information about its workability, and the rheological parameters have several advantages over the slump test results when characterising the workability.

Rheological properties provide information about the workability of cementitious paste. As an example, the yield stress and plastic viscosity indicate the behavior of a specific cement paste composition. As another example, the apparent viscosity indicates what energy is required to move the suspension at a given strain rate. This test may be used to measure the flowability of a cement paste or the influence of a specific material or combination of materials on flowability.

Current empirical and time-independent rheological models (Bingham, Power law, Herschel–Bulkley) represent the shear stress–shear strain rate relationship, yield stress, and apparent (secant) viscosity (Guillot 1990; Mirza 2002; Vipulanandan et al. 2014a, 2015a, d, 2017a). The estimated rheological properties can vary significantly based on the models (Reddy et al. 2004). The Bingham plastic model and the Power law model are widely used in the petroleum industry to describe the flow properties of cement slurries (API 10A Standard). The Bingham plastic model includes both yield stress and a limiting viscosity at finite shear strain rates. The Power law model is a nonlinear relationship, also recommended by the API for cement slurry characterisation. The Herschel–Bulkley model that was developed in 1926, can predict nonlinear shear-thinning and shear-thickening behavior. The new Vipulanandan rheological model that was developed in 2014 is effective in predicting the shear stress–shear strain rate for both shear-thinning and shear-thickening behavior with maximum shear stress tolerance for shear-thinning fluids and maximum shear strain rate tolerance for shear-thickening fluids. The Vipulanandan rheological model also predicted the maximum shear stress tolerance for cement slurry that had a shear-thinning behavior (Vipulanandan et al. 2015d). The Vipulanandan rheological model is used to predict the behavior of various types for drilling muds with and without nanoparticles, and also cement slurries (Tchameni et al. 2019; Afolabi et al. 2017, 2019; Montes et al. 2019; Mohammed 2018).

3.2 STANDARDS

There are ASTM and API standards available to characterise the rheological properties of cement slurries. Some of the ASTM standards are as follows:

ASTM C1749-17a is a Standard Guide for Measurement of the Rheological Properties of Hydraulic Cementitious Paste Using a Rotational Rheometer.

ASTM C1874-20 is a Standard Test Method for Measuring Rheological Properties of Cementitious Materials Using Coaxial Rotational Rheometer.

ASTM D2196 is a Standard Test Methods for Rheological Properties of Non-Newtonian Materials by Rotational Viscometer.

3.3 RHEOLOGICAL MODELS

Rheological models relate the shear stress (τ) to the shear strain rate ($\dot{\gamma}$). The fluids show non-linear shear-thinning or shear-thickening behavior with a yield stress (τ_y). Based on the test results, the following conditions have to be satisfied for the model to represent the observed behavior.

(a) **Shear-Thinning Fluids Conditions Are as Follows:**

$$\tau = \tau_y \quad \text{When } \dot{\gamma} = 0 \tag{3.1}$$

$$\frac{d\tau}{d\dot{\gamma}} > 0 \tag{3.2}$$

$$\frac{d^2\tau}{d\dot{\gamma}^2} < 0 \tag{3.3}$$

$$\dot{\gamma} \to \infty \ then \ \tau = \tau_{max} \tag{3.4}$$

Hence the shear-thinning fluids will have a limit to the shear stress.

(b) **Shear-Thickening Fluids Conditions Are as Follows:**

$$\tau = \tau_y \quad \text{When} \quad \dot{\gamma} = 0 \tag{3.5}$$

$$\frac{d\tau}{d\dot{\gamma}} > 0 \tag{3.6}$$

$$\frac{d^2\tau}{d\dot{\gamma}^2} > 0 \tag{3.7}$$

$$\tau \to \infty \ then \ \dot{\gamma} = \dot{\gamma}_{max} \tag{3.8}$$

Hence the shear-thickening fluids will have a limit to the shear strain rate tolerance.

The rheological models are used for predicating the shear-thinning and shear-thickening behavior of fluids.

3.3.1 Herschel–Bulkley Model (1926)

The Herschel–Bulkley model (Eqn. (3.9)) defines a fluid rheology with three parameters, and the model is as follows:

$$\tau = \tau_o + k * (\dot{\gamma})^n, \tag{3.9}$$

where τ, τ_o, $\dot{\gamma}$, k, and n represent the shear stress, yield stress, shear strain rate, correlation parameter, and flow behavior index respectively. The model assumes that below the yield stress (τ_o), the slurry behaves as a rigid solid, similar to the Bingham plastic model. When $\tau > \tau_o$ the material flows as a Power law fluid. The exponent n describes the shear-thinning and shear-thickening behavior. Fluids are considered as shear thinning when n < 1 and shear thickening when n > 1. A fluid becomes shear thinning when the apparent viscosity decreases with the increase in the shear strain rate.

For shear thinning fluids, the model should satisfy the following conditions (Eqn. (3.2)) and Eqn. (3.3)):

$$\frac{d\tau}{d\dot{\gamma}} = k * n * \dot{\gamma}^{(n-1)} > 0 \to k * n > 0 \tag{3.10}$$

$$\frac{d^2\tau}{d\dot{\gamma}^2} = k * n(n-1) * \dot{\gamma}^{(n-2)} \Rightarrow k*n*(n-1) < 0 \tag{3.11}$$

One condition when both Eqn (3.10) and Eqn. (3.11) will be satisfied is as follows:

0 < n < 1 and k > 0.

From Eqn. (3.9)

When $\dot{\gamma} \to \infty \ then \ \tau = \infty$, as shown in Figure 3.1.

Hence the Herschel–Bulkley model does not satisfy the condition in Eqn. (3.4) for the maximum shear stress limit for the shear-thinning fluid. When $\tau_o = 0$, then the model will represent the Power Law rheological model.

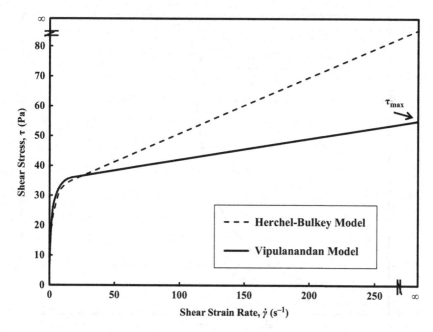

Figure 3.1 Schematics of the rheological models for the shear-thinning fluids.

For shear-thickening fluids, the model should satisfy the following conditions (Eqns. (3.6) and (3.7)):

$$\frac{d\tau}{d\dot{\gamma}} = k*n*\dot{\gamma}^{(n-1)} > 0 \rightarrow k*n > 0 \tag{3.12}$$

$$\frac{d^2\tau}{d\dot{\gamma}^2} = k*n(n-1)*\dot{\gamma}^{(n-2)} \Rightarrow k*n*(n-1) > 0 \tag{3.13}$$

One condition when both Eqn (3.12) and Eqn. (3.13) will be satisfied is as follows:
n > 1 and k > 0.
From Eqn. (3.9)
When $\dot{\gamma} \rightarrow \infty$ *then* $\tau = \infty$
Hence the Herschel–Bulkley model does not satisfy the condition in Eqn. (3.8) for the maximum shear strain limit for the shear-thickening fluid, as shown in Figure 3.2.

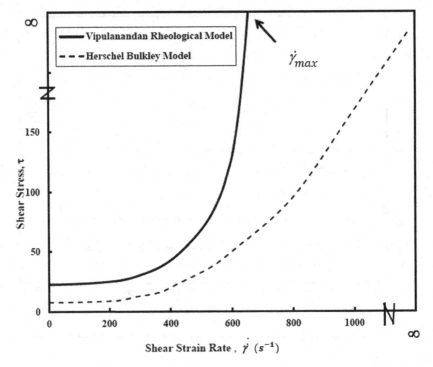

Figure 3.2 Schematics of the rheological models for the shear-thickening fluids.

3.3.2 Vipulanandan Rheological Model (2014)

The relationship between shear stress and the shear strain rate for all types of fluids is represented as follows:

$$\tau = \tau_y + \frac{\dot{\gamma}}{A + B\dot{\gamma}}, \tag{3.14}$$

where τ is shear stress (Pa); τ_y: is yield stress (Pa); A (Pa. s)$^{-1}$ and B (Pa)$^{-1}$: are model parameters; and $\dot{\gamma}$: is the shear strain rate (s^{-1}). The three model parameters τ_y, A, and B are influenced by the composition of the fluids, density, temperature, pressure, curing time, resistivity, and other physical, chemical, and biological properties of the fluids.

For shear-thinning fluids, satisfying Eqns. (3.2) and (3.3) results in the following conditions:

$$\frac{d\tau}{d\dot{\gamma}} = \frac{(A + B\dot{\gamma}) - \dot{\gamma}B}{(A + B\dot{\gamma})^2} = \frac{A}{(A + B\dot{\gamma})^2} > 0 \quad \Rightarrow A > 0 \tag{3.15}$$

$$\frac{d^2\tau}{d\dot{\gamma}^2} = -2(A + B\dot{\gamma})^{-3} * AB = \frac{-2AB}{(A + B\dot{\gamma})^3} < 0, \quad \Rightarrow B > 0 \tag{3.16}$$

$$\dot{\gamma} \to \infty \; then \quad \tau_{max} = \tau_y + \frac{1}{B} \tag{3.17}$$

Hence this model has a limit on the maximum shear stress tolerance or shear strength for shear-thinning fluids, as shown in Figure 3.1, and also the models are compared there.

For shear-thickening fluids, satisfying Eqns. (3.6) and (3.7) results in the following conditions:

$$\frac{d\tau}{d\dot{\gamma}} = \frac{(A + B\dot{\gamma}) - \dot{\gamma}B}{(A + B\dot{\gamma})^2} = \frac{A}{(A + B\dot{\gamma})^2} > 0 \quad \Rightarrow A > 0 \tag{3.18}$$

$$\frac{d^2\tau}{d\dot{\gamma}^2} = -2(A + B\dot{\gamma})^{-3} * AB = \frac{-2AB}{(A + B\dot{\gamma})^3} > 0, \Rightarrow B < 0 \tag{3.19}$$

$$\tau \to \infty \; then \quad \dot{\gamma}_{max} = -\frac{A}{B} \tag{3.20}$$

Hence this model has a limit on the maximum shear strain rate tolerance for shear-thickening fluids, as shown in Figure 3.2, and also the models are compared there.

3.4 MODEL VERIFICATION

Studies have been performed around the world to verify the application of the Vipulanandan rheological model compared to other rheological models for various types of drilling fluids, smart spacer fluids, smart cement, and smart foam cement with and without nanoparticles. All the studies have used statistical parameters such as root-mean-square-error (RMSE) and coefficient of determination (R^2) to compare the various model predictions of the experimental results. Afolabi et al. (2017) investigated the addition of nanosilica to bentonite drilling mud and used several models to compare the predictions, and concluded that the Vipulanandan rheological model had the best prediction of the experimental results. In this study, the maximum shear stress tolerance parameter was also used to compare the effect of adding nanosilica. Montes (2019) used the model to investigate the effect of nanofluids in the viscosity reduction of extra-heavy oils. Tchameni et al. (2019) used the Vipulanandan rheological model to evaluate the thermal effect on the rheological properties of waste vegetable oil biodiesel modified bentonite drilling muds. Afolabi et al. (2019) used the model to predict the rhological properties of nano particle modified drilling muds. Mohammed (2018) used the Vipulanandan rheological model to characterise the rheological properties of cement slurry modified with nano clay. Vipulanandan et al. (2016a, b, 2018a, i) used the model to characterise the behavior of smart foam cement, grouts, and smart spacer fluids, and compared it to the Herschel–Bulkley model.

3.5 EXPERIMENTAL STUDY AND MODEL VERIFICATION

3.5.1 Rheological Test

Using a rheometer, tests were performed on smart cement slurry with different water-to-cement (w/c) ratios at two different temperatures of 25°C and 85°C, and with various additives. The rheometer speed range was 0.3 rpm to 600 rpm (shear strain rate of 0.5 s^{-1} to 1,024 s^{-1}) with a heating chamber. The speed accuracy of this device was 0.001 rpm, and the major components of the instrument are shown in Figure 3.3. The temperature of the slurry was controlled to an accuracy of ±2°C. The viscometer was calibrated using several standard solutions. All the rheological tests were performed after ten minutes of mixing of the cement slurries.

3.5.2 Smart Cement

In this study, Class H smart cement with a water-to-cement ratios of 0.38, 0.44, and 0.54 was used. The samples were prepared according to API standards. To improve the sensing properties and piezoresistive behavior of the cement, varying amounts of up to 0.1% of conductive fillers (CF) were added (by the weight of cement) and mixed with all the samples. After

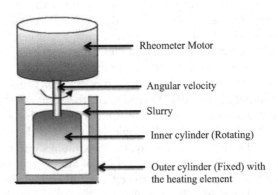

Figure 3.3 Schematic of the rotational viscometer.

mixing, the cement slurries were used for rheological studies at 25°C and 85°C. At least three samples were tested for all the water-to-cement ratio mixes investigated in this study.

(a) Apparent Viscosity

Smart cement slurries with three different w/c ratios at two different temperatures showed significantly different rheological properties. However, regardless of the w/c ratio and temperature, all slurries exhibited non-Newtonian and shear-thinning behavior, as shown in Figure 3.3

Increasing the w/c ratio reduced the apparent viscosity (secant slope; shear stress divided by the shear strain rate) of smart cement slurries. The apparent viscosity of smart cement with a w/c ratio of 0.38, 0.44, and 0.54 at a shear strain rate of 170 s⁻¹ (100 rpm) and at room temperature was 196 cP, 152 cP, and 129 cP respectively. Increasing the temperature to 85°C increased the apparent viscosity of the smart cement slurries. The apparent viscosity of smart cement with a w/c ratio of 0.38, 0.44, and 0.54 at a shear strain rate of 170 s⁻¹ increased by 184%, 105%, and 81% respectively when the temperature was increased from 25°C to 85°C. So increasing the water content in the cement reduced the increase in the apparent viscosity. Also, it is important to monitor the changes in the cement temperature during the installation of wells since it affects the cement rheological properties.

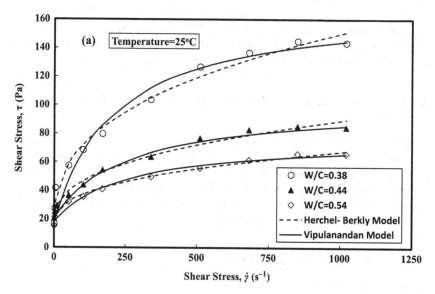

Figure 3.4a Measured and predicted shear stress-shear strain rate relationship for the smart cement slurries with different water-to-cement (w/c) ratios and temperature (a) T = 25°C, (b) T = 85°C.

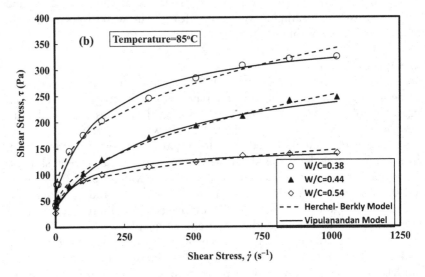

Figure 3.4b

The shear stress–strain rate relationships of smart cement slurries with various w/c ratios at 25°C and 85°C temperatures are shown in Figure 3.4. It was clear that increasing the water-to-cement ratio reduced the shear stress at the same shear strain rates, which indicates the improved flow properties of smart cement with a higher w/c ratio. At 85°C, high shear stress occurred at a very low shear strain rate, which confirmed the formation of a gel structure due to chemical reactions at high temperatures.

3.5.2.1 Comparison of the Constitutive Models for Rheological Properties

Shear stress–shear strain rate relationships for smart cement slurries with w/c ratios of 0.38, 0.44, and 0.54 at two different temperatures were predicted using the Vipulanandan model and compared with the Herschel–Bulkley model, as shown in Figure 3.4. Also all the model parameters are summarised in Tables 3.1 and 3.2 with the root-mean-square error (RMSE) and the coefficient of determination (R^2).

3.5.2.1.1 Herschel–Bulkley Model (1926)

(a) w/c Ratio of 0.38

The shear-thinning behavior of smart cement slurry with a w/c ratio of 0.38 at two different temperatures, 25°C and 85°C, was modeled using the

Table 3.1 Herschel–Bulkley rheological model parameters

w/c	T (°C)	τ_{ol} (Pa)	k (Pa.s^n)	n	RMSE (Pa)	R^2
			Herschel–Bulkley Model (1926)			
0.38	25	21.0 ± 2.08	10.94 ± 0.05	0.37 ± 0.01	2.67	0.99
	85	38.0 ± 2.03	24.26 ± 0.03	0.36 ± 0.01	9.26	0.99
0.44	25	19.2 ± 1.83	5.06 ± 0.06	0.39 ± 0.02	5.89	0.99
	85	33.0 ± 0.97	15.41 ± 0.02	0.37 ± 0.02	11.81	0.97
0.54	25	17.7 ± 1.01	3.89 ± 0.07	0.38 ± 0.01	1.80	0.99
	85	37.5 ± 0.87	14.20 ± 0.1	0.31 ± 0.03	5.26	0.98

Table 3.2 Vipulanandan rheological model parameters

w/c	T (°C)	τ_y (Pa)	A (Pa.s^{-1})	B (Pa)$^{-1}$	$\tau_{max.}$ (Pa)	RMSE (Pa)	R^2
				Vipulanandan Rheological Model (2014)			
0.38	25	20.6 ± 1.13	3.18 ± 0.1	0.008 ± 0.001	146 ± 2.0	2.43	0.99
	85	41.3 ± 1.06	0.62 ± 0.15	0.003 ± 0.001	349 ± 1.7	9.12	0.99
0.44	25	17.3 ± 1.73	1.40 ± 0.1	0.014 ± 0.002	89 ± 3.2	6.38	0.98
	85	40.0 ± 0.75	1.38 ± 0.2	0.004 ± 0.001	312 ± 2.5	5.62	0.99
0.54	25	15.2 ± 0.94	4.25 ± 0.2	0.017 ± 0.001	74 ± 2.0	1.72	0.98
	85	37.5 ± 1.01	1.02 ± 0.1	0.012 ± 0.001	121 ± 1.8	5.10	0.98

Herschel–Bulkley model (Eqn. (3.9)) up to a shear strain rate of 1,024 s^{-1} (600 rpm). The coefficient of determination (R^2) was 0.99, as summarised in Table 3.1. The root-mean-square error (RMSE) was 2.67 Pa and 9.26 Pa for the temperatures of 25°C and 85°C respectively, as summarised in Table 3.1. The yield stress (τ_{o1}) for the cement slurry at a temperature of 25°C was 21 Pa, and with the increase in the temperature of the slurry to 85°C, the yield stress increased to 38 Pa, a 81% increase. The model parameter k for the cement slurry with a w/c ratio of 0.38 at temperatures of 25°C and 85°C was 10.94 Pa.s^n and 24.26 Pa.s^n respectively, as summarised in Table 3.1, a notable increase in the parameter k. The model parameter n for the cement slurry was 0.367 and 0.360 for the temperatures of 25°C and 85°C respectively, as summarised in Table 3.1, a minor reduction in the parameter n.

(b) w/c Ratio of 0.44

The relationships between shear stress with a shear strain rate for smart cement slurry with a w/c ratio of 0.44 at 25°C and 85°C were modeled using the Herschel–Bulkley model (Eqn. (6)). The coefficient of determination (R^2) was greater than 0.96 for the temperatures of 25°C and 85°C, as summarised in Table 3.1. The root-mean-square error (RMSE) was 5.89

Pa and 11.81 Pa respectively, as summarised in Table 3.1. The yield stress (τ_{o1}) for the cement slurry at 25°C was 19.2 Pa and increased to 33 Pa with the increase in the temperature to 85°C, a 72% increase, and the trend was similar to what was observed with a w/c ratio of 0.38. The model parameter k for the cement slurry at 25°C and 85°C were 5.06 Pa.s^n and 15.41 Pa.s^n respectively, a 205% increase with the increase in the temperature to 85°C. The model parameter n decreased by 4% with the increase in the temperature to 85°C, as summarised in Table 3.1.

(c) w/c Ratio of 0.54

Using the Herschel–Bulkley model (Eqn. (3.9)), the relationships between shear stress and the shear strain rate for smart cement slurry mud with w/c ratios of 0.54 at 25°C and 85°C were modeled. The coefficient of determination (R^2) was greater than 0.97, as summarised in Table 3.1. The root-mean-square error (RMSE) varied from 1.80 Pa to 5.26 Pa for the temperatures of 25°C and 85°C, as summarised in Table 3.1. The yield stress (τ_{o1}) of the cement slurry at 25°C was 17.7 Pa and increased to 37.5 Pa with the increase in the temperature to 85°C, a 112% increase. The model parameter k for the cement slurry at 25°C and 85°C were 3.89 Pa.s^n and 14.2 Pa.s^n respectively, a 265% increase, similar to the trend observed with smart cement slurry with a w/c ratio of 0.44. The model parameter n for the cement slurry decreased by 18% with the increase in the temperature to 85°C, as summarised in Table 3.1.

3.5.2.1.2 Vipulanandan Model (2014)

(a) w/c Ratio of 0.38

The shear-thinning behavior of smart cement slurry with a w/c ratio of 0.38 at 25°C and 85°C was modeled using the Vipulanandan rheological model (Eqn. (3.14)). The coefficient of determination (R^2) was 0.99, as summarised in Table 3.2. The root-mean-square error (RMSE) was 2.43 Pa and 9.12 Pa respectively for the temperatures of 25°C and 85°C, as summarised in Table 3.2. The yield stress (τ_y) of the smart cement slurry at 25°C was 20.6 Pa and it increased to 41.3 Pa with an increase in the temperature to 85°C, a 100% increase, as shown in Figure. 3.5. The model parameter A for the smart cement slurry at 25°C and 85°C were 3.18 Pa.s^{-1} and 0.62 Pa.s^{-1} respectively, a 81% reduction. The model parameter B for the smart cement slurry at 25°C and 85°C was 0.008 Pa^{-1} and 0.003 Pa^{-1} respectively, a 63% reduction with the increase in the temperature to 85°C, as summarised in Table 3.2.

(b) w/c Ratio of 0.44

The relationships between shear stress with a shear strain rate of the smart cement slurry with a w/c ratio of 0.44 at 25°C and 85°C were modeled using

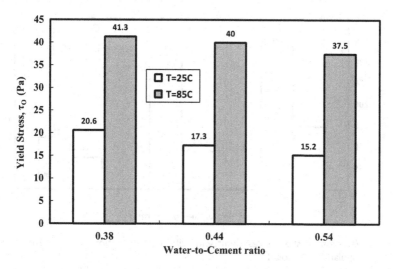

Figure 3.5 Variation of yield stress with different water-to-cement (w/c) ratios predicated using the Vipulanandan rheological model.

the Vipulanandan model (Eqn. (3.10)). The coefficient of determination (R^2) was greater than 0.97, as summarised in Table 3.2. The root-mean-square error (RMSE) was 6.38 Pa to 5.62 Pa respectively for temperatures of 25°C and 85°C, as summarised in Table 3.2. The yield stress (τ_y) at 25°C was 17.3 Pa and increased to 40 Pa with the increase in temperature to 85°C, a 131% increase, as shown in Figure 3.5. The model parameter A for the cement slurry at 25°C and 85°C were 1.4 Pa.s⁻¹ and 1.38 Pa.s⁻¹ respectively, a 1.4% reduction. The model parameter B for the smart cement slurry with a w/c ratio of 0.44 at 25°C and 85°C were 0.014 Pa⁻¹ and 0.004 Pa⁻¹ respectively, a 71% reduction, as summarised in Table 3.2.

(c) w/c Ratio of 0.54

Using the Vipulanandan model (Eqn. (3.9)), the relationships between shear stress with a shear strain rate of smart cement slurry with a w/c ratio of 0.54 at 25°C and 85°C were modeled. The coefficient of determination (R^2) was 0.98. The root-mean-square error (RMSE) was 1.72 Pa and 5.10 Pa respectively for the temperatures of 25°C and 85°C, as summarised in Table 3.2. The yield stress (τ_y) of the cement slurry with a w/c ratio of 0.44 at 25°C and 85°C were 15.2 Pa and 37.5 Pa respectively, a 147% increase, as shown in Figure 3.5. The model parameter A for the cement slurry with w/c ratio of 0.54 at 25°C and 85°C were 4.25 Pa.s⁻¹ and 1.02 Pa.s⁻¹ respectively, a 76% reduction. The model parameter B at 25°C and 85°C were 0.017 Pa⁻¹ and 0.012 Pa⁻¹ respectively, a 29% reduction, as summarised in Table 3.2.

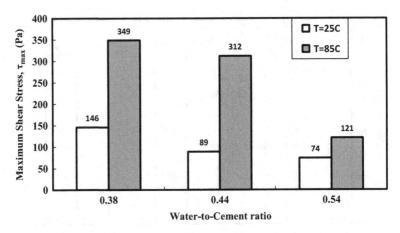

Figure 3.6 Variation of shear stress limit with different w/c ratios predicated using the Vipulanandan model.

(d) Maximum Shear Stress ($\tau_{max.}$)

Based on Eqn. (3.12), the Vipulanandan model has a limit on the maximum shear stress ($\tau_{max.}$) that slurry will produce at a relatively very high rate of shear strains. The τ_{max} for smart cement slurries with a w/c ratio of 0.38, 0.44, and 0.54 at a temperature of 25°C were 102 Pa, 89 Pa, and 74 Pa respectively, as summarised in Table 3.2 and also shown in Figure 3.6. Increasing the temperature of smart cement slurries to 85°C increased the maximum shear stress to 349 Pa, 312 Pa, and 121 Pa for the w/c ratio of 0.38, 0.44, and 0.54 respectively, as shown in Figure 3.6.

3.5.3 Metakaolin and Fly Ash

Based on the application, cement is mixed with various additives to modify its rheological properties. In this study both metakaoline (water-to-cement ratio of 0.40 with a water reducing agent) and fly ash (water-to-cement ratio of 0.40) were used, and the Vipulanandan model was used to characterise the behavior. The amount and type of additive added to the cement will modify the rheological properties and have to be quantified.

3.5.3.1 Metakaolin

Adding metakaolin to cement slurry modified all the rheological behaviors of the cement slurry, as shown in Figure 3.7 and summarised in Table 3.3 with the Vipulanandan rheological model parameters. The reason for these changes in the rheological properties of the cement slurry could be due to

Table 3.3 Rheological model parameters for smart cement slurries modified with metakaolin and fly ash

Amount of Additives	Vipulanandan Rheological Model					
	τ_y (Pa)	A (Pa.s^{-1})	B (Pa)$^{-1}$	$\tau_{max.}$ (Pa)	RMSE (Pa)	R^2
Metakaolin						
0	9.4	2.65	0.0092	142	2.1	0.98
5%	12.0	5.78	0.0067	171	2.4	0.99
10%	37.2	3.74	0.0031	356.5.	3.3	0.99
Fly Ash						
5%	12.5	3.16	0.011	102.9	2.6	0.99
50%	19.7	4.74	0.0079	145.7	2.8	0.99

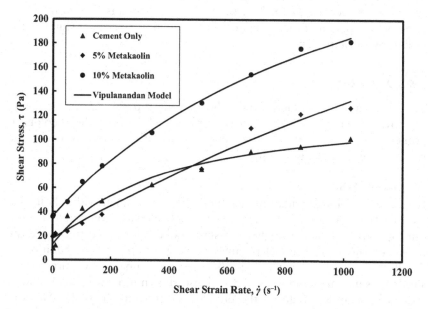

Figure 3.7 Comparing the rheology behavior of the 5% and 10% metakaolin-modified smart cement with the Vipulanandan model.

the fine metakaolin particles minimising the lubricating effect of water in the cement slurry pores.

(a) 5% Metakaolin

Adding 5% metakaolin increased the yield stress to 12 Pa, a 27.5% increase compared to the pure smart cement slurry, as shown in Figure 3.7. Hence, a higher pressure gradient will be needed to initiate the flow. The rheological model parameters were sensitive to the addition of 5% metakaolin, and parameter A increased and parameter B decreased compared to pure cement slurry, as summarised in Table 3.3. Also. modifying smart cement slurry with 5% metakaolin increased the ultimate shear strength by 20.4% from 142 Pa to 171 Pa. as summarised in Table 3.3.

(b) 10% Metakaolin

Adding 10% metakaolin increased the yield stress to 37.2 Pa, a 296%, increase compared to pure cement slurry. Hence, a much higher pressure gradient compared to the 5% metakaolin modification is needed to initiate the flow. The rheological model parameters were sensitive to the addition of 10% metakaolin, and parameter A increased and parameter B decreased compared to pure cement slurry, as summarised in Table 3.3. With the addition of 10% metakaoline, the ultimate shear stress capacity increased to 356.5 Pa, an increase of 151%. Maximum shear stress is also an indicator of potential erosion of the formation during the flow of cement slurry in the well during installation.

3.5.3.2 Fly Ash

The addition of fly ash modified the rheological properties of cement slurry, as shown in Figure 3.8. The reason for these changes in the rheological properties is due to the finer fly ash particles with less reaction with the water in the cement slurry pores.

(a) 5% Fly Ash

Based on the Vipulanandan model, modifying the cement with 5% fly ash increased the yield stress to 12.5 Pa, a 33% increase compared to pure cement slurry. Hence, a higher pressure gradient is needed to initiate the flow. The rheological model parameters were sensitive to the addition of 5% fly ash, and parameters A and B increased compared to pure cement slurry, as summarised in Table 3.3 and also shown in Figure 3.8. The addition of 5% fly ash reduced the ultimate shear strength by 27.5% from 142 Pa to 102.9 Pa, just the opposite trend of what was observed with 5% metakaolin.

(b) 50% Fly Ash

The rheological behavior is shown in Figure 3.8. With the addition of 50% fly ash, the yield stress increased to 19.7 Pa, a 110% increase compared to pure cement slurry. Hence, a higher pressure gradient is needed to initiate the flow. The rheological model parameters were sensitive to the addition of 50% fly ash, and parameter A increased and parameter B decreased

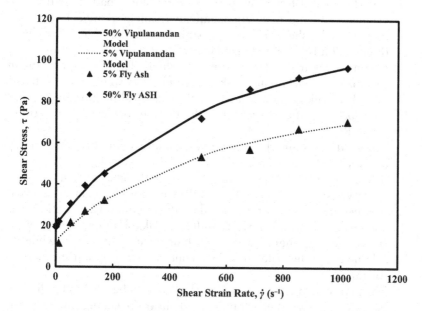

Figure 3.8 Comparing the rheology behavior of the 5% and 50% fly ash modified smart cement with the Vipulanandan model.

compared to pure cement slurry, as summarised in Table 3.3. Adding 50% fly ash increased the ultimate shear stress capacity to 145.7 Pa, an increase of only 2.6% compared to pure cement slurry. Maximum shear stress is also an indicator of potential erosion of the formation during the flow of cement slurry in the well during installation.

3.6 SUMMARY

Characterising the rheological properties of cement slurry is important in order to understand its flowablity, mixing, and pump ability. Also, rheological models are important to better characterise cement slurry and also quantify the effects of various additives related to the pumping pressure gradients and also potential erosion of the formations. Based on the experimental and analytical modeling of the rheological behavior of smart cement with varying water-to-cement ratios (0.38, 0.44, 0.54), temperatures (up to 85°C), and also with additives such as metakaolin (5% and 10%) and fly ash (5% and 50%), the following conclusions are advanced:

1 The rheological test showed that cement slurry had shear-thinning behavior and the new Vipulanandan rheological model was used to predict the shear stress–shear strain rate relationship. For shear-thinning fluids, the Vipulanandan rheological model has a limit to the maximum shear stress tolerance of the fluids. The new Vipulanandan rheological model predicted the test results very well compared to the Herschel–Bulkley model based on the statistical parameters, such as the root-mean-square error (RMSE) and the coefficient of determination (R^2).

2 The yield stress and the maximum shear stress limit of cement slurry reduced with the increase in the water-to-cement ratio, and also increasing the temperature increased these rheological properties.

3 Additives such as metakaolin and fly ash (replacing cement) affected the rheological properties, including the maximum shear stress tolerance of the cement slurries. Adding metakaolin increased the yield stress and maximum shear stress limit of cement slurry, while fly ash increased the yield stress but reduced the maximum shear stress tolerance.

4 The new Vipulanandan rheological model can be used to predict the behavior of both shear-thinning and shear-thickening fluids. It can also be used to predict the behavior of Newtonian fluids. Moreover, the Vipulanandan rheological model predicted the shear-thinning behavior of cement slurries with and without additives very well based on the root-mean-square error (RMSE).

Chapter 4

Cement Curing Resistivity and Thermal Properties

4.1 BACKGROUND

Once cement is mixed with water, chemical reactions are initiated and cement slurry will hydrate with time to transform into a solid with enhanced chemical, physical, and mechanical properties. There are a few methods to investigate the curing of cement. The Vicat needle method was developed in 1854, and it is still used to determine the initial setting time and final setting time of cement (ASTM C 191). In this test, based on the penetration depths of a needle into hydrating cement slurry placed in a cone-shaped mold, the initial and final setting times are determined. This physical method is all about determining the penetration depth of the needle from the curing cement surface, which will be influenced by not only the bulk cement hydration but also the environmental condition surrounding the exposed surface. Different types of penetration methods are used to determine the setting time of concrete, mortar, and grouts (ASTM C403 / 403 M). Also, tests have been performed using thermal calorimeters to determine the heat of hydration and also the changes in temperature during the curing of cement.

It is important to develop testing methods to better characterise bulk cement hydration not only in the laboratory but also in the field. Also, the hydration of cement will very much depend on the mixing method and the admixtures added to the mix (Zhang et al. 2010; Vipulanandan et al. 2014a through 2018c). In this study, the electrical resistivity (inverse of conductivity) of the bulk materials has been identified as the monitoring parameter for hydrating and curing cement, based on the material characterisation for smart cement in Chapter 2. There are many laboratory devices available to measure the resistivity and conductivity of cement slurry.

It is important to investigate the sensitivity of the resistivity parameter to indicate the hydration changes in all types of smart cements with and without additives. Since cement is also used as a thermal insulator in some applications, quantifying the thermal conductivity of smart cement with and without additives is of interest and also can be compared to the electrical resistivity, two material properties.

4.2 THREE-PHASE MODEL

Three phase (solid (S), liquid (L) and gas (G)) representation is the most general case for all natural and artificial materials that are used in the field of engineering, medicine and science. Currently, three-phase models are very much focused on quantifying the changes in liquid content. In this study, cement mixed with water and additives is represented as a chemically reactive three-phase material model with nine time-dependent variables representing the chemical reactions occurring between the solid, liquid, and gas in the three-phase material, as shown in Figure 4.1. These reactions between the three phases will change the composition and density of hydrating cement and are represented in Eqns. (4.1), (4.2), and (4.3).

In order to determine the nine reactive parameters (a, b, c, p, q, r, x, y, and z), theoretically it is essential to have at least six independent measurements. Hence, for hydrating cement, the changes in the weight, volume, and moisture content with time, and the developing correlation between the few selected parameters (based on the information available in the literature) and the six independent material parameters have to be quantified. These six independent reactive parameters can be used to verify the influence on the changes in the physical (shrinkage, porosity) and electrical properties of the cement.

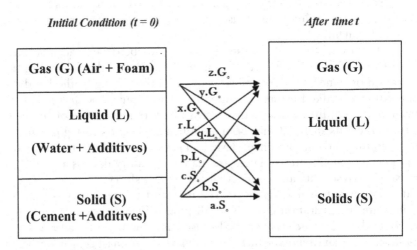

Figure 4.1 Three-phase diagram with the nine reaction parameters.

4.2.1 Reactive Parameters

Three-phase material such as cement slurry includes solid (S) (cement + additives), liquid (L) (water and additives), and gas (G) (air + foam) interacting over time (t) under various curing environments such as pressure (P), temperature (T), and relative humidity (RH).

(a) Solid Phase (S)
There will be reaction between the solid phase, the liquid phase, and the gas phase. Due to the reaction, part of the solid phase will become the liquid phase, and the fractional parameter based on weight is represented as 'b(t,T,P,RH)'. The unreacted fraction of the solid phase will be represented by a fractional parameter 'a(t,T,P,RH)', and the remaining fraction of the solid phase will be transferred to the gas phase 'c(t,T,P,RH)'. Hence, based on the conservation of mass, the solid reactive parameters can be represented as follows:

$$a(t,T,P,RH) + b(t,T,P,RH) + c(t,T,P,RH) = 1. \tag{4.1}$$

(b) Liquid Phase (L)
Due to another reaction, a portion of the liquid phase will become solidified, for which the fractional parameter is 'p (t,T,P,RH)'. The unreacted fraction of the liquid phase is represented as 'q (t,T,P,RH)', and the remaining portion evaporates 'r (t,T,P,RH)'. Hence, based on the conservation of mass, the liquid reactive parameters can be represented as

$$p(t,T,P,RH) + q(t,T,P,RH) + r(t,T,P,RH) = 1. \tag{4.2}$$

(c) Gas Phase
Due to two reactions, one portion of the gas becomes solid 'x(t,T,P,RH)', one portion becomes liquid 'y(t,T,P,RH)', and the remaining portion is 'z(t,T,P,RH)'. The gas reactive parameters can be represented as

$$x(t,T,P,RH) + y(t,T,P,RH) + z(t,T,P,RH) = 1. \tag{4.3}$$

In Figure 4.1, the three phases and the reaction parameters are shown.

After curing time t, the weight of the solid phase, , $W_S(t,T,P,RH)$ the weight of the liquid phase, $W_L(t,T,P,RH)$ and the weight of the gas phase $W_G(t,T,P,RH)$ can be represented as portions of the initial weights of solid, liquid, and gas as, W_{S_0}, W_{L_0} and W_{G_0} as following:

$$W_S(t,T,P,RH) = a(t,T,P,RH).W_{S_0} + p(t,T,P,RH).W_{L_0}$$
$$+ x(t,T,P,RH).W_{G_0}$$

$$W_L(t,T,P,RH) = b(t,T,P,RH).W_{S_0} + q(t,T,P,RH).W_{L_0}$$
$$+ y(t,T,P,RH).W_{G_0} \tag{4.4}$$

$$W_G(t,T,P,RH) = c(t,T,P,RH).W_{S_0} + r(t,T,P,RH).W_{L_0}$$
$$+ z(t,T,P,RH).W_{G_0}$$

Hence Eqn. (4.4) can be represented in a matrix form as

$$
\begin{bmatrix} W_S(t,T,P,RH) \\ W_L(t,T,P,RH) \\ W_G(t,T,P,RH) \end{bmatrix} =
\begin{bmatrix} a(t,T,P,RH) & p(t,T,P,RH) & x(t,T,P,RH) \\ b(t,T,P,RH) & q(t,T,P,RH) & y(t,T,P,RH) \\ c(t,T,P,RH) & r(t,T,P,RH) & z(t,T,P,RH) \end{bmatrix} .
\begin{bmatrix} W_{S_0} \\ W_{L_0} \\ W_{G_0} \end{bmatrix}.
$$
$$\tag{4.5}$$

The composition of hydrating cement will very much depend on the nine parameters a, b, c, p, q, r, x, y, and z, with six independent variables based on Eqns. (4.1), (4.2), and (4.3). Hence, the monitoring method and the selected parameter must be sensitive to the changes in hydrating cement products with time, pressure, and curing conditions (relative humidity and temperature).

4.3 HYDRATION OF CEMENT

4.3.1 Calorimetric Study

A high-performance semi-adiabatic calorimeter (P-CAL 1000), as shown in Figure 4.2 (insert), was used to record the temperature changes during the hydration (up to four days) of a Class H oil-well cement with a water-to-cement (w/c) ratio of 0.38 and with and without carbon fiber additive. The volume of the sample tested was 1.64 L. The peak temperature was 171.7°F for cement only after 20.5 hours of curing in the calorimeter. As shown in Figure 4.2 and summarised in Table 4.1, the addition of 0.075% carbon fiber slightly reduced the peak temperature to 169.8°F in 20.7 hours.

4.3.2 Setting Time

Based on the Vicat needle test (ASTM C191) at room condition, the initial setting time was determined to be after six hours and the final setting time

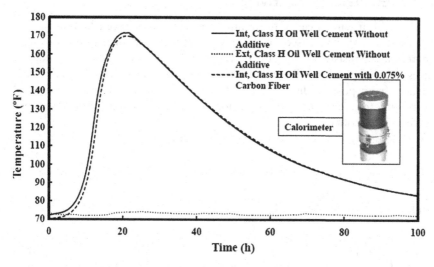

Figure 4.2 Hydration temperature with time for the Class H oil-well cement with and without carbon fibers in the calorimeter study (Int: Internal temperature or hydration temperature; Ext: External room temperature).

Table 4.1 Summary of peak temperature and corresponding time for a Class H oil-well cement with and without carbon fibers in the calorimeter study

Additives	Peak Temperature (°F)	Peak Time (h)
With no Additive	171.7	20.5
With 0.075% Carbon Fiber	169.8	20.7

was about eight hours. As shown in Figure 4.2, the cement continued to hydrate beyond the experimentally determined setting time based on the Vicat needle test (determined by the needle penetration). The hydration process, reactions between the cement, water, air, and admixtures will continue through the entire service life of cement, and hence it is important to select a highly sensitive parameter to monitor it.

Fibers
Carbon Fiber: The addition of 0.075% carbon fiber to the cement with a w/c ratio of 0.4 reduced the peak temperature by about 2°F and increased the time to reach the peak temperature by about 0.2 hr. Hence, adding 0.075% carbon fiber (amount used in piezoresistive smart cement studies) had a minimal effect on the hydration process of the cement. This was also reflected in the resistivity and changes in resistivity during the initial 24 hours of curing.

4.3.3 Electrical Resistivity

In the field, it will be easier to monitor electrical resistance (R), but it is not a material property. Based on the type of material, including piezoresistive material, the resistance and resistivity will influenced by the curing time (t), stress (σ), temperature (T), and also the concentration of contamination (C). Hence, from past studies and also theoretical understanding of electrical resistivity (ρ), the following conditions between resistivity and resistance have to be satisfied:

$$\frac{d\rho}{dR} > 0 \tag{4.6}$$

and

$$\frac{d^2\rho}{dR^2} < 0 \tag{4.7}$$

or

$$\frac{d^2\rho}{dR^2} = 0 \tag{4.8}$$

or

$$\frac{d^2\rho}{dR^2} > 0 \tag{4.9}$$

In order to satisfy the above conditions, the following resistivity–resistance relationship is proposed:

$$\rho(t,\sigma,T,C) = \frac{R(t,\sigma,T,C)}{K + GR(t,\sigma,T,C)} \tag{4.10}$$

And

$$d\rho = \frac{K\,dR}{(K+GR)^2} \tag{4.11}$$

Hence $\qquad \dfrac{d\rho}{\rho} = \dfrac{K}{(K+GR)}\dfrac{dR}{R}$ $\tag{4.12}$

Both material parameters K and G have to be determined using experiments where ρ and R are measured independently to determine the parameters. For conductive materials such as metals, it has been shown that the parameter G is equal to zero.

Electrical resistivity is a unique property of the material that can be monitored from the time of mixing to the entire service life of solidified cement in oil wells, concrete, and other materials where cement is part of the constituent. The electrical resistivity (sensing property) of cement slurry was measured using a digital resistivity meter and a conductivity meter. Also, electrical resistance changes in the curing time were measured using an LCR (inductance (L), capacitance (C), resistance (R)) meter employing the two-probe method (Vipulanandan et al. 2013a). Each specimen was first calibrated to obtain the parameters K and G using the resistivity (ρ) and the measured electrical resistance (R) based on Eqn. (4.10). Several experimental studies were performed to determine the parameters K and G for cement. Based on the studies, it was proven that parameter G was zero for cement. Hence, the changes in the resistivity (dρ) can be related to the change in resistance (dR) represented in Eqn. (4.12) as follows (parameter G = 0):

$$\frac{d\rho}{\rho_0} = \frac{dR}{R_0} \qquad\qquad (4.13)$$

where ρ_0 and R_0 are the measured initial resistivity and resistance. The nominal parameter K_n is determined as the ratio of the measuring distance between the two measuring electrical probes (L) and the cross-sectional area (A) of the specimen and is equal to L/A. The parameter K (effective K) determined experimentally from this study on cement by directly measuring the resistivity (ρ) and resistance (R) was more than double the nominal parameter K_n. This is because the actual current flow path through the cement, a semiconducting material, will be higher than a conductive material such as metal, and also the current flow area for cement will be less than the nominal cross-sectional area A, resulting in a higher value for parameter K (effective K) for cement determined in this study.

4.3.3.1 Initial Resistivity of Smart Cement Slurry

Two different methods were used to measure the electrical resistivity of cement slurries. To assure the repeatability of the measurements, the initial resistivity was measured at least three times for each cement slurry, and the average resistivity is reported. The electrical resistivity of the cement slurries was measured using a conductivity probe and digital resistivity meter.

(a) Conductivity Probe

A commercially available conductivity probe was used to measure the conductivity (inverse of resistivity) of the slurries. In the case of cement, this meter was used during the initial curing of the cement. The conductivity measuring range was from 0.1 μS/cm to 1,000 mS/cm, representing a resistivity of 0.1 Ω.m to 10,000 Ω.m.

(b) Digital Resistivity Meter

A digital resistivity meter (used in the oil industry) was used measure the resistivity of fluids, slurries, and semi-solids directly. The resistivity range for this device was 0.01 Ω-m to 400 Ω-m.

The conductivity probe and the digital electrical resistivity device were calibrated using a standard solution of sodium chloride (NaCl).

4.3.3.2 Two-Probe Method

In this study, high-frequency alternative current (AC) measurement was adopted to overcome interfacial problems and minimise contact resistances. Electrical resistance (R) was measured using an LCR meter (measures the inductance (L), capacitance (C), and resistance (R)) during the curing time. This device has a least count of 1 μΩ for electrical resistance and measures the impendence (resistance, capacitance, and inductance) in the frequency range of 20 Hz to 300 kHz. Based on the impedance (z)–frequency (f) response, it was determined that smart cement is a resistive material (Vipulanandan et al. 2013a, 2014c, 2016c, d). Hence the resistance was measured at 300 kHz using the two-probe method during the entire testing time.

4.4 TESTING AND MODELING

The electrical resistivity (ρ) changes with time during the hydration of cement are shown in Figure 4.3. Electrical conductivity is the inverse of resistivity, as shown in Figure 4.3. Electrical resistivity was reduced to a minimum value, and then gradually increased with time. Initially after mixing the cement with water, resistivity decreased to a minimum value (ρ_{min}), and the corresponding time to reach the minimum resistivity was (t_{min}), as shown in Figure 4.3. The t_{min} can be used as an index for the speed of chemical reactions and cement setting times. Also, electrical resistivity is predominated by the conductivity of the pore solution and the connectivity of the pores. Immediately after mixing, the pores are connected and more electrical conduction paths are formed within cement slurry. After hours of curing, the hydration products block the conduction path and tortuosity increases. The decrease in the connectivity of pores results in the increase in resistivity. There is very limited information in the literature about

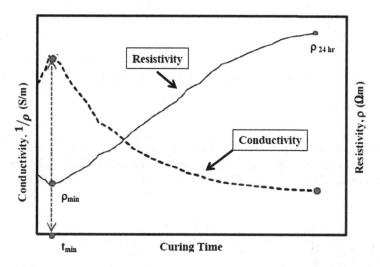

Figure 4.3 Typical changes in the resistivity during the hydration of cement with time.

quantification of electrical resistivity during the curing of cement, including the minimum resistivity and the corresponding time.

4.4.1 Vipulanandan Curing Model

It has been observed that the relationship between resistivity (ρ) and curing time (t) for various types of cement slurries followed similar trends, as shown in Figure 4.3. Also, note that the inverse of electrical resistivity is conductivity ($1/\rho$). Based on experimental results, the following conditions were observed:

(i) Initial condition; *when* t = 0 $\dfrac{1}{\rho} = \dfrac{1}{\rho_o}$ (4.14a)

(ii) when $0 < t < t_{min}$

$$\dfrac{d\left(\dfrac{1}{\rho}\right)}{dt} > 0 \quad \text{and} \quad \dfrac{d^2\left(\dfrac{1}{\rho}\right)}{dt^2} < 0 \qquad (4.14b)$$

(iii) *When* $t = t_{min}$ $\dfrac{1}{\rho} = \dfrac{1}{\rho_{min}}$ (4.14c)

$$\frac{d\left(\dfrac{1}{\rho}\right)}{dt} = 0 \tag{4.14d}$$

(iv) When $t > t_{min}$
$$\frac{d\left(\dfrac{1}{\rho}\right)}{dt} < 0 \quad \frac{d^2\left(\dfrac{1}{\rho}\right)}{dt^2} > 0 \tag{4.14e}$$

Hence, the model proposed by Vipulanandan et al. (1990) was modified and used to predict the electrical resistivity of cement during hydration the entire curing. The proposed Vipulanandan curing model is as follows:

$$\frac{1}{\rho} = \left(\frac{1}{\rho_{min}}\right) \left[\frac{\left(\dfrac{t + t_o}{t_{min} + t_o}\right)}{q_1 + (1 - p_1 - q_1) * \left(\dfrac{t + t_o}{t_{min} + t_o}\right) + p_1 * \left(\dfrac{t + t_o}{t_{min} + t_o}\right)^{\frac{q_1 + p_1}{p1}}} \right] \tag{4.15}$$

where ρ is the electrical resistivity (Ω-m); ρ_{min} is the minimum electrical resistivity (Ω-m); t_{min} is the time corresponding to minimum electrical resistivity (ρ_{min}); parameters $p_1(t)$, t_o, and q_1 (t) are model parameters; and t is the curing time (min). The parameter q_1 represents the initial rate of change in resistivity, and parameter p_1 influences the prediction of the changes in resistivity with time. Also, the parameter ratio q_1/p_1 influences the long-term prediction of the resistivity and also the type of additives used in the cement. The model will also predict the initial resistivity (ρ_0) when the time t = 0.

4.4.2 Initial Resistivity (ρ_0)

It is important to develop methods to characterise the mixing of cement with water and various additives. This monitoring method also can help with the quality control of the mixing of cement in the field. At present there is no method to determine the quality control of mixing since there are many variables influencing the mixing. In this study, the sensitivity of resistivity, a material property, in quantifying the mixing of cement was investigated. The initial resistivity of cement slurry will be affected by not only the type of cement and also the ratio but also the method of mixing and the composition of the mixture with various additives. It is important to test and quantify the initial resistivity of cement with temperature, w/c ratio during mixing, and also various additives.

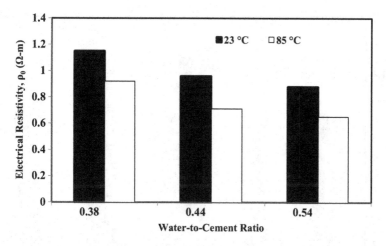

Figure 4.4 Effects of temperature and water-to-cement (w/c) ratio on the initial resistivity of the smart cement.

(a) Temperature and Water-to-Cement Ratio

The initial resistivity of smart cement is sensitive to the temperature of mixing and also to the w/c ratio. The initial resistivity of smart cement reduced with an increase in temperature for all the w/c ratios tested, as shown in Figure 4.4. For a w/c ratio of 0.38, when the mixing temperature was increased from 23°C to 85°C, the resistivity reduced from about 1.15 Ω.m to 0.92 Ω.m, a -20% reduction. Also, the initial resistivity reduced with the increase in the w/c ratio, as shown in Figure 4.4. The initial resistivity reduced from 1.15 Ω.m to 0.88 Ω.m when the w/c ratio was increased from 0.38 to 0.54, over a -23% reduction. This clearly indicates the sensitivity of resistivity initial mixing conditions. Also, the total and the changes in resistivity can be related to the temperature and the w/c ratio using the nonlinear model (NLM) (Joseph et al. 2010; Demircan et al. 2011).

(b) Effect of Additives

The effect of modifying smart cement with different additives on the initial resistivity was investigated by partially replacing the cement with fly ash (FA) and metakaolin (MK). Class H cement with a w/c ratio of 0.38 was chosen as the base slurry, and the percentage of these additives was reported based on the weight of cement (BWOC). The average initial resistivity of the cement was 1.12 Ωm. As shown in Figure 4.5 and summarised in Table 4.2, replacing oil-well cement with 5% fly ash (FA) decreased the resistivity to 1.05 Ω.m, and 50% fly ash (FA) increased it to 1.53 Ω.m, which represented a -7% reduction and a 36.6% increment based on neat smart cement, respectively. By modifying smart cement with 5% metakaolin (MK), the

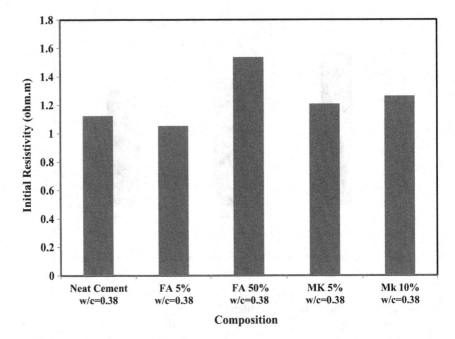

Figure 4.5 Comparing the initial resistivity of the smart cement with various additives.

Table 4.2 Initial resistivity of smart cement with additives

Mix Proportions		ρinitial (Ω-m)
W/C ratio = 0.38	Smart Cement (C)	1.12
	C+5%FA	1.05
	C+50%FA	1.53
	C+5%MK	1.21
	C+10%MK	1.26

initial resistivity was 1.21 Ω.m, representing an 8% increase in resistivity, as compared to the base OWC slurry. The initial resistivity increased by 12% to 1.26 Ω.m when 10% metakaolin was mixed with oil-well cement Class H. The initial resistivity was sensitive to the type and amount of additives.

4.5 CURING

In order to quantify the changes in resistivity with curing time, three types of cements, Portland Cement (Type I), Class G, and Class H, were investigated

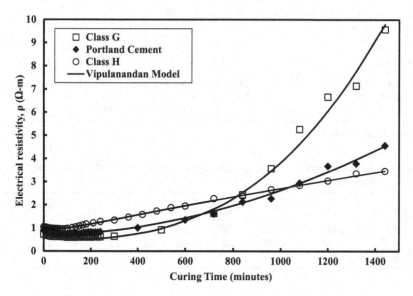

Figure 4.6 Variation of electrical resistivity of the Portland cement and oil-well smart cements with curing time up to 1 day of curing.

with a w/c ratio of 0.38. The samples were cured under room condition, 23°C and 50% relative humidity. Test results show the differences in resistivity during cement curing, as shown in Figure 4.6.

4.5.1 Resistivity Index (RI$_{24}$)

A new parameter has been defined to better quantify the maximum resistivity change in the first 24 hours. The resistivity index (RI$_{24}$) is represented as follows:

$$RI_{24} = \frac{\rho_{24} - \rho_{min}}{\rho_{min}} \, x100\% \tag{4.16}$$

It has been proven that the RI$_{24}$, obtained in one day, can be used to predict the long-term properties of smart cement.

(a) Portland Cement (Type 1)

The initial resistivity was 0.90 Ωm, the lowest of the three cements tested (Table 4.3 and Figure 4.6). The time taken to reach the minimum resistivity was 150 minutes, the highest of the three cements tested. The initial setting time using the Vicat needle (ASTM C-191) was 270 minutes, and it

Table 4.3 Initial resistivity of smart cement with additives

| Cement Type | Setting time (min.)* | | Resistivity, ρ (Ωm) | | Time (min) |
	Initial	Final	Initial	Minimum	t_{min}
Portland (Type I)	270 ± 4	305 ± 5	0.90 ± 0.02	0.75 ± 0.03	150 ± 4
Class G	310 ± 4	415 ± 5	0.73 ± 0.04	0.59 ± 0.04	120 ± 4
Class H	330 ± 4	420 ± 5	1.05 ± 0.03	0.96 ± 0.03	80 ± 4

Note: *Vicat Needle Test.

Table 4.4 Resistivity index and curing model parameters for Portland, Class G, and Class H smart cements

Cement Type	ρ_{24h} (Ωm)	RI_{24} (%)	t_0 (min.)	p_1	q_1	Ratio q_1/p_1
Portland (Type I)	4.54 ± 0.02	505	100 ± 4.5	0.15	0.099	0.67
Class G	9.57 ± 0.02	1,522	170 ± 3.5	0.16	0.112	0.70
Class H	3.46 ± 0.02	260	98 ± 5.0	0.65	0.281	0.43

was 120 minutes (2 hours) after reaching the minimum resistivity. The minimum resistivity was 0.75 Ωm, and the resistivity reduced by 17% in 150 minutes. The final setting time (Vicat needle) was 305 minutes, 35 minutes after the initial setting time, and it was the fastest setting cement. The resistivity after 24 hours of curing was 4.54 Ωm, a change of over 400% in 24 hours, a clear indication of the sensitivity of the resistivity to curing, as shown in Figure 4.6. The parameter RI_{24} was 505%, as summarised in Table 4.4. Also, the Vipulanandan curing model parameters p_1, q_1, and t_o are summarised in Table 4.4.

(b) Class G Cement

The initial resistivity of class G cement mixed with a w/c ratio of 0.38 was 0.73 Ωm, the lowest of the three types of cement tested. The time taken to reach the minimum resistivity was 120 minutes. The initial setting time (Vicat needle) was 310 minutes, higher than Portland cement, as summarised in Table 4.3. The initial setting was 190 minutes (over three hours) after reaching the minimum resistivity. The minimum resistivity was 0.59 Ωm, the lowest of the three cements tested, and the resistivity reduced by 19% in 310 minutes. The final setting time was 415 minutes, 105 minutes after the initial setting time. The resistivity after 24 hours of curing was 9.57 Ωm, change of over 1,200% in 24 hours, a clear indication of the sensitivity of the resistivity to curing, as shown in Figure 4.6. The parameter RI_{24} was 1,522%, the highest value for the three cements investigated, as summarised in Table 4.4. Also, the Vipulanandan curing model parameters p_1, q_1, and t_o

are summarised in Table 4.4. The curing parameter ratio (q1/p1) was 0.70, the highest for the three cements tested.

(c) Class H Cement

The initial resistivity of Class H cement mixed with a w/c ratio of 0.38 was 1.05 Ωm, the highest of the three cements tested. The percentage difference in the initial resistivity between class G cement and Class H cement was 44%, a clear indication of the sensitivity of the measuring parameter resistivity. The time taken to reach the minimum resistivity was 80 minutes, the lowest of the three cements tested. The initial setting time (Vicat needle) was 330 minutes, the highest of the three cements tested (Table 4.3), and it was 250 minutes (over four hours) after reaching the minimum resistivity. The minimum resistivity was 0.96 Ωm and the resistivity reduced by 8.6% in 330 minutes. The final setting time was 420 minutes, 90 minutes after the initial setting time, and it had the longest final setting time. The resistivity after 24 hours of curing was 3.46 Ωm, a change of 230% in 24 hours, and a clear indication of the sensitivity of the resistivity to curing. The parameter RI_{24} was 260%, the lowest value for the three cements investigated, as summarised in Table 4.4. Also, the Vipulanandan curing model parameters p_1, q_1, and t_o are summarised in Table 4.4. The curing parameter ratio (q1/p1) was 0.43, the lowest for the three cements tested.

There was no direct correlation between the setting times and the electrical resistivity parameters, summarised in Table 4.3.

4.5.2 Effect of Water-to-Cement Ratios

In this study, the API method of mixing was used to mix Class H cement with varying amounts of water. The w/c ratios investigated were 0.38, 0.44, and 0.54. The cement initial resistivity was measured after mixing. The test results are summarised in Table 4.5 and also shown in Figure 4.7, and the initial resistivity decreased with an increased w/c ratio.

(a) w/c = 0.38

The unit weight of smart cement with a w/c of 0.38 was 1.98 g/cc (16.48 ppg). The initial electrical resistivity (ρ_o) of smart cement with a w/c ratio of 0.38 modified with 0.1% CF was 1.15 Ω-m, and the electrical resistivity reduced to reach the ρ_{min} of 1.02 Ω-m after 78 minutes (t_{min}), as summarised in Table 4.6 and shown in Figure 4.7a. The 24-hour electrical resistivity (ρ_{24hr}) of the cement was 4.15 Ω.m. Hence, the maximum change in electrical resistivity after 24 hours (RI_{24hr}) was 306%, as summarised in Table 4.6. The seven-day electrical resistivity (ρ_{7days}) of smart cement was 7.75 Ω.m; hence, the maximum change in electrical resistivity after seven days (RI_{7days}) was 683%.

Table 4.5 Density and initial resistivity of smart cement with varying water-to-cement ratios

Water-to-Cement Ratio	Density (g/cc)	Initial Resistivity (Ωm)
0.38	1.98 (16.48 ppg)	1.15
0.44	1.93 (16.12 ppg)	1.00
0.54	1.89 (15.78 ppg)	0.90

Note: ppg – pounds per gallon.

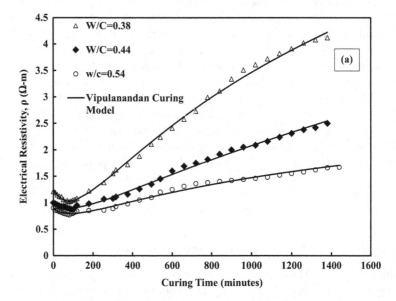

Figure 4.7a Curing electrical resistivity development for the smart cement with various W/C ratios (a) 1 day, (b) 7 days.

Table 4.6 Resistivity indices of smart cement with varying water-to-cement ratios

				1 Day and 7 Days			
w/c ratio	Initial Resistivity, ρ_o (Ω-m)	ρ_{min} (Ω-m)	t_{min} (min)	ρ_{24h} (Ω-m)	ρ_{7days} (Ωm)	RI_{24} (%)	RI_{7days} (%)
0.38	1.15	1.02	78	4.15	7.75	306	683
0.44	1.00	0.87	90	2.55	5.00	193	462
0.54	0.90	0.77	102	1.67	4.60	117	490

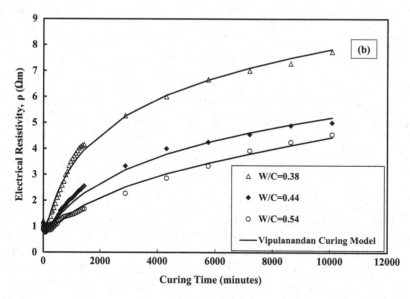

Figure 4.7b

The curing model parameters q_1 and p_1 for one day and seven days are summarised in Table 4.6 and were dependent on the curing time. The parameter ratio (q_1/p_1) decreased with an increase in curing time. The Vipulanandan curing model predicted the measured resistivity very well, as shown in Figure 4.7. The coefficient of determination (R^2) varied from 0.98 to 0.99, and the root-mean-square error (RMSE) varied from 0.034 1Ω.m to 0.158 Ω.m for one day and seven days of curing respectively.

(b) w/c = 0.44

The unit weight of smart cement with a w/c of 0.44 was 1.93 g/cc (16.12 ppg). The initial electrical resistivity (ρ_o) of smart cement with a w/c ratio of 0.44 and modified with 0.1% CF was 1.00 Ω-m. The electrical resistivity reduced to reach the ρ_{min} of 0.87 Ω-m after 90 minutes (t_{min}), as summarised in Table 4.6. The 24-hour electrical resistivity (ρ_{24hr}) of the sample was 2.55 Ω.m. Hence, the maximum change in electrical resistivity after 24 hours (RI_{24hr}) was 193%. The seven-day electrical resistivity (ρ_{7days}) of the sample was 5.00 Ω.m; hence, the maximum change in electrical resistivity after seven days (RI_{7days}) was 475%.

The curing model parameters q_1 and p_1 for one day and seven days are summarised in Table 4.7 and were dependent on the curing time. The parameter ratio (q_1/p_1) decreased with an increase in curing time. The coefficient of determination (R^2) varied from 0.98 to 0.99, and the root-mean-square error (RMSE) varied from 0.036 Ω.m to 0.14 Ω.m for one day and seven days of curing respectively.

Table 4.7 Vipulanandan curing model parameters for smart cement with varying water-to-cement ratios

Curing Time					1 Day			7 Days
w/c ratio	t_o (min)	Parameter p_l	Parameter q_l	Parameter ratio (q_l/p_l)	Parameter p_l	Parameter q_l	Parameter ratio (q_l/p_l)	
0.38	85	0.506	0.493	0.974	0.601	0.398	0.662	
0.44	100	0.532	0.467	0.878	0.619	0.380	0.614	
0.54	115	0.661	0.338	0.511	0.628	0.330	0.525	

(c) w/c = 0.54

The unit weight of smart cement with a w/c of 0.38 was 1.89 g/cc (15.78 ppg). The initial electrical resistivity (ρ_o) of smart cement with a w/c ratio of 0.54 modified with 0.1% CF was 0.90 Ω-m (Table 4.7), and the electrical resistivity reduced to reach the ρ_{min} of 0.77 Ω-m after 102 minutes (t_{min}), as summarised in Table 4.6. The 24-hour electrical resistivity (ρ_{24hr}) of the sample was 1.67 Ω.m. Hence, the maximum change in electrical resistivity after 24 hours (RI_{24hr}) was 117%, as summarised in Table 4.6. The seven-day electrical resistivity (ρ_{7days}) of the sample was 4.6 Ω.m; hence, the maximum change in electrical resistivity after seven days (RI_{7days}) was 497%.

The curing model parameters q_1 and p_1 for one day and seven days are summarised in Table 4.7 and were dependent on the curing time. The parameter ratio (q_1/p_1) decreased with an increase in curing time. The coefficient of determination (R^2) varied from 0.97 to 0.99, and the root-mean-square error (RMSE) varied from 0.042 Ω.m to 0.08 Ω.m for one day and seven days of curing respectively.

Overall, with the increase in the w/c ratio, the density reduced and also the initial resistivity. Also, the curing model parameters were sensitive to the w/c ratio and the curing time.

Summary: The initial electrical resistivity (ρ_o) of smart cement decreased by 13% and 22% when the w/c ratio increased from 0.38 to 0.44 and 0.54 respectively, as summarised in Table 4.5. The minimum electrical resistivity (ρ_{min}) of smart cement also decreased by 15% and 25% when the w/c ratio was increased from 0.38 to 0.44 and 0.54 respectively, as summarised in Table 4.6. The time to reach the minimum electrical resistivity (t_{min}) increased by 15% and 31% when the w/c ratio increased from 0.38 to 0.44 and 0.54 respectively, as summarised in Table 4.6.

Also, the model parameter ratio (q_1/p_1) decreased with the w/c ratio. The resistivity parameters can be correlated to the w/c ratio and used in various applications, including verifying the quality control of the cement mixing and curing conditions.

Table 4.8 Resistivity changes with curing time for oil-well cement modified with fly ash and metakaolin

Mix Proportions		$\rho_{initial}$ (Ω-m)	ρ_{min} (Ω-m)	t_{min} (min)	ρ_{24} (Ω-m)	RI_{24} (%)
W/C ratio	C+5%FA	1.05	0.96	48	3.02	215%
0.38 to	C+50%FA	1.53	1.36	164	2.62	93%
0.40	C+5%MK	1.21	1.09	85	4.48	312%
	C+10%MK	1.26	1.17	95	5.45	366%

4.6 EFFECT OF ADDITIVES

(a) Fly Ash and Metakaolin

Resistivity with varying amounts of additives are tabulated in Table 4.8. The time to reach the minimum resistance was 48 minutes with 5% fly ash (FA) content. The time to reach the minimum resistivity was 164 minutes with 50% fly ash. The RI_{24} was 215% and 93% with 5% and 50% fly ash respectively. The time to reach the minimum resistance was 85 minutes with 5% metakaolin (MK), higher than with 5% FA. The time to reach the minimum resistivity was 95 minutes with 10% metakaolin. The RI_{24} was 312% and 366% with 5% and 10% metakaolin respectively. This clearly demonstrates the sensitivity of resistivity to the type of additive and also the amount of additive.

(b) Sodium Meta Silicate (SMS)

Sodium meta silicate was added to cement slurry in varying amounts based on the applications to modify the physical and mechanical properties of the cement slurry and also the long-term properties. In this study, SMS was first added to the water and then the cement slurry was prepared. Also, the percentage of SMS added to the water is based on the weight of water.

With the addition of SMS to the cement slurry, the initial resistivity and the minimum resistivity were reduced based on the concentration of SMS added to the cement slurry, as summarised in Table 4.9. Part of the reason for the reduction in the resistivity is due to the SMS in the pore solution. Also, with time SMS will react with cement and also the calcium hydroxide produced during the cement hydration. Due to these reactions, the time to reach the minimum resistivity also increased with the amount of SMS added, as summarised in Table 4.9. Also, adding SMS reduced the 24-hour resistivity and the RI_{24}.

4.7 LONG-TERM CURING

The changes in electrical resistivity with the curing time for smart cement with and without 1% SMS were tested for one year. The unit weight of

Table 4.9 Resistivity change with curing time for smart oil-well cement with different percentages of SMS

Mix Proportions		ρ_o (Ω-m)	ρ_{min} (Ω-m)	t_{min} (min)	ρ_{24} (Ω-m)	$RI_{(24)}$ (%)
W/C = 0.4	Cement (C)	1.00±0.02	0.90±0.01	87	3.23	259
	C + 0.2% SMS	0.90±0.01	0.80±0.02	130	2.70	238
	C + 1 % SMS	0.80±0.02	0.60±0.01	240	1.65	175

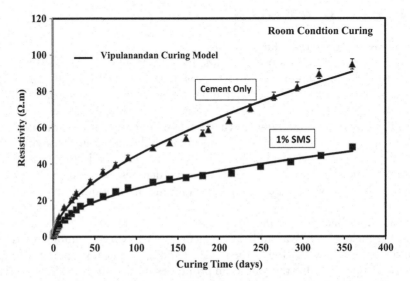

Figure 4.8 Variation of resistivity of the smart cement cured at room temperature with and without 1% sodium meta silicate (SMS) up to 12 months modeled with the Vipulanandan curing model.

smart cement with a w/c of 0.4 and 0.075% carbon fiber was 16.2±0.12 ppg (19.04 kN/m³). With the addition of 1% SMS, the unit weight of the smart cement increased to 16.4±0.10 ppg (19.27 kN/m³). In addition to measuring the resistance, the weight of the test specimens was also measured.

4.7.1 Room Condition

In this test, specimens were cured under the room condition. The room temperature was 23°C and the relative humidity was 50%. The test results are shown in Figures 4.8 and 4.9, and the resistivity and the weight loss were low with the addition of 1% SMS. Also, the curing model predicted the

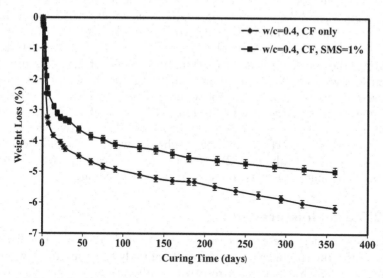

Figure 4.9 Percentage weight loss in the smart cement cured at room temperature with and without 1% sodium meta silicate (SMS) up to 12 months of curing.

resistivity with curing time very well, and the model parameter ratio (q_1/p_1) reduced with the addition of 1% SMS.

(a) Zero SMS

After one week of room condition curing, the resistivity increased to 9.98 Ω.m from the initial resistivity of 1.00 Ω.m, a 898% increase, and the weight loss was 3.44%. The weight loss was due to moisture loss in the test specimens, and the moisture content loss was about 5.73%. After 28 days of curing, the resistivity was 21.93 Ω.m, an increase of 2,093%, and the weight loss was 4.27%. In 60 days of curing, the resistivity was 35.84 Ω.m, an increase of 3,584%, and the weight loss was 4.68%. After 90 days of curing, the resistivity was 43.51 Ω.m, an increase of 4,251%, and the weight loss was 4.92%. After 180 days of curing, the resistivity was 56.88 Ω.m, an increase of 5,588%, and the weight loss was 5.31%. After 360 days of curing, the resistivity was 93.81 Ω.m, an increase of 9,281%, and the weight loss was 6.17% and the moisture content loss was about 10.28%. The test results clearly showed the sensitivity of changing resistivity to monitor the cement condition.

(b) 1% SMS

After one week of room condition curing, the resistivity increased to 7.33 Ω.m from the initial resistivity of 0.85 Ω.m, a 762% increase, and the weight loss was 2.48%. The weight loss was due to moisture loss in the test specimens, and the moisture content loss was about 4.13%, lower than the cement without SMS. After 28 days of curing, the resistivity was 14.62 Ω.m, an increase of 1,620%, and the weight loss was 3.32%. In 60 days of curing, the resistivity was 22.17 Ω.m, an increase of 2,508%, and the weight loss was 3.85%. After 90 days of curing, the resistivity was 25.18 Ω.m, an increase of 2,862%, and the weight loss was 4.12%. After 180 days of curing, the resistivity was 34.43 Ω.m, an increase of 3,951%, and the weight loss was 4.53%. After 360 days of curing, the resistivity was 46.86 Ω.m, an increase of 5,413%, and the weight loss was 4.99% and the moisture content loss was about 8.32%. The test results clearly showed the sensitivity of changing resistivity to monitor the cement condition.

4.7.2 Zero Moisture Loss

The test specimens were cured in containers with relative humidity of 100%. Also, the specimens were weighed regularly to ensure there was no weight change. The changes in resistivity with curing time are shown in Figure 4.10. After one week of curing, the resistivity increased to 6.82 Ω.m from the initial resistivity of 1.00 Ω.m, a 582% increase, and was 68.3% of the room-cured cement, with the difference being due to the moisture loss of 5.73%. After 28 days of curing, the resistivity was 12.03 Ω.m, an increase

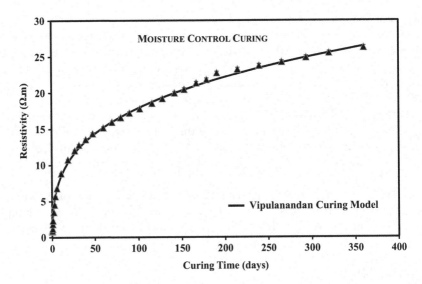

Figure 4.10 Variation of resistivity of the curing smart cement with the curing time up to 12 months with no moisture loss and modeled with the Vipulanandan curing model.

of 1,103%. In 60 days of curing, the resistivity was 15.19 Ω.m, an increase of 1,419%, and was 42.4% of the room-cured cement due to higher moisture loss. After 90 days of curing, the resistivity was 17.23 Ω.m, an increase of 1,623%. After 180 days of curing, the resistivity was 21.81 Ω.m, an increase of 2,081%. After 360 days of curing, the resistivity was 26.27 Ω.m, an increase of 2,527%, and was 27.7% of the room-cured cement, with the difference being due to the moisture loss of about 10.28%. The test results clearly showed the sensitivity of changing resistivity to monitor the cement condition.

4.7.3 Underwater Curing

The change in resistivity with curing time is shown in Figure 4.11, and the weight gain with time is shown in Figure 4.12. After one week of room-condition curing, the resistivity increased to 5.24 Ω.m from the initial resistivity of 1.00 Ω.m, a 424% increase, and the weight gain was 0.84%. The weight gain was due to moisture gain in the test specimens, and the moisture content increase was about 1.40%. After 28 days of curing, the resistivity was 8.14 Ω.m, an increase of 714%, and the weight gain was 1.38%. The weight gain was due to moisture gain in the test specimens, and the moisture content increase was about 2.30%. After 60 days of curing, the resistivity was 11.84 Ω.m, an increase of 1,084%, and the weight gain was

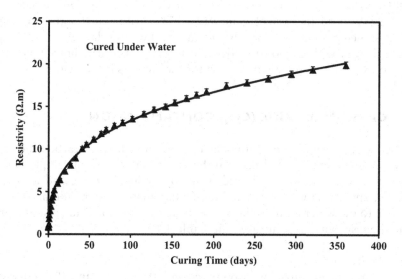

Figure 4.11 Variation of resistivity of curing cement specimen with time up to 12 months for specimens cured under water modeled with the Vipulanandan curing model.

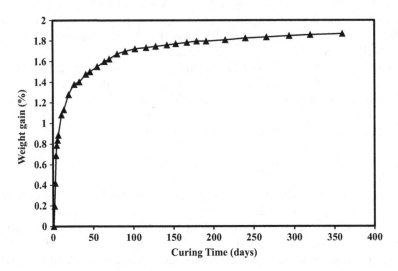

Figure 4.12 Percentage weight gain in the smart cement with the curing time up to 12 months for specimens cured under water.

1.60%. After 90 days of curing, the resistivity was 13.13 Ω.m, an increase of 1,213%, and the weight gain was 1.70%. After 180 days of curing, the resistivity was 16.34 Ω.m, an increase of 1,534%, and the weight gain was 1.80%. After 360 days of curing, the resistivity was 19.93 Ω.m, an increase of 1,893%, and the weight gain was 1.87%, with the moisture content increase being about 3.12%. The resistivity after 360 days was 75.9% of the resistivity with no moisture loss. The test results clearly showed the sensitivity of changing resistivity to monitor the changes in the cement.

4.8 CARBON DIOXIDE (CO₂) CONTAMINATION

There is increasing interest in understanding the effects of carbon dioxide (CO_2) contamination of cement. In this study, smart cement samples were prepared with CO_2-contaminated water by adding dry ice, and also the cement specimens were cured in CO_2-contaminated water. Adding 3% of dry ice to the water reduced the temperature by about 3°C and also the pH and the resistivity, as summarised in Table 4.10.

Sample Preparation
The smart cement slurries were prepared with 0.1%, 1%, and 3% of dry ice (CO_2) in water. The test specimens were prepared following API standards. API Class H cement was used with a water-cement ratio of 0.38. For all the

Table 4.10 Water with and without CO_2 contamination

Solutions	pH	$\rho(\Omega m)$
Pure Water	7.4	25.1
3% CO_2 in Water	4.2	22.3

samples, 0.04% (based on weight of cement) of carbon fiber (CF) was added to the slurry in order to enhance the piezoresistivity of the cement and to make it more sensing. After mixing, the slurries were casted into cylindrical molds with a height of 100 mm and a diameter of 50 mm, with two conductive wires embedded 50 mm apart vertically to monitor the resistivity development of the specimens during the curing time. After one day, all the specimens were demolded and were cured for 28 days under water.

4.8.1 Results and Discussion

(a) Density
The average density of smart cement was 1.95 g/cc (16.28 ppg). With 0.1% of CO_2-contaminated water, the density reduced by 0.06%. With 1% of CO_2-contaminated water, the density was reduced by 0.49% to 1.94 g/cc (16.20 ppg), and with 3% of CO_2-contaminated water, it reduced to 1.93 g/cc (16.14 ppg), a 0.86% reduction.

(b) Electrical Resistivity

Initial Resistivity
The initial resistivity of smart cement slurries with varying CO_2 (dry ice) concentrations was investigated.

(a) Smart Cement: The average initial resistivity of the cement slurry was 1.10 Ω.m.
(b) Smart Cement with CO_2 Contamination: Smart cement with 0.1%, 1%, and 3% of CO_2 contamination resulted in a reduction in the initial resistivity to 1.03 Ω.m, 0.93 Ω.m, and 0.90 Ω.m respectively, as shown in Figure 4.13. Hence, CO_2 contamination with concentrations of 0.1%, 1%, and 3% resulted in a resistivity reduction of 6%, 15%, and 18% respectively. The main reason for the reduction in the initial electrical resistivity of the contaminated cement slurries was due to the existence of carbonic acid (H_2CO_3) in the slurries.

4.8.2 Curing

During the initial period of curing, resistivity will reduce with time. Also, the time to reach the minimum resistivity is also a good monitoring parameter,

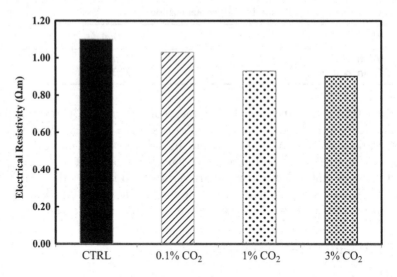

Figure 4.13 Initial electrical resistivity of the smart cement slurries contaminated with different CO_2 concentrations.

and it is important to quantify the sensitivity of these parameters due to CO_2 contamination.

(a) **Smart Cement:** The minimum resistivity of the smart cement slurry was 0.85 Ω.m and was reached in 85 minutes (t_{min}) after mixing the sample.

(b) **CO_2-Contaminated Smart Cement:** CO_2 contamination decreased the ρ_{min} of the smart cement slurry by 7%, 15%, and 17% from 0.85 Ω.m to 0.79 Ω.m, 0.72 Ω.m, and 0.70 Ω.m respectively with 0.1%, 1%, and 3% of CO_2 contamination. CO_2 exposure also delayed the hydration process. With 0.1%, 1%, and 3% of CO_2 contamination, it delayed t_{min} by 15 minutes, 35 minutes, and 45 minutes respectively.

One Day
The test results are shown in Figure 4.14, and the Vipulanandan curing model predicted the test results very well. The model parameter ratios (q_1/ p_1) reduced with the CO_2 contamination.

(a) **Smart Cement:** After one day of curing, the smart cement resistivity was 4.8 Ω.m and the resistivity index was 465%

(b) **CO_2-Contaminated Smart Cement:** CO_2 contamination reduced the development of resistivity during the one day of curing. The

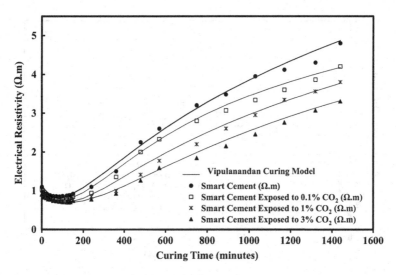

Figure 4.14 Development of electrical resistivity of the smart cement contaminated with different CO_2 concentrations up to 1 day of curing.

Table 4.11 Electrical resistivity parameters of the smart cement slurries exposed to different CO_2 concentrations

Smart Cement	ρ_0 $(\Omega.m)$	ρ_{min} $(\Omega.m)$	t_{min} (minute)	ρ_{24} $(\Omega.m)$	$\frac{\rho_{24} - \rho_{min}}{\rho_{min}}$%
Uncontaminated cement	1.10	0.85	85	4.80	465%
0.1% CO$_2$-Contaminated Smart Cement	1.03	0.79	100	4.20	432%
1% CO$_2$-Contaminated Smart Cement	0.93	0.72	120	3.80	428%
3% CO$_2$-Contaminated Smart Cement	0.90	0.70	130	3.30	371%

0.1%, 1%, and 3% of CO_2-contaminated resistivity after one day of curing were 4.20 $\Omega.m$, 3.80 $\Omega.m$, and 3.30 $\Omega.m$ respectively. Also, the contamination reduced the resistivity indices, as summarised in Table 4.11

28 Days

The test results are shown in Figure 4.15, and the Vipulanandan curing model predicted the test results very well. The model parameter ratios (q_1/ p_1) reduced with the CO_2 contamination.

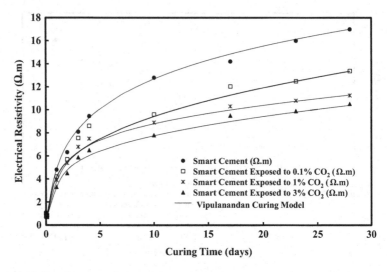

Figure 4.15 Development of electrical resistivity of the smart cement contaminated with different CO_2 concentration up to 28 days of curing.

(a) **Smart Cement:** After 28 days of curing, the smart cement resistivity was 17.0 Ω.m.

(b) **CO_2-Contaminated Smart Cement:** CO_2 contamination reduced the development of resistivity during the 28 days of curing. The 0.1%, 1%, and 3% of CO_2-contaminated cement reduced the resistivity of the cement by 21%, 34%, and 38% to 13.4 Ω.m, 11.3 Ω.m, and 10.5 Ω.m respectively after 28 days of curing.

4.9 THERMAL CONDUCTIVITY

A commercially available thermal properties analyzer, which complied with the ASTM D5334-08, was used. It also had a controller that recorded the data, as shown in Figure 4.16 (a). It had three different probes, as shown in Figure 4.16 (b). Two of the probes (KS-1 and TR-1) were single-needle sensors that are used to measure thermal conductivity and resistivity. The probe SH-1 was a dual-needle sensor used to measure specific heat and diffusivity.

The operating concept of the instrument is based on a hot wire where the thermal conductivity was calculated by monitoring the heat dissipation from a linear heat source at a given voltage (ASTM D 5334).

The probe consisted of a heating wire (60 mm in length and 1.28 mm in diameter) and a thermistor (a resistor whose resistance is dependent on

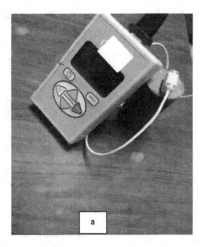

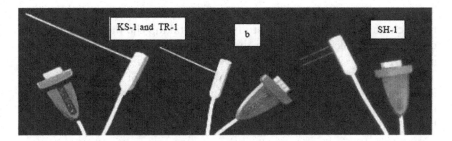

Figure 4.16 (a) and (b)   Commercially available thermal conductivity measuring device (a) controller, (b) three different probes.

the temperature) in the middle of the wire. First, the controller heats the probe for 30 seconds and then calculates the thermal properties. For solid materials, two holes 1.3 mm in diameter and 30 mm in depth were drilled in the middle of the specimen for providing better thermal contact between the sample and the probe. Also, the probe was coated with a thin gray layer of thermal grease. Figure 4.17 shows the setup for thermal conductivity test.

The device produced constant current and recorded the voltage and current to an accuracy of 0.01 V and 0.01 amperes. The probe SH-1 can measure the thermal capacity and thermal diffusivity, and is mostly used for solids. Sensors operate in the temperature range of -50°C to +150°C.

In this study, the KS-1 probe was used for measuring the cement slurry and the SH-1 probe was used for the hardened cement. Before taking the measurements, the probe was calibrated using the specific calibration material provided by the manufacturer. There were different calibration materials

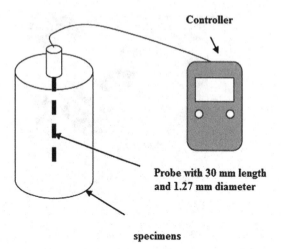

Figure 4.17 Test setup for thermal conductivity measurements.

for each of the probes. In this study, before taking the measurements, the device was double-checked with distilled water, and the results are shown in Figure 4.18.

As shown in Figure 4.18, the thermal conductivity of distilled water was in the range of 0.6 Watts/(meter. Kelvin) and it agreed with results in the literature. The initial readings had some fluctuation until the temperature became constant.

4.9.1 Cement

There is very limited data available in the literature on the thermal properties of cements. Thermal conductivity is responsible for temperature distribution inside the cement sheath. The changes in thermal conductivity of Class H cement are shown in Figure 4.19. Thermal conductivity decreased from 0.87 to 0.68 W/m.K after 50 days of curing under room condition, a 21.8% reduction. The changes in thermal conductivity for modified cements with metakaolin and fly ash are shown in Figure 4.20 and Figure 4.21 respectively.

4.9.2 Additives

(a) Metakaoline
As shown in Figure 4.20, partial replacement of cement with 10% metakaolin had a minor effect on the thermal conductivity of cement, and the initial thermal conductivity increased to 0.89 W/m.K. With time it reduced below

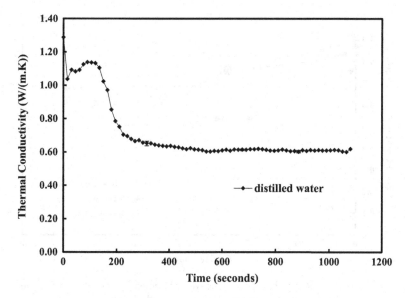

Figure 4.18 Thermal conductivity of the distilled water.

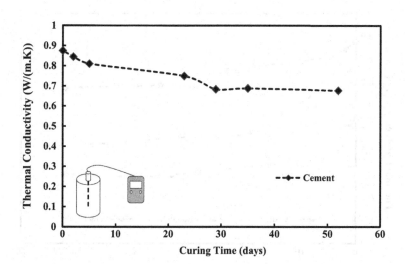

Figure 4.19 Variation of the thermal conductivity of Class H cement (water-to-cement ratio = 0.38) with the curing time.

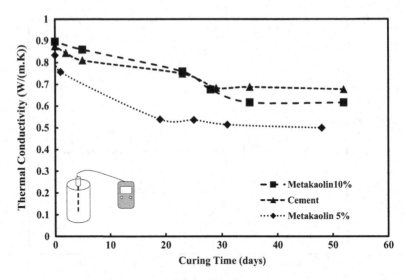

Figure 4.20 Variation of the thermal conductivity of cement modified with 5% and 10% metakaolin (water-to-cement ratio = 0.4) with the curing time.

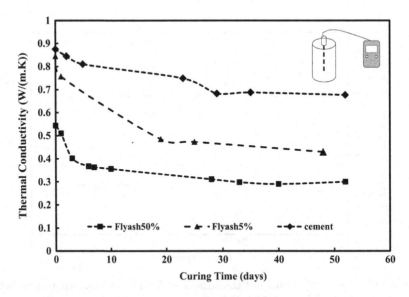

Figure 4.21 Variation of the thermal conductivity of modified cement with 50% fly ash (water-to-cement ratio = 0.38) with the curing time.

Table 4.12 Thermal conductivity of Class H oil-well cement with additives

Cement Type	Initially ($\frac{W}{m.K}$)	After 50 Days ($\frac{W}{m.K}$)
Class H	0.87	0.68
5% Metakaolin	0.83	0.50
5% Fly ash	0.84	0.43
10% Metakaolin	0.89	0.61
50% Fly ash	0.54	0.30

the cement, and after 50 days it was 0.62 W/m.K. With a 5% addition of metakaolin, the initial thermal conductivity was 0.84 W/m.K, and after 50 days of curing it reduced to 0.50 W/mK, as summarised in Table 4.12.

(b) Fly Ash
As shown in Figure 4.21, cement modifying with 5% and 50% fly ash decreased the thermal conductivity of Class H cement. The thermal conductivity of cement with 50% fly ash was 0.54 W/m.K initially and then decreased to 0.3 W/m.K after 50 days of curing (Table 4.12). Increasing fly ash content reduced the thermal conductivity of cement, which could be used in cement to make it a thermally isolating material.

The effect of the w/c ratio on the thermal conductivity of cement was also investigated. As shown in Figure 4.22, when the w/c ratio (w/c) was increased, thermal conductivity decreased for specimens cured up to 50 days. Also, adding carbon fibers up to 0.5% slightly increased the thermal conductivity of cement cured for three days, as shown in Figure 4.23.

4.10 SUMMARY

Based on the experimental studies on curing resistivity and the thermal properties of various types of cements with and without additives and also the modeling of the curing behavior, the following conclusions are advanced:

1 Resistivity was a highly sensing parameter to monitor the mixing of the cement with various additives and also the curing of the cement, and the experimental results clearly demonstrated this. Differences in the initial resistivity clearly indicated the addition of the amount and type of additives and also the w/c ratios. Initial resistivity can be used as a quality control measure in the field.

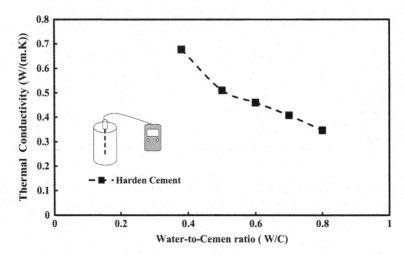

Figure 4.22 The effect of water-to-cement ratio (w/c) on the thermal conductivity of Class H cement cured for 50 days.

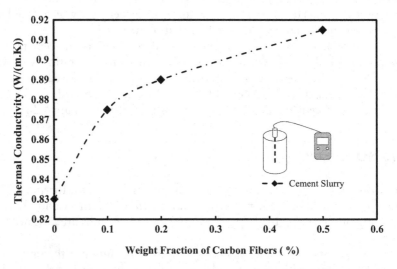

Figure 4.23 The effect of carbon fiber content on the thermal conductivity of Class H cement cured for 3 days.

2 Cement slurry is a three-phase material. Its curing using the calorimeter has been investigated and also compared to Vicat needle test results.

3 Also during the initial 24 hours of curing, several parameters such as the minimum resistivity (ρ_{min}), time to reach minimum resistivity (t_{min}), resistivity after 24 hours of curing (ρ_{24}), and the resistivity index (RI_{24}) can be used to characterise the cement.

4 The continuous changes in the resistivity of curing cement under different conditions were measured and modeled using the new Vipulanandan curing model up to one year. The model also captured all the important initial 24-hour curing parameters. The resistivity changes were highly sensitive to the curing environment of the cement. Also, the effects of additives, including carbon dioxide (CO_2) contamination, on the curing characteristics of cement have been measured and quantified using the curing model.

5 The Vipulanandan curing model predicted the resistivity changes with the curing time very well, based on the root-mean-square error (RMSE).

6 The thermal conductivity of cement with and without additives has been quantified. The thermal conductivity decreased with the increase in the curing time. Also, increasing the ratio decreased the thermal conductivity of smart cement cured under room condition.

7 The electrical resistivity increased with the curing time after reaching the minimum resistivity, and the percentage changes were much higher than the changes in the thermal conductivity.

Chapter 5

Piezoresistive Smart Cement

5.1 BACKGROUND

It is important to develop innovative methods to make cement highly sensing without any buried sensors in it. Also, highly sensing monitoring parameters have to be identified that can be easily adopted in the field. Past studies have investigated changes in electrical resistivity with applied stress, referred to as the piezoresistive behavior, of modified cement-based and polymer composites (Sett et al. 2004; Han et al. 2007; Vipulanandan et al. 2008; Pakeerathan et al. 2012; Mangadlao et al. 2015). These studies have shown that changes in resistivity with applied stress were 30 to 50 times higher than the strain in the materials. Hence, the change in resistivity has the potential to be used to determine the integrity of materials. Past studies have also reported that interfacial factors are important in obtaining electrical resistivity from electrical resistance (Chung 2001). Due to the voltage present during electrical resistance measurement, electric polarisation occurs as the resistance measurement is made continuously. Polarisation results in an increase in the measured resistance. The conventional methods of measuring the electrical resistivity of cementitious materials can be categorised into direct-current (DC) methods and alternating-current (AC) methods, both of which require electrodes for their measurement. Therefore, there is the potential for contact problems between the electrodes and the matrix, which can completely affect the accuracy of the measurement. Recent studies have suggested that replacing the DC measurement with the AC measurement can eliminate the polarisation effect (Zhang et al. 2010, Vipulanandan et al. 2013a, 2014a–2016e).

The compressive stress-strain behavior of strain-softening materials such as concrete, glass fiber-reinforced polymer concrete, fine sands grouted with sodium silicate grout, and cement mortar have been predicted using the Vipulanandan stress-strain p-q model (Mebarkia et al. 1992; Vipulanandan et al. 1990–2020). Also, the stress-strain behavior of Portland cement-stabilised sand has been modeled using the Vipulanandan p-q stress-strain model (Usluogullari et al. 2011). As well, the Vipulanandan p-q stress-strain model was used to predict the compressive stress-strain behavior of sulfate-contaminated clay soils treated with polymer and lime (Mohammed et al. 2013 and 2015 and Vipulanandan et al. 2020c). In addition, there is very limited information on the tensile testing and modeling of cement-based materials (Vipulanandan et al. 1985 and 1987)

There is very limited information in the literature about the mechanical properties and also the constitutive stress-strain modeling of cements with various additives. The overall objective was to characterise the mechanical behavior of smart cement under different loading conditions (compression, direct tension, splitting tension, and bending) and model the behavior and also compare the responses to classical cements.

5.2 MODELS

5.2.1 Vipulanandan p-q Stress-Strain Model

Cement compressive, tensile, and bending stress (σ)-strain (ε) behaviors are nonlinear and also strain softening after the peak stress (σ_f), and hence the stress-strain model has to satisfy the following conditions:

(i)   For $\sigma \leq \sigma_f$

$$\frac{d\sigma}{d\varepsilon} > 0 \tag{5.1}$$

$$\frac{d^2\sigma}{d\varepsilon^2} < 0 \tag{5.2}$$

(ii) At peak stress, when $\sigma = \sigma_f$

$$\frac{d\sigma}{d\varepsilon} = 0 \tag{5.3}$$

For $\sigma > \sigma_f$

$$\frac{d\sigma}{d\varepsilon} < 0 \tag{5.4}$$

Based on how the strength is lost with the increase in strain (low or high), it can be as follows:

$$\frac{d^2\sigma}{d\varepsilon^2} \geq 0 \tag{5.5a}$$

for a lower strength drop with an increase in strain and for a higher strength drop with an increase in strain it will be as follows:

$$\frac{d^2\sigma}{d\varepsilon^2} \leq 0 \tag{5.5b}$$

In order to satisfy the above five conditions, the following stress-strain relationship was developed (Vipulanandan and Paul 1990; Mebarkia et al. 1992) and the updated relationship is as follows:

$$\sigma = \left[\frac{\dfrac{\varepsilon}{\varepsilon_f}}{q_o + (1 - p_o - q_o)\dfrac{\varepsilon}{\varepsilon_f} + p\left(\dfrac{\varepsilon}{\varepsilon_f}\right)^{\frac{(p_o+q_o)}{p_o}}} \right] * \sigma_f \tag{5.6}$$

where σ is compressive/tension/bending stress; and σ_f, and ε_f are the peak stress and the corresponding strain. The model material parameters p_o and q_o are related to material properties such as composition, density, resistivity and porosity, and also the mixing process, curing time, and environmental conditions (temperature, pressure, and relative humidity).

5.2.2 Vipulanandan p-q Piezoresistivity Models

(a) Resistive Strain Softening: The addition of less than 0.1% conductive filler (CF) substantially improved the nonlinear stress (σ)-resistivity strain response of smart cement, satisfying the conditions related to piezoresistive axial strain ($\Delta\rho/\rho$) instead of the strain (ε) in Eqns. 5.1 to 5.5. From the experimental results, a constitutive model developed by Vipulanandan et al. (1990) was used to predict the changes in the piezoresistive axial strain of smart cement with applied stress. The new piezoresistive stress-strain model is as follows:

$$\frac{\sigma}{\sigma_f} = \left[\frac{\left(\dfrac{x}{x_f}\right)}{q_2 + (1 - p_2 - q_2)*\left(\dfrac{x}{x_f}\right) + p_2 * \left(\dfrac{x}{x_f}\right)^{\frac{p_2+q_2}{p_2}}} \right] \tag{5.7}$$

where ρ is the resistivity and σ_f is the strength (MPa); $x = \left(\dfrac{\Delta\rho}{\rho_o}\right) * 100 =$ percentage of change in resistive strain due to the stress, which is the piezoresistive axial strain; $x_f = \left(\dfrac{\Delta\rho}{\rho_o}\right)_f * 100 =$ percentage of piezoresistive axial strain at failure; and $\Delta\rho$ is the change in the ρ. The initial electrical resistivity (ρ_o) (at σ=0 MPa) and the model material parameters p2, q_2 and $\dfrac{q_2}{p_2}$ are related to the material properties and testing environments (temperature, pressure, relative humidity).

(b) **Resistive Strain Hardening:** If the material is resistivity strain hardening, the conditions is modified as follows:

For $\sigma \leq \sigma_f$

$$\frac{d\sigma}{dx} > 0 \tag{5.8}$$

$$\frac{d^2\sigma}{dx^2} > 0 \tag{5.9}$$

The new piezoresistive model is represented as follows:

$$\sigma = \left[\frac{\dfrac{x}{x_f}}{q_2 + \left(1 - p_2 - q_2\right)\dfrac{x}{x_f} + p_2\left(\dfrac{x}{xf}\right)^{\left(\frac{p_2-q_2}{p_2}\right)}} \right] \sigma_f \tag{5.10}$$

or

$$= \left[\frac{\dfrac{x}{x_f}}{q_2 + \left(1 - p_2 - q_2\right)\dfrac{x}{x_f} + p_2\left(\dfrac{x}{xf}\right)^{\left(\frac{p_2}{p_2-q_2}\right)}} \right] \sigma_f \tag{5.11}$$

where σ_f is the failure strength (MPa); $x = \left(\dfrac{\Delta\rho}{\rho_o}\right)*100$ = percentage of

piezoresistive axial strain due to the stress; $x_f = \left(\dfrac{\Delta\rho}{\rho_o}\right)_f *100$ = percentage

of piezoresistive axial strain at failure; and $\Delta\rho$ is the change in the ρ. The initial electrical resistivity (ρ_o) (at $\sigma=0$ MPa) and the model parameters p_2, q_2 and q_2/p_2 are related to the material properties and testing environments.

5.3 MATERIALS AND METHODS

In this study, all types of cements (Portland and oil well) were investigated. After mixing the cement with varying water-to-cement (w/c) ratios, the samples were prepared in cylindrical plastic molds (50 mm dia.*100 mm height) for a compression test and specially designed tension and bending molds, and were cured at room condition of 23°C and relative humidity of 50%. The specimens were demolded and capped before testing.

(a) Compression Test (ASTM C39)
The cylindrical specimens (50mm dia.*100 mm height) were capped and tested at a predetermined controlled displacement rate of 0.01 mm/min. Compression tests were performed on sulfur-capped cement samples after 1, 7, and 28 days, and longer curing times using a hydraulic compression machine. At least three specimens were tested under each testing condition, and average results are reported. Also, the capping electrically insulated the testing specimen from the testing machine. The dimension of the specimen was measured using a Vernier caliper accurate to 0.0001 mm. In order to measure the strain, commercially available extensometers (accuracy of 0.001% strain) were used. The extensometers were also calibrated using 12 mm strain gauges with 120 Ω resistance and glued to the cement specimens.

(b) Direct Tension Test
Direct tension testing is important to understand the tensile behavior and failure mechanism. Special molds were designed and built in the CIGMAT laboratory and used to prepare the testing specimens, as shown in Figure 5.1. The direct tension test set up with the instrumentation and data collection system is shown in Figure 5.1 and Figure 5.2. Two electrode probes (wires) were used to monitor the changes in electrical resistance during the curing and loading.

The axial strain was measured along the stress axis by using an extensometer and a 12 mm strain gauge simultaneously. The strain gauge was glued

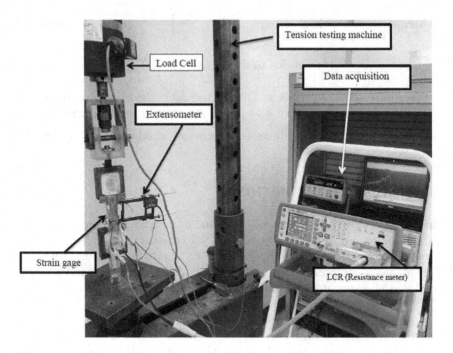

Figure 5.1 Direct tension test setup.

to the specimen directly. During testing, an LCR meter was used to monitor electrical resistance changes along the stress axis between two known points, as shown in Figure 5.2. Also, the specimen was electrically isolated from the testing machine using insulator connectors.

(c) Bending Test

Three-point bending tests were performed to determine the flexural properties of cement. Three-point bending tests were conducted according to ASTM C293/C293M. The configuration of the wooden molds used for casting the specimen is shown in Figure 5.3. Wires (electric probes) were used to measure the change in the resistance during the loading, not only in the compression (top) and tension (bottom) sides but also along the neutral axis.

The cement samples were placed on two lower static knife-edges, and the upper moveable knife-edge was moved downward until the cement failed. Strains and loads were recorded using the data acquisition system. Strain gauges were glued on the bottom (tension) and top (compression) sides close to the midpoint of the beam to monitor the strains at the outer edges from

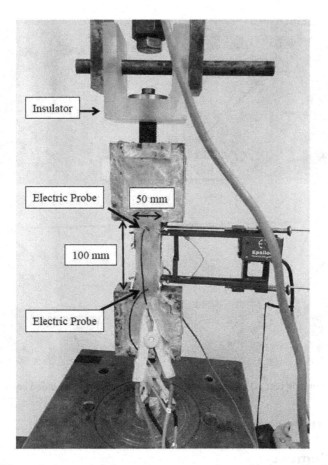

Figure 5.2 Direct tension test on the smart cement specimen with the monitoring probes.

a neutral axis. The experimental setup is shown in Figure 5.4. Also, the specimen was electrically isolated from the testing machine.

The stresses at the outer edges were calculated using the following relationship:

$$\sigma_b = \frac{3PL}{2bd^2},$$

(5.12)

where σ_b is flexural stress, P is the applied load, L is the span length, b is the average width of specimen, and d is the average depth of the specimen.

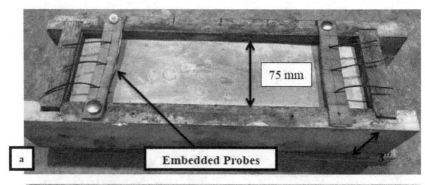

Figure 5.3 Bending test (a) mold used to cast the specimen, (b) demolded beam specimen.

5.4 RESULTS AND ANALYSES

5.4.1 Cement

5.4.1.1 Compressive Behavior

(a) Strength

One Day of Curing
The average compressive strengths (σ_{cf}) of the cement with a w/c ratio of 0.38, 0.44, and 0.54 for one day of curing were 10.6 MPa, 8.4 MPa, and 4.6 MPa respectively, a 21% and 57% reduction when the w/c ratio was increased from 0.38 to 0.44 and 0.54 respectively, as summarised in Table 5.1.

Seven Days of Curing
The average compressive strengths (σ_{cf}) of the cement with a w/c ratio of 0.38, 0.44, and 0.54 after seven days of curing were 15.8 MPa, 13.0 MPa, and 8.9 MPa respectively. The compressive strength (σ_{cf}) of the cement with

Table 5.1 Compressive strength of Class H cement with different w/c ratios

Material	w/c	Curing Time (day)	Compressive Strength σ_{cf} (MPa)
Cement only	0.38	1	10.6
		7	15.8
		28	17.3
	0.44	1	8.4
		7	13.0
		28	15.1
	0.54	1	4.6
		7	8.9
		28	11.3

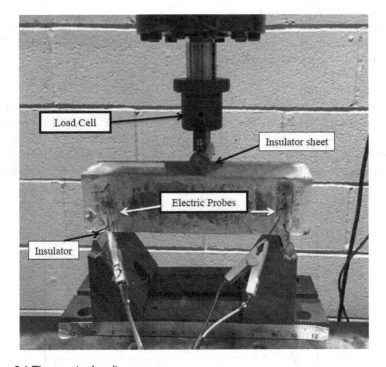

Figure 5.4 Three-point bending test setup.

a w/c ratio of 0.38, 0.44, and 0.54 after seven days of curing increased by 49%, 55%, and 93% respectively compared with the compressive strength (σ_{cf}) of the cement after one day of curing, as summarised in Table 5.1. The compressive strength (σ_{cf}) of the cement reduced by 18% and 44% when the w/c ratio was increased from 0.38 to 0.44 and 0.54 respectively, as summarised in Table 5.1.

Twenty-eight Days of Curing

The average compressive strengths (σ_{cf}) of the cement with a w/c ratio of 0.38, 0.44, and 0.54 for 28 days of curing were 17.3 MPa, 15.1 MPa, and 11.3 MPa respectively, as summarised in Table 5.1. The compressive strength (σ_{cf}) of the cement with a w/c ratio of 0.38, 0.44, and 0.54 for 28 days of curing increased by 9%, 16%, and 27% respectively compared with the seven-day compressive strengths, as summarised in Table 5.1. The average compressive strength (σ_{cf}) of the cement reduced by 13% and 35% when the w/c ratio was increased from 0.38 to 0.44 and 0.54 respectively, as summarised in Table 5.1.

(b) Stress-Strain Behavior

It is important to characterise the compressive stress-strain behavior of class H cement with curing time to quantify the changes in modulus, failure strain, and also model parameters.

One Day of Curing

The compressive strength (σ_{cf}) and failure strain (ε_{cf}) of class H cement with a w/c ratio of 0.38 after one day of curing were 10.6 MPa and 0.30% respectively, as shown in Figure 5.5 and also summarised in Table 5.2. The secant modulus at peak stress was 3,500 MPa (0.51×10^6 psi). The Vipulanandan p-q stress-strain model was used to model the strain-softening behavior of

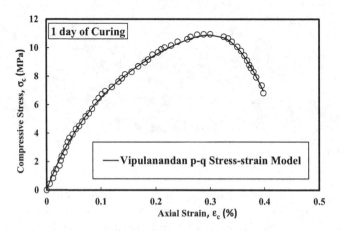

Figure 5.5 Measured and predicted compressive stress-strain behavior of the Class H oil-well cement after one day curing.

Table 5.2 Compressive stress-strain behavior and model parameters for smart cement

Curing Time (day)	σ_{cf} (MPa)	ε_{cf} (%)	Ei (MPa)	p_o	q_o	RMSE (MPa)	R^2
1	10.6 ± 1	0.30 ± 0.02	7,850	0.034 ± 0.03	0.45 ± 0.02	0.128	0.99
28	17.3 ± 0.5	0.19 ± 0.02	9,950	0.070 ± 0.01	0.91 ± 0.03	0.110	0.99

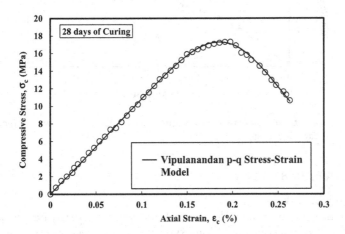

Figure 5.6 Measured and predicted compressive stress-strain behavior of Class H oil-well cement after 28 days of curing.

the cement, and model parameter p_0 and q_0 are summarised in Table 5.2. The parameter q_0 represent the nonlinearity of the stress-strain curve up to the peak stress. The initial modulus was 7,850 MPa (1.14 x 10^6 psi).

Twenty-eight Days of Curing

The compressive strength (σ_{cf}) and failure strain (ε_{cf}) of class H cement with a w/c ratio of 0.38 after 28 days of curing were17.3 MPa and 0.19% respectively, as shown in Figure 5.6 and also summarised in Table 5.2. The compressive strength increased by about 63% compared to the one day of curing. The compressive failure strain at peak stress reduced by about 37% compared to the one day of curing. The secant modulus at peak stress was 9,100 MPa (1.32 x 10^6 psi). The Vipulanandan p-q stress-strain model was used to predict the strain-softening behavior of the cement, and model parameters p_0 and q_0 are summarised in Table 5.2. Both parameters increased with the curing time. The model parameter q_0 represent the nonlinearity of the stress-strain relationship up to the peak stress. The initial modulus was 9,950 MPa (1.44 x 10^6 psi), which increased by 27% compared to the one day of curing.

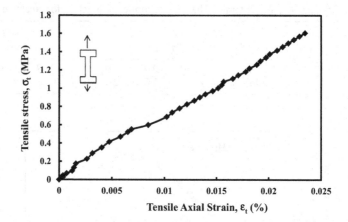

Figure 5.7 Direct tensile stress-strain response of the Class H oil-well cement after 28 days of curing.

5.4.1.2 Tensile Behavior

Twenty-eight Days

Uniaxial direct tension tests (Figures 5.1 and 5.2) were performed on cement samples cured at room condition. Typical tensile stress-strain relationship for the 28-day cured cement sample is shown in Figure 5.7. The tensile failure stress (σ_{tf}) was 1.6 MPa, 9.3% of the compressive strength. The tensile failure strain (ε_{tf}) was 0.023%, 12.1% of the compressive failure strain. The initial tangent modulus was 8,300 MPa, about 83% of the compressive initial modulus. The secant modulus at failure was 6,880 MPa, about 75% of the compressive secant modulus. The Vipulanandan p-q stress-strain model parameter q_0 was 0.83, which indicates and also quantifies the nonlinearity of the tensile response.

5.4.1.3 Bending Behavior

Bending tests (Figure 5.4) were performed on cement samples cured at room condition. The typical bending stress-strain relationship for the 28-day cured cement sample is shown in Figure 5.8. The bending tensile failure stress (σ_{btf}) was 3.35 MPa, which doubled the direct tensile strength. The tensile failure strain (ε_{btf}) was 0.045%, double the direct tensile failure strain. The initial flexural tangent modulus was about 20,000 MPa, double the compressive initial modulus. The secant modulus at failure was about 7,400 MPa, about 82% of the compressive secant modulus at peak stress. The Vipulanandan p-q stress-strain model parameter q_0 was 0.37, which indicates and also quantifies the nonlinearity of the bending response.

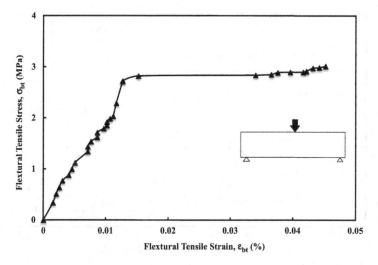

Figure 5.8 Measured flexural stress-strain response of the Class H oil-well cement after 28 days of curing.

Under the uniaxial unconfined compressive, uniaxial tensile, and three-point bending loading conditions, the failure strains of the cements cured for 28 days were about 0.2% or less as demonstrated by the test results. Most of the past cement performance-monitoring methods, such as using fiber optics and strain gauges, are based on the strain, and hence there is a need to develop new technology to make the cement more sensing to better monitor cement performance.

5.5 NEW THEORY FOR PIEZORESISTIVE CEMENT

Cement is not an insulator, and its initial resistivity is about 1 Ωm based on the composition and method of mixing. When conductive fibers (carbon fibers) are added to cement, based on the dispersion of the fibers they may not be in contact (region A) or may be in contact (region B) to make the material piezoresisitive (resistivity will change with applied stress) (Figure 5.9). A new theory is being proposed by Vipulanandan to make bulk cement be piezoresisitive by dispersing the fibers, as shown in region A (Figure 5.9). Adding conductive fibers (carbon fibers) to cement will affect its resistivity (ρ), and the relationship is as follows:

$$\rho = \rho_0 - \frac{X_f}{H + JX_f} \tag{5.13}$$

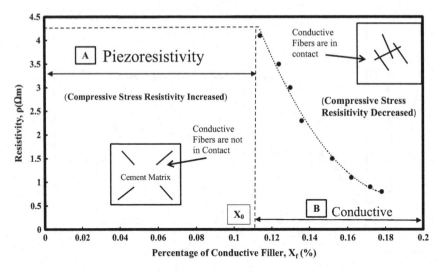

Figure 5.9 Fiber distribution configuration for making the cement highly piezoresistive without changing the initial cement resistivity (no stress applied).

where X_f is the fiber content, and ρ_0, H, and J are three material parameters that are influenced by the curing time. The general trend observed for the cement cured for seven days with the conductive fiber content is shown in Figure 5.9.

5.5.1 Region A

When the fiber content is less than X_0, there will be no change in the resistivity of the cement since the fibers are dispersed well $(X_f < X_0)$ and are not in contact. Hence, the current will keep flowing though the cement matrix (Region A) and pick up the changes in the cement matrix due to stress, strain, temperature, cracks, and contamination. When stress is applied and the resistivity is measured, the current will flow through the conductive fiber with the changes in the orientation of the fibers (3D) and cement matrix. Also, with any type of stress applied, the resistivity will increase because the deviatoric stress (shear stress) will cause shape changes in the microstructural configuration with the reorientation of the fibers and increase the resistivity, since the current will flow through not only the conductive fiber but also the cement matrix. This will make the cement a bulk 3D sensor, since the resistivity is a second-order tensor. The amount of increase in resistivity will depend on the type and magnitude of the stress and how well the fibers are dispersed within the cement matrix. Also, the increase in resistivity will make the material more stable and minimise the eddy currents, corrosion, and degradation of the cement.

5.5.2 Region B

With the addition of conductive fibers (carbon fibers), if the resistivity is reduced (Xf >Xo), then it will represent region B. In this region the conductive fibers (carbon fibers) are in contact and measuring current will flow though the conductive fibers. When compressive stress is applied to the cement, the fibers will move closer and reduce the resistivity because the current will flow through the fibers and not flow through the cement. This is what has been done in the past to make piezoresistive cement.

It must be noted that X_o will vary with the curing time, mixing process, and the composition of the cement, and only a schematic is shown in Figure 5.9. The basic concept is to not to change the initial resistivity with the addition of conductive or semiconductive fibers (carbon, metal, basaltic, and others) to make the cement be chemo-thermo-piezoresisitive smart cement.

5.6 MATERIALS AND METHODS

5.6.1 Resistivity of Slurry

Two different methods were used for electrical resistivity measurements of Class H oil-well cement slurries. To assure the repeatability of the measurements, the resistivity of the cement slurry was measured at least three times and the average resistivity is reported. The electrical resistivity of the cement slurries was measured using the following instruments:

(a) Conductivity Probe
A commercially available conductivity probe was used to measure the conductivity (inverse of resistivity) of the slurries. In the case of cement, this meter was used during the initial curing of the cement. The conductivity measuring range was from $0.1 \mu S/cm$ to $1,000$ mS/cm, representing a resistivity of $0.1 \Omega.m$ to $10,000$ $\Omega.m$.

(b) Digital Resistivity Meter
A digital resistivity meter (used in the oil industry) was used measure the resistivity of fluids, slurries, and semisolids directly. The resistivity range for this device was 0.01Ω-m to 400Ω-m.

The conductivity probe and the digital electrical resistivity device were calibrated using a standard solution of sodium chloride (NaCl).

5.6.2 Resistivity of Solidified Smart Cement

In this study, high frequency AC measurement was adopted to overcome interfacial problems and minimise the contact resistances in the solid cement measurements. Electrical resistance (R) was measured using an LCR meter (measures the inductance (L), capacitance (C), and resistance (R)) during the curing time. This device has a least count of 1 μΩ for electrical resistance and measures the impedance (resistance, capacitance, and inductance) in the frequency range of 20 Hz to 300 kHz. Based on the impedance (z)–frequency (f) response, it was determined that the smart cement was a resistive material (Vipulanandan et al. 2013a and 2015d). Hence, the resistance measured at 300 kHz using the two-probe method was correlated to the resistivity (measured using the digital resistivity device) to determine the effective parameter K factor with parameter G = 0 (Eqn. (4.10)) for a time period of an initial five hours of curing. This effective K factor was used to determine the resistivity of the cement with the curing time.

5.6.3 Piezoresistivity Test

Piezoresistivity describes the change in the electrical resistivity of a material under stress. Since oil-well cement serves as a pressure-bearing part of oil and gas wells in real applications, the piezoresistivity of smart cement (stress-resistivity strain relationship) with different w/c ratios was investigated under compressive loading at different curing times. During the compression test, electrical resistance was measured in the direction of the applied stress. To eliminate the polarisation effect, AC resistance measurements were made using an LCR meter at a frequency of 300 kHz (Vipulanandan et al. 2013a).

5.7 SMART CEMENT

Tests have been performed on Portland cements and oil-well cements to demonstrate the piezoresistivity behavior of smart cement (Vipulanandan 2014a–2019a; U.S. Patent Number 10,481,143 (2019)). In this study, oil-well cements with w/c ratios of 0.38, 0.44, and 0.54 were used. The samples were prepared according to API standards. To improve the sensing properties and piezoresistive behavior of the cement modified with less than 0.1% of carbon fiber (CF) by the weight of cement was mixed with all the samples. After mixing, specimens were prepared using cylindrical molds with a diameter of 50 mm and a height of 100 mm, and four conductive wires were placed in all of the molds to measure the changes in electrical resistivity. At least three specimens were prepared for each mix.

5.7.1 Compressive Behavior

(a) Strength

One Day of Curing
For the smart cement with w/c ratios of 0.38, 0.44, and 0.54, the compressive strengths compared to the unmodified cement (Table 5.2) increased to 10.9

Table 5.3 Model parameters of piezoresistive smart cement with different w/c ratios

Material	w/c	Curing Time (day)	$(\Delta\rho/\rho_0)_f$ (%)	σ_f (MPa)	q_2	p_2	RMSE (MPa)	R^2
Smart	0.38	1	583	10.9	0.30	0.16	0.01	0.99
cement		7	432	17.2	0.14	0.09	0.03	0.99
		28	401	19.4	0.05	0.03	0.03	0.99
	0.44	1	531	9.8	1.59	0.85	0.02	0.99
		7	405	13.7	0.33	0.07	0.02	0.99
		28	389	16.8	0.41	0.06	0.02	0.99
	0.54	1	355	5.3	1.37	0.0	0.04	0.99
		7	325	9.2	0.41	0.0	0.03	0.99
		28	289	12.6	0.39	0.0	0.02	0.99

MPa, 9.8 MPa, and 5.3 MPa respectively. In comparison, the pure cement (Table 5.1), with the addition of less than 0.1% carbon fibers, increased its compressive strength by 3%, 17%, and 15% of the cement with w/c ratios of 0.38, 0.44, and 0.54 respectively, as summarised in Table 5.3.

Seven Days of Curing
The addition of 0.1% carbon fibers to the cement (smart cement) with w/c ratios of 0.38, 0.44, and 0.54 increased its compressive strength to 17.2 MPa, 13.7 MPa, and 9.2 MPa respectively, as summarised in Table 5.3. Compared to the pure cement-only case in Table 5.1, the addition of less than 0.1% carbon fibers to the cement increased its compressive strength by 9%, 5%, and 4% for cement with w/c ratios of 0.38, 0.44, and 0.54 respectively.

Twenty-eight Days of Curing
With the addition of 0.1% carbon fibers to the Class H cement, the compressive strengths (σ_{cf}) with w/c ratios of 0.38, 0.44, and 0.54 were 19.4 MPa, 16.8 MPa, and 12.6 MPa, as summarised in Table 5.3, and they increased by 12%, 11%, and 12% respectively compared to the unmodified cement strengths summarised in Table 5.1 after 28 days of curing.

Hence the addition of 0.1% carbon fibers increased the compressive strength of cement by varying amounts based on the w/c ratio and curing time. The increase in the compressive strength varied from 3% to 17%.

(b) Piezoresistivity

One Day of Curing
With the addition of less than 0.1% carbon fibers (CF), the piezoresistive axial strains at failure $\left(\dfrac{\Delta\rho}{\rho_o}\right)_f$ for the smart cement with w/c ratios of 0.38, 0.44, and 0.54 were 583%, 531%, and 355% respectively, as summarised in

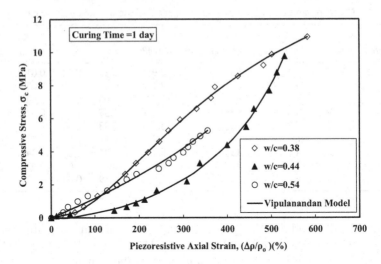

Figure 5.10 Compressive piezoresistive behavior of the smart cement after 1 day of curing.

Table 5.3 and also shown in Figure 5.10. The electrical resistivity increased with compressive stress, representing Region A in Figure 5.9, an important demonstration of the piezoresistive cement theory. The addition of less than 0.1% carbon fiber to the cement substantially enhanced the change in the electrical resistivity of oil-well cement at failure $\left(\dfrac{\Delta\rho}{\rho_o}\right)_f$ (piezoresistive axial strain) compared to the 0.2% failure strain. Compared to the compressive failure strain, the sensing parameter piezoresistive axial strain at failure was magnified by over 1,500 times (150,000%) with w/c ratios of 0.38, 0.44, and 0.54.

Using the Vipulanandan p-q piezoresistive strain-softening model (Eqn. 5.7)), the relationships between compressive stress and the change in electrical resistivity $\left(\dfrac{\Delta\rho}{\rho_o}\right)$ (piezoresistive axial strain) of the cement with different w/c ratios of 0.38, 0.44, and 0.54 for one day of curing were modeled. The piezoresistive model (Eqn. (5.7)) predicted the measured stress change in the resistivity relationship very well, as shown in Figure 5.10. The model parameters q_2 and p_2 are summarised in Table 5.3. The coefficients of determination (R^2) were 0.98 and 0.99. The root-mean-square error (RMSE) varied between 0.02 MPa and 0.04 MPa, as summarised in Table 5.3.

Seven Days of Curing

With the addition of less than 0.1% carbon fibers to the cement (smart cement), the piezoresistive axial strains at failure $\left(\dfrac{\Delta\rho}{\rho_o}\right)_f$ for the smart cement with w/c ratios of 0.38, 0.44, and 0.54 were 432%, 405%, and 325% respectively, as shown in Figure 5.11. The addition of 0.1% CF, compared to the compressive failure strain of 0.2%, increased the piezoresistive axial strain at failure $\left(\dfrac{\Delta\rho}{\rho_o}\right)_f$ with different w/c ratios of 0.38, 0.44, and 0.54 by 2,160 (216,000%), 2,025 (202,500%), and 1,625 (162,500%) respectively.

Twenty-eight Days of Curing

With the addition of less than 0.1% CF to the cement (smart cement), the piezoresistive axial strains at failure $\left(\dfrac{\Delta\rho}{\rho_o}\right)_f$ for the smart cement with a w/c of 0.38, 0.44, and 0.54 were 401%, 389%, and 289% respectively, as shown in Figure 5.12 and summarised in Table 5.3. The addition of 0.1% CF, compared to the compressive failure strain of 0.2%, increased the piezoresistive axial strains at failure $\left(\dfrac{\Delta\rho}{\rho_o}\right)_f$ with different w/c ratios of 0.38, 0.44, and 0.54 by 2,005 (200,500%), 1,945 (194,500%), and 1,445 (144,500%) respectively

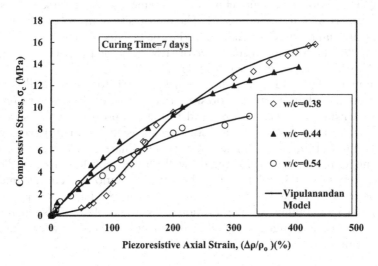

Figure 5.11 Compressive piezoresistive behavior of the smart cement after 7 days of curing.

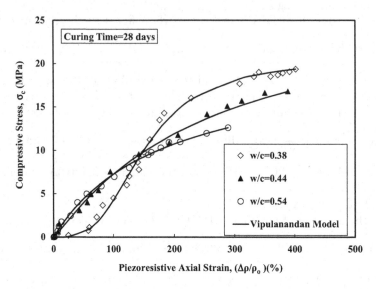

Figure 5.12 Compressive piezoresistive behavior of the smart cement after 28 days of curing.

The measured compressive stress and the piezoresistive axial strain $\left(\dfrac{\Delta\rho}{\rho_o}\right)$ of the cement with different w/c ratios of 0.38, 0.44, and 0.54 after 28 day of curing were modeled using the Vipulanandan p-q piezoresistive model (Eqn. (5.7)). The piezoresistive model predicted the measured experimental results very well, as shown in Figure 5.12. The piezoresistive model parameters q_2 and p_2 are summarised in Table 5.3. The coefficients of determination (R^2) were 0.99. The root-mean-square error (RMSE) varied between 0.02 MPa and 0.04 MPa, as summarised in Table 5.3.

Summary: The addition of less than 0.1% CF to the oil-well cement substantially enhanced the piezoresistivity behavior of the cement to make it very sensing and smart. Compared to the compressive failure strain of 0.2%, the piezoresistive axial strains (changes in resistivity) were over 1,400 times (140,000%) higher, making the smart cement very sensing. The model parameters q_2 for the oil-well cement with less than 0.1% CF varied between 0.05 and 1.59 based on the w/c ratio and curing time, as summarised in Table 5.3. For the smart cement, the parameter p_2 varied from 0 to 0.85 (Table 5.3). The addition of less than 0.1% CF also increased the compressive strength of the oil-well cement based on the w/c ratios and curing times.

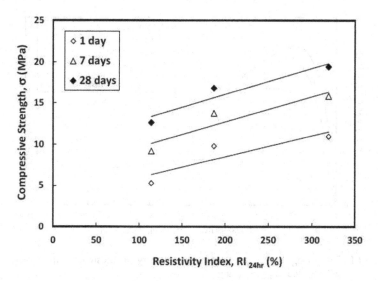

Figure 5.13 Relationship between resistivity index ($RI_{24\,hr}$) and compressive strength of the smart cement for ($0.54 \geq$ w/c ≥ 0.38).

(c) Compressive Strength–Resistivity Index (RI_{24hr}) Relationship

During the entire cement hydration process, both the electrical resistivity and the compressive strength of the cement increased gradually with the curing time. For cement pastes with various w/c ratios, the change in resistivity was varied during the hardening. The cement paste with the lowest w/c ratio had a higher electrical resistivity change (RI_{24hr}) than the cement with a higher w/c ratio, as summarised in Table 4.6.

The linear relationship between (RI_{24hr}) and one-day, 7-day, and 28-day compressive strengths (MPa) of the smart cement with varying w/c ratios, as shown in Figure 5.13, are as follows:

$$\sigma_{1day} = 0.0.3 \times RI_{24hr} + 3.3 \qquad R^2 = 0.81 \qquad (5.14a)$$

$$\sigma_{7days} = 0.031 \times RI_{24hr} + 6.5 \qquad R^2 = 0.89 \qquad (5.14b)$$

$$\sigma_{28days} = 0.03 \times RI_{24hr} + 9.7 \qquad R^2 = 0.94 \qquad (5.14c)$$

Hence, the compressive strength of the smart cement after various curing times was linearly related to the electrical resistivity index, RI_{24hr}. Since RI_{24hr} can be determined in one day, it can be used to predict the compressive strength of smart cement up to 28 days.

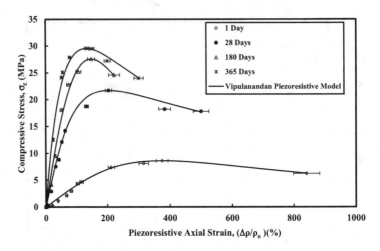

Figure 5.14 Comparing the compressive piezoresistivity of the smart cement after 1, 28, 180 and 365 days of curing.

(d) Long-term Piezoresistivity

In another study, the piezoresistive behavior of smart cement (Portland cement Type 1) was evaluated up to 365 days of curing, as shown in Figure 5.14, using the piezoresistivity strain-softening model (Eqn. (5.7)). After one day of curing, the piezoresistivity of the smart cement was 375%. The parameters p_2 and q_2 for the model were 0.13 and 0.55 respectively. After 28 days of curing, the piezoresistivity of the smart cement was 204%, a 46% reduction compared to the one day-curing smart cement. The parameters p_2 and q_2 for the model were 0.20 and 0.45 respectively, as summarised in Table 5.4. After 180 days of curing, the piezoresistivity of the smart cement reduced by 61% to 148% compared to the one day-curing smart cement. Also, the parameters p_2 and q_2 for the model were 0.30 and 0.49 respectively. After 365 days of curing, the piezoresistivity of the smart cement was 133%, a 65% reduction compared to the one day-curing smart cement. Also, the parameters p_2 and q_2 for the model were 0.55 and 0.42 respectively.

The variation of the piezoresistive axial strain for the smart cement at the peak compressive stress with curing time is shown in Figure 5.15 and was modeled using the Vipulanandan correlation model as follows:

$$\frac{\Delta\rho}{\rho} = \left(\frac{\Delta\rho}{\rho}\right)_0 - \frac{t}{A_1 + B_1 t}, \tag{5.15}$$

Table 5.4 Piezoresistive model parameters, strengths, and piezoresistive axial strain with curing time for smart cement

Curing Time	p_2	q_2	R^2	RMSE (MPa)	Compressive Strength (MPa)	Piezoresistivity (%)
I Day	0.13	0.55	0.99	0.36	8.6	375 ± 25
28 Days	0.20	0.45	0.99	0.56	21.7	204 ± 21
180 Days	0.30	0.49	0.99	0.44	27.6	148 ± 15
365 Days	0.55	0.42	0.99	0.62	29.5	133 ± 7

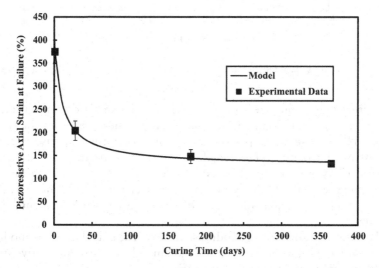

Figure 5.15 Variation of piezoresistive axial strain at failure for the smart cement over one year.

where $\left(\dfrac{\Delta\rho}{\rho}\right)_0 (\%)$ is the piezoresistive axial strain after one day of curing and it is 375%. Model parameters A_1 and B_1 were 0.0394 days/% and 0.0037/% respectively, and the R^2 was 0.99 and the RMSE was 2.81%. Using the model, at infinite time, the limiting value of piezoresistivity at failure was 104%, 520 times (52,000%) higher than the compressive failure strain of 0.2%.

5.7.2 Tensile Behavior

(a) Direct Tension

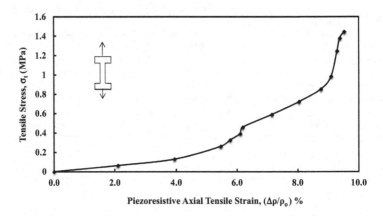

Figure 5.16 Direct tensile piezoresistivity of the smart cement after 28 days of curing.

The direct tensile piezoresisitive behavior of the smart cement cured for 28 days is shown in Figure 5.16. The resistivity increased with the tensile stress. The tensile stress-piezoresisitive tensile strain response is resistivity strain hardening. The tensile strength was about 1.50 MPa, and the resistivity strain change was 9.52%. The tensile strain failure for the cement was 0.023% (Figure 5.7); hence, the change in resistivity was over 414 times (41,400%) higher, making the smart cement highly sensing.

The initial piezoresisitive modulus was 5 MPa, and the secant modulus at failure was 15.1 MPa. Hence, the Vipulanandan p-q piezoresisitive strain hardening model parameter q_2 was 3.02.

(b) Splitting Tension

Cement samples cured for 28 days were tested using splitting tension. Figure 5.17 shows the percentage change in resistivity tensile strain with applied tensile stress. The change in resistivity was measured across the section in a horizontal direction (along the diameter perpendicular to the loading direction). The percentage change in resistivity was positive up to the failure. The failure tensile stress was 1.8 MPa. The piezoresistive failure strain was 13%, and compared to the direct tensile failure strain of 0.023% it was 565 times (56,500%) higher. The initial piezoresistive modulus was 20 MPa, and at failure the secant modulus was 13.8 MPa. The Vipulanandan p-q piezoresistive strain softening model (Eqn. (5.7)) parameter q_2 was 0.69, representing piezoresistivity softening material.

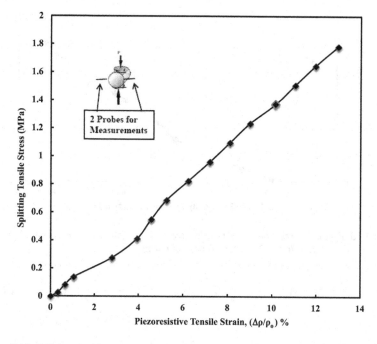

Figure 5.17 Splitting tensile stress-resistivity tensile strain relationship of the smart cement after 28 days of curing.

5.7.3 Bending Behavior

Cement samples cured for 28 days were tested in bending. Figure 5.18 shows the percentage change in tensile resistivity strain with applied bending tensile stress (σ_{bt}). The change in resistivity was measured across the section in a horizontal direction. The percentage change in the resistivity was positive up to the failure. The failure tensile stress was 3.35 MPa. The piezoresistive failure strain was 8.2%, compared to the cement flexural tensile failure strain of 0.045%, a magnification of 182 times (18,200%) higher. The initial piezoresistive bending modulus was 5.5 MPa, and at failure the secant modulus was 41 MPa. The Vipulanandan p-q piezoresistive strain hardening model (Eqn. (5.10)) parameter q_2 was 7.3, representing piezoresistivity strain hardening material, similar to the direct tension response in Figure 5.16.

5.8 VIPULANANDAN FAILURE MODEL (2018) (3D MODEL)

It is important to identify the critical stress parameters for the 3D failure model. The first stress invariant (I_1) and the second deviatoric stress invariant) (J_2) are defined as follows:

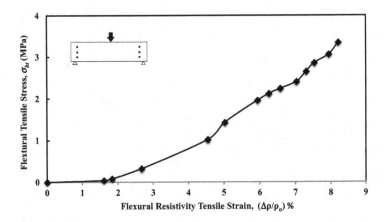

Figure 5.18 Flexural stress resistive tensile strain relationship for the Class H oil-well cement (w/c = 0.38) after 28 days of curing.

$$I_1 = \sigma_1 + \sigma_2 + \sigma_3 \tag{5.16}$$

$$J_2 = \frac{1}{6}\left[(\sigma_1 - \sigma_2)^2 + (\sigma_2 - \sigma_3)^2 + \left(\sigma_3 - \sigma_1\right)^2\right] \tag{5.17}$$

where σ_1, σ_2, *and* σ_3 are the major principal stresses at failure, the intermediate principal stress and the minor principal stress respectively.

Based on the understanding of material behavior, there is a limit to the maximum shear stress tolerance for all the materials, including cement and concrete, and hence the following conditions are proposed using stress invariants:

$$\frac{d\sqrt{J_2}}{dI_1} > 0 \tag{5.18}$$

$$\frac{d^2\sqrt{J_2}}{dI_1^2} < 0 \tag{5.19}$$

$$I_1 \to \infty \quad \sqrt{J_2} = \sqrt{J_2}_{\,max}$$

The Vipulanandan failure model is represented as follows (Vipulanandan et al. 2018h; Mayooran 2018):

$$\sqrt{J_2} = (\sqrt{J_2})_o + \frac{I_1}{L + N * I_1} \qquad (5.20)$$

$$\frac{d\sqrt{J_2}}{dI_1} > \frac{(L + NI_1) - NI_1}{(L + NI_1)^2} = \frac{L}{(L + NI_1)^2} > 0 \qquad \Rightarrow L > 0$$

$$\frac{d^2\sqrt{J_2}}{dI_1^2} = -2(L + NI_1)^{-3}LN = \frac{-2LN}{(L + NI_1)^3} < 0 \qquad \Rightarrow N > 0$$

When $I_1 \rightarrow \infty$

$$(\sqrt{J_2})_{max} = (\sqrt{J_2})_o + \frac{1}{N} \qquad (5.21)$$

The Drucker-Prager model is represented as follows:

$$\sqrt{J_2} = (\sqrt{J_2})_o + \propto I_1 \qquad (5.22)$$

where α is the slope of the linear relationship in Eqn. 5.22. When parameter $N = 0$ in the Vipulanandan failure model, it will represent the Drucker-Prager model. Based on the limited experimental data on smart cement cured for 28 days, as summarised in Table 5.5, the relationship between I_1 and $\sqrt{J_2}$ is shown in Figure 5.19 for the Vipulanandan model and the Drucker-Prager linear model. The RMSE value for the Vipulanandan model was lower than for the Drucker-Prager model, an indication of better prediction of the test results. The model parameters are summarised in Table 5.6.

Based on the Vipulanandan model, $(\sqrt{J_2})_{max}$ for the smart cement will be equal to 144.7 MPa, and for the Drucker-Prager model it will be infinity. It has been shown that for shale rock and limestone rock, the maximum shear stress limit was 103 MPa and 102 MPa respectively (Vipulanandan et al. 2018j). It is interesting to note that cement is produced using constituents from shale rock (clay) and limestone and that the resulting strength is better than basic rock constituents. The, $(\sqrt{J_2})_o$ representing the pure shear strength of cement, was 1.81 MPa, lower than the pure shear strength of shale rock and limestone rock of 3.0 MPa and 2.6 MPa respectively (Vipulanandan et al. 2018j). Also the pure tension strength (tension in three perpendicular directions) for the cement will be 1.14 MPa, using the Vipulanandan model, about 10% of the compressive strength of cement. The Drucker-Prager model pure tension strength was 1.25 MPa, 11% higher.

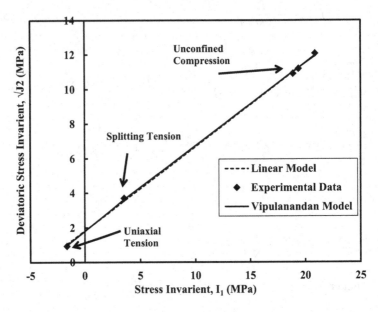

Figure 5.19 Failure models for the piezoresistive smart cement.

Table 5.5 Smart cement strength data (28 days)

I_1 (MPa)* (X-axis)	$\sqrt{J2}$ (MPa) (Y-axis)	Remarks
19.4	11,19	Compression
20.9	12.07	Compression
18.9	10.91	Compression
3.56	3.70	Splitting Tension
-1.61	0.93	Direct Tension

*Sign Convention used: Compressive stress is positive.

Table 5.6 Failure model parameters for the smart cement strength data (28 days)

				Model Parameters				
Drucker-Prager Linear Model					Vipulanandan Model			
Slope	$(\sqrt{J_2})_0$ (MPa)	R^2	RMSE (MPa)	L	N (MPa)$^{-1}$	$(\sqrt{J_2})_0$ (MPa)	R^2	RMSE (MPa)
0.485	1.82	0.99	0.11	1.92	0.007	1.81	1.0	0.09

Hence, more cement strength data is needed to verify the uniqueness of the new Vipulanandan failure model.

5.9 SUMMARY

Based on the experimental study, the development and verification of the piezoresistive cement concept, and behavior modeling, the following conclusions are advanced:

(1) The mechanical properties of standard cement (without any carbon fiber addition) under compression, direct tension, and three-point bending loading have been tested and quantified. The uniaxial compression and direct tension failure strains after 28 days of curing were about than 0.2% and 0.023% respectively.

(2) The new Vipulanandan piezoresistivity theory has been developed for cement and verified with experiments under compression, direct and splitting tension, and bending loading. With loading, the piezoresistive strain increased and the resistivity change was positive under both the compression and the tension loading, which also verified the piezoresistive cement theory.

(3) Electrical resistivity has been identified as the monitoring parameter. Also, electrical resistivity is a second-order tensor and the changes can be monitored in 3D. The long-term piezoresistive compressive axial strain was over 500 times (50,000%) higher than the compressive failure strain of 0.2%. Also, the direct piezoresistive tensile axial strain was over 400 times (40,000%) higher than the direct tensile failure strain of 0.023%.

(4) Vipulanandan p-q models were used to characterise the stress-strain and stress-piezoresisitive strain of standard cement and smart cement respectively.

(5) The Vipulanandan failure model was used to characterise the failure of smart cement with limited data.

Chapter 6

Chemo-Thermo-Piezoresistive Smart Cement

6.1 BACKGROUND

Based on the applications and the environments, all types of standard cements are modified with various types of inorganic and organic additives (Salib et al. 1990; Vipulanandan et al. 1992 and 2012c; Plank et al. 2010 and 2013; Pakeetharan et al. 2016; Amani et al. 2020). Also, during the construction and the service life of structures produced using cement-based materials, contamination is a possibility and hence investigating the sensitivity of smart cement to detect chemical, temperature, and stress changes for real-time monitoring must be investigated. To minimise delays during construction, failures, and also safety issues, it is important to quantify the changes in chemo-thermo-piezoresistive cement.

Portland cement slurries are not only used in the construction but also in the repairing applications related to slurry walls, piles, other foundations, pipelines, tunnels, wells (oil, gas, and water), bridges, buildings, and highways (McCarter et al. 2000; Fuller et al. 2002; Vipulanandan et al. 2000b, 2005, and 2014b; Wilson 2017). Based on the applications, cement slurries are made with additives and water-to-cement (w/c) ratios varying from 0.3 to over 1 (Nelson 1990; Vipulanandan et al. 2017a). Construction of deep foundations, near surface structures and underground structures, that requires drilling in the ground using drilling muds and placing the cementitious materials in the boreholes, may result in various types of clay soil contamination. Clay soil contamination will impact the cement hydration and long-term properties (Vipulanandan 1995; Kim et al. 2003 and 2006;; Vipulanandan et al. 2018k). Unfortunately, there are no real-time monitoring methods to detect the clay soil contamination of cementitious materials during construction and also the effects of clay contaminations during the service life of infrastructures (Mohammed 2018; Vipulanandan et al. 2018k and 2020c).

Clay soils are mainly characterised as montmorillonite, kaolinite, illite, or a mixture of these clay constituents with particle sizes less than 2 μm (Vipulanandan 1995a, b and 2016f). Chemically the main constituent of clay is aluminum silicates with vary amounts of cations such as sodium (Na), potassium (K), magnesium (Mg), and calcium (Ca) (Vipulanandan 1995; Mohammed et al. 2013 and 2015). Clays are hydrophilic inorganic materials that can react with both hydrating cement particles and the pore fluid. During the drilling of boreholes to install water, oil, and gas wells and drilled shafts to support bridges and buildings, water-based drilling muds are used with varying amounts of bentonites clay contents (Vipulanandan 2014a). If the bore holes are not cleaned before placing the cement or concrete to construct drilled shafts, the cement will get contaminated (Vipulanandan et al. 2018k). When installing oil and gas wells, after drilling is finished the metal casing is placed inside the well bore and then the cement slurry is pumped through the casing so that it comes from the bottom, pushing the drilling mud and mud cake up, and fills the gap between the casing and the formation (Wilson 2017). Also, the construction of tunnels in clay soils and shale rock formations can also contaminate cement and concrete with clays. Flooding on construction sites will also result in contaminating the surfaces of the cementitious construction materials in-place by depositing transported clay sediments, which will significantly impact the construction. Hence, there is potential for the cement to be contaminated with clays from the drilling muds, mud cakes, flooding, and the geological formations. Based on the type and the amount contamination, it will affect the performance of the cement and concrete (Vipulanandan 1995, 2014b, 2015c, and 2018k).

During oil, gas, and water well installations, cement can also get contaminated by the oil-based drilling mud (OBM) and also the salty formations. Oil-based drilling muds are used in both onshore and offshore applications when clay shale rocks are encountered to minimise fluid losses and also degradations of the formations. Also, OBM can contaminate the cement during well installations, affecting the hydration and performance of the cement and impacting the integrity of the well. Cement will be contaminated as well with carbon dioxide in sequestration wells, which will impact the integrity of the wells. Cement-based composites are used in storage facilities where there is potential for contamination. Based on the environmental and operational impacts, there is a need for developing real-time monitoring systems to not only monitor the quality of the cement during the installation with quality cement but also the performance of the cement during the entire service life of infrastructures.

The potential applications of smart cement with various types of chemical additives have been investigated under different curing conditions (temperatures, and saturated sand simulating the water-saturated conditions in the bore holes), and the results are analyzed in this chapter to demonstrate the sensitivity of the chemo-thermo-piezoresistivity of smart cements. Also, the effects of clay (inorganic), oil-based mud (organic), and carbon dioxide (CO_2) contaminations on chemo-thermo-piezoresistive smart cement sensing characteristics were investigated.

6.2 CURING METHODS

(a) Room Condition
Specimens were cured in plastic molds at room temperature (23°C) and a relative humidity of 50%, and the specimens were demolded just before testing.

(b) Oven Cured
Specimens were kept in the plastic mold and cured in the oven at elevated temperature. Also, specimens were placed in saturated sand in a closed bottle (Figure 6.1) to simulate the field condition under water and groundwater, and cured at room temperature and elevated temperatures, and were demolded just before testing.

Also, water was added regularly to keep the sand saturated.

6.3 SODIUM META SILICATE (SMS)

From the initial use in the late 1800s, sodium silicate–based compounds have been used in a number of applications including cementing, grouting, emulsifying, and in cleaning agents (Mbaba et al. 1983). Of the various forms of sodium silicate–based compounds, sodium meta-silicates (anhydrous) have been used in the oil and gas industry and infrastructure-repairing applications. Sodium meta-silicate (Na_2SiO_3; SMS) is a water-soluble powder, which is produced by fusing silica sand with sodium carbonate at 1400°C (Nelson 1990). Because of its emulsification and interfacial tension reduction characteristics, SMS has been used in alkaline flooding, a chemical recovery method to recover oil from various types of geological formations and sand. The overall objective of the study was to investigate the effects of adding varying amounts of SMS and higher temperature (80°C/176°F) curing on the piezoresisitive behavior of smart cement with and without SMS.

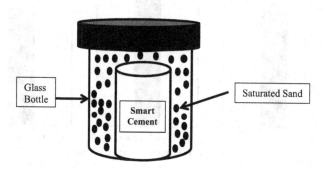

Figure 6.1 Bottle filled with saturated sand and the smart cement sample was placed for curing.

6.3.1 Results and Discussion

SMS solution was characterised by determining the pH and the resistivity of the water solutions. With the addition of 0.1% SMS, the pH of water increased from 7.7 to 11.8, a 50% change in the pH. With the addition of 0.3% SMS, the pH of the solution was 12.4, a 60% change. The resistivity of tap water decreased from 27.0 Ω.m to 4.15 Ω.m with the addition of 0.1% SMS, a 85% reduction in resistivity. With the addition of 0.3% SMS, the resistivity reduced to 2.0 Ω.m, a 93% reduction.

(a) Density

Adding SMS powder to cement slurry (w/c ratio of 0.40) slightly increased the density of the cement mixtures. Adding 0.3% SMS (by weight of water) to cement slurry increased the density from 1.94 g/cm³ (16.2 ppg) to 1.95 g/cm³ (16.3 ppg) at room condition curing, a 0.6% increase.

(b) Initial Resistivity

The electrical resistivity of cement slurry with and without SMS was measured immediately after mixing. The initial resistivity of the smart cement slurry was 0.97 Ω.m, and it decreased with the addition of SMS, as shown in Figure 6.2. With the addition of 0.1% SMS, the resistivity decreased to 0.92 Ω.m, a 5% reduction. With the addition of 0.2% and 0.3% SMS, the resistivity were 0.9 Ω.m and 0.88 Ω.m. Hence, the resistivity was sensitive to the concentration of SMS in the cement. The resistivity was decreased to 0.8 Ω.m with 1% SMS, which is a 17% decrease. Hence, the resistivity is a

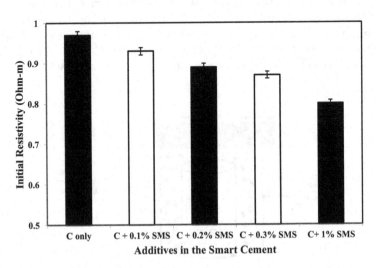

Figure 6.2 Changes in the initial resistivity of the smart cement slurry with varying amounts of sodium meta silicate (SMS).

highly sensitive material property and will be a good monitoring and quality control parameter in the field.

(c) Compressive Piezoresistivity Behavior

With the addition of up to 0.3% SMS (inorganic additive), the tests showed that smart cement cured under high temperature and in different environments (dry and saturated sand) was a highly sensitive chemo-thermo-piezoresistive material.

One Day of Curing

The compressive strength (σ_{cf}) of smart cement after one day of curing at 80°C in the oven was 15.81 MPa, which increased to 18.00 MPa when cured in saturated sand at 80°C, a 14% increase. Smart cement with 0.3% SMS cured at 80°C had a compressive strength of 14.93 MPa, which increased to 16.91 MPa when oven cured in saturated sand at 80°C, a 13% increase (Table 6.1).

The piezoresistive axial strain at failure $\left(\dfrac{\Delta\rho}{\rho_o}\right)_f$ for smart cement air cured at 80°C was 433%, and it increased to 475% for smart cement cured in saturated sand at 80°C. Smart cement with 0.3% SMS cured in the oven showed the piezoresistive axial strain at failure $\left(\dfrac{\Delta\rho}{\rho_o}\right)_f$ was 331%, and it

Table 6.1 Piezoresistivity model parameters for smart cement with and without SMS cured at 80°C for one day of curing

Composition and Curing Conditions	Curing Time (day)	Strength σ_{cf} (MPa)	Piezoresistive Strain at Peak Stress, $(\Delta\rho/\rho_o)_{cf}$ (%)	Model Parameter p_2	Model Parameter q_2	R^2	RMSE (MPa)
w/c = 0.4, (Oven cured)	1 day	15.81	433	0.010	0.673	0.99	0.12
w/c = 0.4, (Cured in Saturated Sand)		18.00	475	0.030	0.802	0.98	0.38
w/c = 0.4, SMS = 0.3% (Oven cured)		14.93	331	0.048	0.730	0.99	0.16
w/c = 0.4, SMS = 0.3% (Cured in Saturated Sand)		16.91	345	0.081	0.897	0.99	0.11

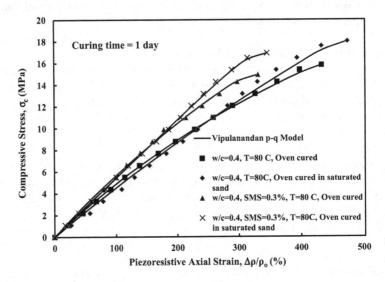

Figure 6.3 Piezoresistive response of the smart cement with and without SMS after 1 day of curing at 80°C.

increased to 345% when cured in saturated sand at 80°C (Table 6.1). The piezoresistivity at the peak compressive stress varied from 1,555 to 2,375 times the compressive strain of the smart cement.

Using the p-q Piezoresistive model (Eqn. (5.12)), the relationships between compressive stress and the piezoresistive axial strain $\left(\dfrac{\Delta \rho}{\rho_o} \right)$ of smart cement with and without 0.3% SMS for one day of curing at 80°C in air and saturated sand were modeled. The piezoresistive model (Eqn. (5.7) predicted the measured stress-piezoresistivity strain relationship very well, as shown in Figure 6.3. The model parameters q_2 and p_2 are summarised in Table 6.1. The coefficients of determination (R^2) were 0.98 to 0.99. The root-mean-square error (RMSE) varied between 0.11 MPa and 0.38 MPa, as summarised in Table 6.1.

Seven Days of Curing
The compressive strength (σ_{cf}) of smart cement after seven days of curing at 80°C in the oven was 29.65 MPa, which increased to 30.08 MPa when cured at 80°C in saturated sand, a 2% increase. Smart cement with 0.3% SMS air cured at 80°C had a compressive strength of 27.88 MPa, which increased to 28.54 MPa when cured in saturated sand at 80°C, a 2.3% increase, as summarised in Table 6.2.

Table 6.2 Piezoresistivity model parameters for smart cement with and without SMS cured at 80°C for seven days of curing

Composition and Curing Conditions	Curing Time (day)	Strength σ_{cf} (MPa)	Piezoresistive Strain at Peak Stress, $(\Delta\rho/\rho_o)_{cf}$ (%)	Model Parameter p_2	Model Parameter q_2	R^2	RMSE (MPa)
w/c = 0.4, (Oven cured)	7 days	29.65	360	0.070	0.768	0.99	0.37
w/c = 0.4, (Cured in Saturated Sand)		30.08	420	0.040	0.612	0.98	0.47
w/c = 0.4, SMS = 0.3% (Oven cured)		27.88	250	0.010	0.813	0.99	0.23
w/c = 0.4, SMS = 0.3% (Cured in Saturated Sand)		28.54	290	0.038	0.668	0.99	0.18

The piezoresistive axial strain at failure $\left(\dfrac{\Delta\rho}{\rho_o}\right)_f$ for smart cement air cured at 80°C was 360%, and it increased to 420% for smart cement cured in saturated sand at 80°C. For smart cement with 0.3% SMS and cured at 80°C in the oven, the piezoresistive axial strain at failure $\left(\dfrac{\Delta\rho}{\rho_o}\right)_f$ was 250%, which increased to 290% when cured in saturated sand at 80°C, as summarised in Table 6.2. The piezoresistivity at the peak compressive stress varied from 1,250 to 2,040 times the compressive strain of the smart cement.

Using the p-q piezoresistive model (Eqn. (5.7)), the relationships between compressive stress and the piezoresistive axial strain $\left(\dfrac{\Delta\rho}{\rho_o}\right)$ of smart cement with and without 0.3% SMS after seven days of curing in air and in saturated sand at 80°C were modeled. The piezoresistive model (Eqn. (5.7)) predicted the measured stress-change in the resistivity relationship very well, as shown in Figure 6.4. The model parameters q_2 and p_2 are summarised in Table 6.2. The coefficients of determination (R^2) were 0.98 to 0.99. The root-mean-square error (RMSE) varied between 0.18 MPa and 0.47 MPa, as summarised in Table 6.2.

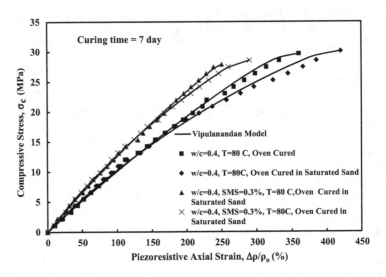

Figure 6.4 Piezoresistive response of the smart cement with and without SMS after 7 days of curing at 80°C.

Twenty-eight Days of Curing

The compressive strength (σ_{cf}) of smart cement after 28 days of curing at 80°C in the oven was 33.56 MPa, which increased to 37.98 MPa when cured in saturated sand at 80°C, a 13% increase. Smart cement with 0.3% SMS cured in the oven had a compressive strength of 33.15 MPa, which increased to 38.42 MPa when cured in saturated sand at 80°C, a 16% increase, as summarised in Table 6.3.

The piezoresistive axial strain at failure $\left(\dfrac{\Delta\rho}{\rho_o}\right)_f$ for smart cement cured at 80°C in the oven was 245%, which increased to 302% when cured in saturated sand at 80°C. Smart cement with 0.3% SMS cured in the oven at 80°C showed the piezoresistive axial strain at failure $\left(\dfrac{\Delta\rho}{\rho_o}\right)_f$ was 160%, which increased to 220% for specimens cured in saturated sand at 80°C, as summarised in Table 6.3. The piezoresistivity at the peak compressive stress varied from 900 to 1,510 times the compressive strain of the smart cement.

Using the p-q piezoresistive model (Eqn. (5.7)), the relationships between the compressive stress and the piezoresistive axial strain $\left(\dfrac{\Delta\rho}{\rho_o}\right)$ of smart cement with and without 0.3% SMS for 28 days of curing were modeled. The piezoresistive model (Eqn. 5.7) predicted the measured stress-change in the resistivity relationship very well, as shown in Figure 6.5. The model

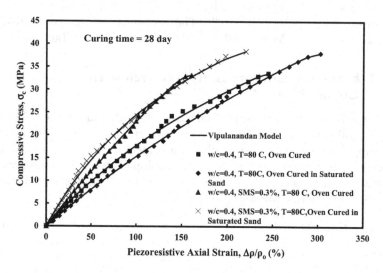

Figure 6.5 Piezoresistive Response of the Smart Cement with and without SMS after 28 days of curing at 80°C.

Table 6.3 Piezoresistivity model parameters for smart cement with and without SMS cured at 80°C for 28 days of curing

Composition and Curing Conditions	Curing Time (day)	Strength σ_{cf} (MPa)	Piezoresistive Strain at Peak Stress, $(\Delta\rho/\rho_o)_{cf}$ (%)	Model Parameter p_2	Model Parameter q_2	R^2	RMSE (MPa)
w/c = 0.4, (Oven cured)	28 days	33.59	245	0.010	0.626	0.98	0.39
w/c = 0.4, (Cured in Saturated Sand)		37.98	302	0.019	0.744	0.99	0.26
w/c = 0.4, SMS = 0.3% (Oven cured)		33.15	160	0.050	0.825	0.99	0.30
w/c = 0.4, SMS = 0.3% (Cured in Saturated Sand)		38.42	220	0.010	0.487	0.97	0.46

parameters q_2 and p_2 are summarised in Table 6.3. The coefficients of determination (R^2) were 0.97 to 0.99. The root-mean-square error (RMSE) varied between 0.26 MPa and 0.46 MPa, as summarised in Table 6.3.

6.3.2 Modeling Compressive Strength with Curing Time

The compressive strength of the cement made with and without SMS and cured at 80°C in the oven and saturated sand was measured up to 28 days of curing. The compressive strength of the cement increased with the curing time in a nonlinear manner. The relationship between the compressive strength of the cement and curing time was modeled with the Vipulanandan correlation model as follows:

$$\sigma_{cf} = t/(C_1 + D_1 t), \tag{6.1}$$

where

σ_{cf} = Compressive strength of the smart cement (MPa)
t = Curing time (day)

Parameters C_1 (day/MPa) and D (MPa^{-1}) are model parameters, and parameter C represents the initial rate of change and parameter D_1 determines the ultimate strength. For the cement cured at 80°C in the oven, experimental results matched very well, as shown in Figure 6.6 with the proposed model with coefficient of determination (R^2) of 0.99. For smart cement only, parameters C_1 and D_1 were found as 0.035 day/MPa and 0.028 MPa^{-1}. For smart cement with 0.3% SMS, parameters C_1 and D_1 were found as 0.039 day/MPa and 0.029 MPa^{-1}. For cement cured in saturated sand at 80°C, the experimental results also matched very well, as shown in Figure 6.6 with the proposed model with coefficient of determination (R^2) of 0.94–0.95. For smart cement only, parameters C_1 and D_1 were found as 0.032 day/MPa and 0.026 MPa^{-1}. For smart cement with 0.3% SMS, parameters C_1 and D_1 were found as 0.038 day/MPa and 0.026 MPa^{-1}.

6.3.3 Modeling Piezoresistive Strain at Failure with Curing Time

Piezoresistivity at failure for smart cement made with and without SMS oven cured and cured up to 28 days was investigated. With curing time increases, the piezoresistivity at failure of the cement specimen changes. The relationship between the piezoresistivity at failure of the cement grout and curing time has been modeled using the Vipulanandan correlation model as follows:

$$\Delta\rho/\rho_o = (\Delta\rho/\rho_o)_1 - t/(E_1 + F_1 t), \tag{6.2}$$

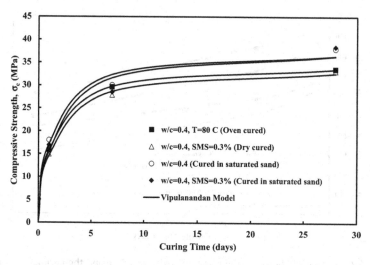

Figure 6.6 Relationship between the compressive strength and the curing time for the smart cement with and without SMS.

where

$\Delta\rho/\rho_o$ = piezoresistivity at failure (%)
$(\Delta\rho/\rho_o)_1$ = piezoresistivity at failure after 1 day (%)
t = Curing time (day)

Parameters E_1 (day/Ω.m) and F_1 (Ω.m)$^{-1}$ are model material parameters, and the parameter E_1 represent the initial rate of change, and the parameter F determines the ultimate piezoresistivity. For cement cured at 80°C in the oven, the model predicted the experimental results very well, as shown in Figure 6.7 with the coefficient of determination (R^2) in the range of 0.98 to 0.99. For the oven dry cured smart cement, parameters E_1 and F_1 were 0.083 (day/Ω.m) and 0.0033 (Ω.m)$^{-1}$ respectively, and the estimated ultimate piezoresistivity (infinite time) was over 130%. For the oven dry cured smart cement with 0.3% SMS, parameters E_1 and F_1 were found as 0.067 (day/Ω.m) and 0.0044 (Ω.m)$^{-1}$ respectively, and the estimated ultimate piezoresistivity (infinite time) was over 100%. For the cement specimens cured in saturated sand at 80°C, experimental results also matched very well, as shown in Figure 6.7 with the model with coefficient of determination (R^2) of 0.98–0.99. For smart cement cured in saturated sand at 80°C, parameters E_1 and F_1 were 0.121 (day/Ω.m) and 0.0033 (Ω.m)$^{-1}$ respectively, and the estimated ultimate piezoresistivity (infinite time) was over 175%. For smart cement with 0.3% SMS cured in saturated sand at 80°C,

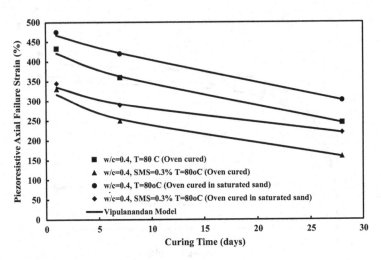

Figure 6.7 Relationship between the piezoresistivity at failure and the curing time for the smart cement with and without SMS.

parameters E_1 and F_1 were found as 0.104 (day/Ω.m) and 0.0043 $(\Omega.m)^{-1}$ respectively, and the estimated ultimate piezoresistivity (infinite time) was over 120%.

6.4 METAKAOLIN

6.4.1 Tensile Piezoresistivity Behavior

Tensile piezoresistive behaviors were experimentally investigated by performing direct tension, splitting tension, and bending tests. With the addition of up to 10% metakaolin (inorganic additive), the tests showed that smart cement cured at room condition was a highly sensitive chemo-piezoresistive material.

(a) Direct Tension

The direct tensile piezoresisitive behavior of smart cement with 10% metakaolin cured for 28 days is shown in Figure 6.8 and also summarised in Table 6.4. The resistivity increased with applied tensile stress. The tensile stress-piezoresisitive tensile strain response was resistivity strain hardening. The tensile strength was about 1.82 MPa, and the resistivity strain change was 9.9.%. Adding 10% metakaolin increased the tensile strength of smart cement of 1.59 MPa by about 14%. Also, with the addition of 10% metakaolin, the piezoresistive tensile strain at failure of smart cement of

Table 6.4 Comparing the direct tensile strengths and piezoresistive tensile strains at failure for smart cement modified with 10% metakaolin and 50% fly ash

| Smart Cement | 28 days of Curing | | Remarks |
	Strength (MPa)	Piezoresistive Axial Strain (%)	
Class H*	1.59	10 9.5	Strain hardening. The parameter q_2 (modulus ratio) was 3.02.
10% Metakaolin	1.82	9.9	Strain hardening. The strength and piezoresistive failure strain increased. The parameter q_2 (modulus ratio) was 3.54.
50% Fly ash	1.26	8.1	Strain softening. The strength and piezoresistive failure strain decreased. The parameter q_2 (modulus ratio) was 0.29.

*Chapter 5

9.5% was increased by about 4%. The failure tensile strain of smart cement was 0.023% (Figure 5.7); hence, the change in resistivity was over 430 times (43,000%) higher, making smart cement highly sensing.

The initial piezoresisitive modulus (Ei) was 5 0 MPa, and the secant modulus at failure was 18.4 MPa. Hence, the Vipulanandan piezoresisitive strain hardening model parameter q_2, representing the ratio of secant modulus at peak stress to the initial modulus was 3.54.

(b) Splitting Tension

The splitting tensile piezoresisitive behavior of smart cement with 10% metakaolin cured for 28 days is shown in Figure 6.9 and also summarised in Table 6.5. The change in resistivity was measured across the section in a horizontal direction (along the diameter perpendicular to the loading direction). The percentage change in resistivity was positive up to the failure. The tensile stress-piezoresisitive tensile strain response was resistivity strain hardening. The tensile strength was 1.92 MPa, and the resistivity strain change was 13.1%. Adding 10% metakaolin increased the tensile strength of smart cement of 1.78 MPa by about 8%. Also, with the addition of 10% metakaolin the piezoresistive tensile strain at failure of smart cement of 13% was increased by about 0.8%. The failure tensile strain of smart cement was 0.023% (Figure 5.7); hence, the change in resistivity was over 570 times (57,000%) higher, making the smart cement modified with 10% metakaolin highly sensing.

The initial piezoresisitive modulus (Ei) was 2.86 MPa, and the secant modulus at failure was 14.66 MPa. Hence, the Vipulanandan piezoresisitive

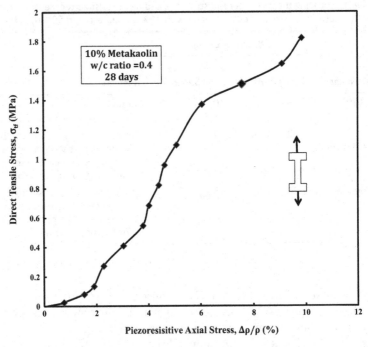

Figure 6.8 Direct tensile stress – piezoresistive axial strain relationship for the smart cement modified with 10% metakaolin and cured for 28 days.

Table 6.5 Comparing the splitting tensile strengths and piezoresistive tensile strains at failure for smart cement modified with 10% metakaolin and 50% fly ash

| Smart Cement | 28 days of Curing | | Remarks |
	Strength (MPa)	Piezoresistive Axial Strain (%)	
Class H*	1.78	13	Strain softening. The parameter q_2 (modulus ratio) was 0.69.
10% Metakaolin	1.92	13.1	Strain hardening. The strength and piezoresistive failure strain increased. The parameter q_2 (modulus ratio) was 5.12
50% Fly ash	1.8	8.3	Strain softening. The strength decreased but the piezoresistive failure strain increased. The parameter q_2 (modulus ratio) was 0.41.

*Chapter 5

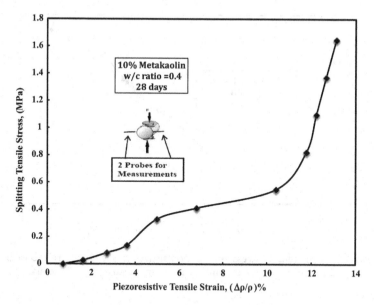

Figure 6.9 Splitting tensile stress versus piezoresistive lateral diagonal strain for the smart cement modified with 10% metakaolin after 28 days of curing.

strain hardening model parameter q_2, representing the ratio of secant modulus at peak stress to the initial modulus was 5.12.

(c) Bending Behavior

The bending tensile piezoresisitive behavior of smart cement with 10% metakaolin cured for 28 days is shown in Figure 6.10 and also summarised in Table 6.6. The change in resistivity was measured along the length close to the bottom of the beam. The percentage change in resistivity was positive up to the failure. The bending tensile stress-piezoresisitive tensile strain response was resistivity strain hardening, as shown in Figure 6.10. The bending tensile strength was 3.81 MPa, and the piezoresistive tensile strain change was 9.0%. Adding 10% metakaolin increased the bending tensile strength of smart cement of 3.35 MPa by about 14%. Also, with the addition of 10% metakaolin, the piezoresistive tensile strain at failure of smart cement of 8.2% was increased by about 10%. The failure bending tensile strain of smart cement was 0.045% (Figure 5.8); hence, the change in resistivity was over 200 times (20,000%) higher, making smart cement modified with 10% metakaolin highly sensing.

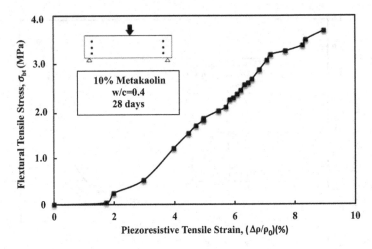

Figure 6.10 Bending tensile stress and piezoresistive tensile strain relationship after 28 days of curing for the smart cement modified with 10% metakaolin.

Table 6.6 Comparing the flexural strengths and piezoresistive tensile strains at failure for smart cement modified with 10% metakaolin and 50% fly ash

Smart Cement	28 days of Curing		Remarks
	Strength (MPa)	Piezoresistive Axial Strain (%)	
Class H*	3.35	8.2%	Strain hardening. The parameter q_2 (modulus ratio) was 7.3.
10% Metakaolin	3.81	9.0%	Strain hardening. The strength and piezoresistive failure strain increased. The parameter q_2 (modulus ratio) was 3.39
50% Fly ash	3.15	11.4%	Strain hardening. The strength decreased, but the piezoresistive tensile failure strain increased. The parameter q_2 (modulus ratio) was 2.72.

*Chapter 5

The initial piezoresisitive modulus (E_i) was 12.5 MPa, and the secant modulus at failure was 42.3 MPa. Hence, the Vipulanandan piezoresisitive strain hardening model parameter q_2, representing the ratio of secant modulus at peak stress to the initial modulus was 3.39

6.5 FLY ASH

6.5.1 Tensile Piezoresistive Behavior

Tensile piezoresistive behaviors were experimentally investigated by performing direct tension, splitting tension, and bending tests. With the addition of 50% fly ash (inorganic additive), the tests showed that smart cement cured under room condition was a highly sensitive chemo-piezoresistive material.

(a) Direct Tension

The direct tensile piezoresisitive behavior of smart cement with 50% fly ash cured for 28 days is shown in Figure 6.11. The resistivity increased with applied tensile stress. The tensile stress-piezoresisitive tensile strain response was resistivity strain softening. The tensile strength was about 1.24 MPa, and the resistivity strain change was 8.5%. Adding 50% fly ash decreased the tensile strength of smart cement of 1.59 MPa by about 22%. Also, with the addition of 50% fly ash the piezoresistive tensile strain at failure of smart cement of 9.5% was decreased by about 10.5%. The failure tensile strain of smart cement was 0.023% (Figure 5.7); hence, the change in resistivity was over 370 times (37,000%) higher, making smart cement highly sensing.

The initial piezoresisitive modulus (E_i) was 50 MPa, and the secant modulus at failure was 14.6 MPa. Hence, the Vipulanandan piezoresisitive strain softening model parameter q_2, representing the ratio of secant modulus at peak stress to the initial modulus was 0.29.

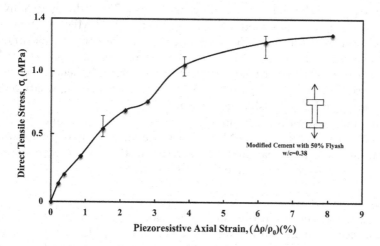

Figure 6.11 Direct tensile stress versus piezoresistive axial strain relationship for the smart cement modified with 50% fly ash.

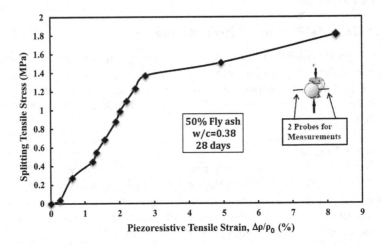

Figure 6.12 Splitting tensile stress versus piezoresistive lateral diagonal strain for the smart cement modified with 50% fly ash after 28 days of curing

(b) Splitting Tension

The splitting tensile piezoresisitive behavior of smart cement with 50% fly ash cured for 28 days is shown in Figure 6.12. The change in resistivity was measured across the section in the horizontal direction (along the diameter perpendicular to the loading direction). The percentage change in resistivity was positive up to the failure. The tensile stress-piezoresisitive tensile strain response was resistivity strain softening. The tensile strength was 1.80 MPa, and the resistivity strain change was 8.3%. Adding 50% fly ash increased the tensile strength of smart cement of 1.78 MPa by about 1%. Also, with the addition of 50% fly ash, the piezoresistive tensile strain at failure of smart cement of 13% was decreased by about 36%. The failure tensile strain of smart cement was 0.023% (Figure 5.7); hence, the change in resistivity was over 361 times (36,100%) higher, making smart cement modified with 50% fly ash highly sensing.

The initial piezoresisitive modulus (Ei) was 53.33 MPa, and the secant modulus at failure was 21.69 MPa. Hence, the Vipulanandan piezoresisitive strain softening model parameter q_2, representing the ratio of secant modulus at peak stress to the initial modulus was 0.41.

(c) Bending Behavior

The bending tensile piezoresisitive behavior of smart cement with 50% fly ash cured for 28 days is shown in Figure 6.13. The change in resistivity was measured along the length close to the bottom of the beam. The percentage change in resistivity was positive up to the failure. The bending tensile stress-piezoresisitive tensile strain response was resistivity strain hardening (Figure 6.13). The bending tensile strength was 3.15

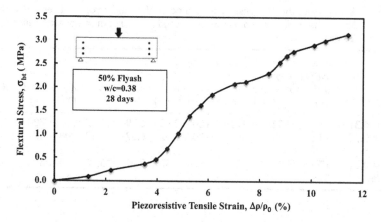

Figure 6.13 Bending tensile stress and piezoresistive tensile strain relationship after 28 days of curing for the smart cement modified with 50% fly ash.

MPa, and the piezoresistive tensile strain change was 11.4%. Adding 50% fly ash decreased the bending tensile strength of smart cement of 3.35 MPa by about 6%. Also, with the addition of 50% fly ash, the 8.2% piezoresistive tensile strain in bending at failure of smart cement was increased by about 39%. The failure bending tensile strain of smart cement was 0.045% (Figure 5.8); hence, the change in resistivity was over 253 times (25,300%) higher, making smart cement modified with 50% fly ash highly sensing.

The initial piezoresisitive modulus (E_i) was 10 MPa, and the secant modulus at failure was 27.2 MPa. Hence, the Vipulanandan piezoresisitive strain hardening model parameter q_2, representing the ratio of secant modulus at peak stress to the initial modulus was 2.72.

6.6 CONTAMINATION

6.6.1 Clay

In this study, the effects of up to 5% montmorillonite clay soil contamination on the initial properties and piezoresistive behavior of smart Portland cement was investigated. Smart Portland cement was made by mixing the cement (Type I) with 0.1% carbon fibers to enhance the sensing properties. Based on the type of construction, cement might get contaminated with varying amounts of clay soils. Hence, a series of experiments were performed to evaluate the cement behavior with and without up to 5% of montmorillonite clay contamination to determine the effects on the initial properties and the piezoresistivity with strength up to 28 days under room condition.

Table 6.7 Summary of the bulk resistivity parameters for smart Portland cement with
and without clay soil contamination cured under room temperature up to
28 days

Mix Type	Unit Weight (kN/ m³)	Initial resistivity, ρ_o (Ω.m)	ρ_{min} (Ω.m)	t_{min} (min)	ρ_{24hr} (Ω.m)	$\rho_{7\,days}$ (Ω.m)	$\rho_{28\,days}$ (Ω.m)	$RI_{24\,hr}$ (%)	$RI_{7\,days}$ (%)	$RI_{28\,days}$ (%)
w/c = 0.38	19.8	0.92	0.84	180	2.48	6.79	11.37	195	708	1,253
w/c = 0.38 Clay = 1%	19.5	0.94	0.85	180	2.62	6.57	12.30	208	673	1,347
w/c = 0.38 Clay = 5%	17.8	1.15	1.07	180	2.82	8.17	15.10	164	664	1,311

(a) Density

The initial unit weight of smart Portland cement with a w/c ratio of 0.38 was
19.8 kN/m³, as summarised in Table 6.7. The initial unit weight decreased
with montmorillonite clay contamination. With 5% clay soil contamination, the unit weight decreased to 17.8 kN/m³, a 10% reduction.

(b) Initial Resistivity

The initial resistivity of modified Portland cement slurry was 0.92 Ω.m, and
it increased with the clay soil contamination, as summarised in Table 6.7.
With 1% clay soil contamination, the initial resistivity was increased to 0.94
Ω.m, and with 5% clay soil contamination the initial resistivity was 1.15
Ω.m, a 25% increase. Hence, the initial resistivity increases were more than
two times more sensitive than the density changes of clay soil contamination
in the cement.

(c) Piezoresistivity and Strength

One Day of Curing

The compressive strengths (σ_{cf}) of smart Portland cement with 0%, 1%, and
5% clay soil contamination for one day of curing were 9.88 MPa, 9.44 MPa,
and 8.12 MPa, a 4% and 18% reduction when the clay content increased
about 1% and 5% respectively, as summarised in Table 6.8.

The piezoresistive axial strain at failure $\left(\dfrac{\Delta\rho}{\rho_o}\right)_f$ for modified Portland

cement was 432%, which was reduced to 411% and 230% respectively
with 1% and 5% clay, as summarised in Table 6.8. With 5% clay soil
contamination to smart Portland cement, the piezoresistive axial strain at

failure $\left(\dfrac{\Delta\rho}{\rho_o}\right)_f$ was reduced about 45% from that of smart Portland cement.

Table 6.8 Compressive strength, piezoresistivity, model parameters p_2 and q_2 for smart Portland Cement after 1 day and 28 days of curing

Mix Type	Curing Time (day)	Strength σ_f (MPa)	Piezoresistivity at peak stress, $(\Delta\rho/\rho_o)_f$ (%)	p_2	q_2	R^2	RMSE (MPa)
w/c = 0.38	1 day	9.88	432	0.047	2.77	0.99	0.21
w/c = 0.38 Clay = 1%		9.44	411	0.025	1.48	0.98	0.43
w/c = 0.38 Clay = 5%		8.12	230	0.031	1.64	0.97	0.43
w/c = 0.38	28 days	31.40	270	0.062	0.75	0.98	0.44
w/c = 0.38 Clay = 1%		30.08	209	0.052	0.75	0.99	0.34
w/c = 0.38 Clay = 5%		27.44	158	0.125	0.78	0.99	0.34

Using the p-q Piezoresistive model (Eqn. 5.7)), the relationships between compressive stress and the piezoresistive axial strain $\left(\dfrac{\Delta\rho}{\rho_o}\right)$ of smart Portland cement with different clay content of 0%, 1%, and 5% for one day of curing were modeled. The piezoresistive model (Eqn. (5.7) predicted the measured stress-change in the resistivity relationship very well, as shown in Figure 6.14. The model parameters q_2 and p_2 are summarised in Table 6.8. The coefficients of determination (R^2) were 0.97 to 0.99. The root-mean-square error (RMSE) varied between 0.21 MPa and 0.43 MPa, as summarised in Table 6.8.

Twenty-eight Days of Curing

The compressive strengths (σ_{cf}) of modified Portland cement with 0%, 1%, and 5% clay soil contamination for one day of curing were 31.40 MPa, 30.08 MPa, and 27.44 MPa, a 4% and 13% reduction when the clay content increased about 1% and 5% respectively, as summarised in Table 6.8.

The piezoresistive axial strain at failure $\left(\dfrac{\Delta\rho}{\rho_o}\right)_f$ for modified Portland cement was 270%, which was reduced to 209% and 158% respectively with 1% and 5% clay, as summarised in Table 6.8. With 5% clay soil contamination to the modified Portland cement, the piezoresistive axial strain at failure $\left(\dfrac{\Delta\rho}{\rho_o}\right)_f$ was reduced about 40% from that of modified Portland cement.

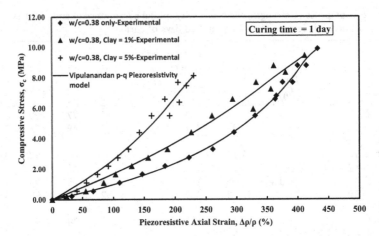

Figure 6.14 Piezoresistive response of the smart cement with and without montmorillonite clay soil contamination after 1 day of curing and predicted using the
Vipulanandan p-q piezoresistivity model.

Using the p-q Piezoresistive model (Eqn. (5.7)), the relationships between compressive stress and the piezoresistive axial strain $\left(\dfrac{\Delta\rho}{\rho_o}\right)$ of modified Portland cement with different clay content of 0%, 1%, and 5% for one day of curing were modeled. The piezoresistive model (Eqn. (5.7)) predicted the measured stress-change in the resistivity relationship very well, as shown in Figure 6.15. The model parameters q_2 and p_2 are summarised in Table 6.8. The coefficients of determination (R^2) were 0.98 to 0.99. The root-mean-square error (RMSE) varied between 0.34 MPa and 0.44 MPa, as summarised in Table 6.8.

6.6.1.1 Relationship between Curing Time and Strength and Piezoresistive Strain at Failure

The strength of the smart cement specimen made with and without clay soil contamination was measured up to 28 days of curing. With curing time increase, the compressive strength of the cement specimen increased. The relationship between the compressive strength of the cement and the curing time has been modeled with the Vipulanandan property correlation (two parameters since the initial condition is zero) model used for over two decades, and the relationship is as follows:

$$\sigma_c = t/(C_1 + D_1 t), \tag{6.3}$$

where

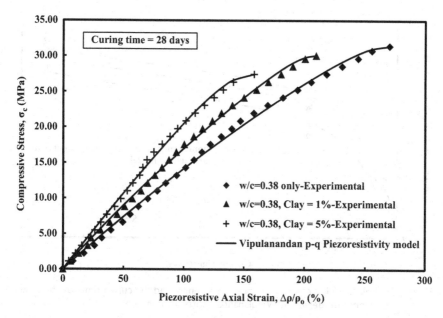

Figure 6.15 Piezoresistive response of the smart cement with and without montmorillonite clay soil contamination after 28 days of curing and predicted using the Vipulanandan p-q piezoresistivity model.

σ_c = Compressive strength of the grout (MPa)
t = Curing time (day)

Parameters C_1 and D_1 are model parameters, and parameter C_1 represents the initial rate of change and parameter D_1 determines the ultimate strength. The experimental results matched very well, as shown in Figure 6.16, with the proposed model with coefficient of determination (R^2) varied from 0.95 to 0.96. For smart Portland cement only, parameters C_1 and D_1 were found as 0.098 MPa^{-1}day and 0.029 MPa^{-1}. For smart Portland cement with 1% clay, parameters C_1 and D_1 were 0.119 MPa^{-1}day and 0.030 MPa^{-1}. For smart Portland cement with 5% clay, parameters C_1 and D_1 were 0.151 MPa^{-1}day and 0.032 MPa^{-1}.

The piezoresistivity at the peak failure stress for the cement specimen made with and without clay soil contamination was measured up to 28 days of curing. With the increase in curing time, the piezoresistivity at the peak failure stress for the cement reduced. The relationship between the piezoresistivity at failure of the cement and curing time has been modeled with the Vipulanandan property correlation model as follows:

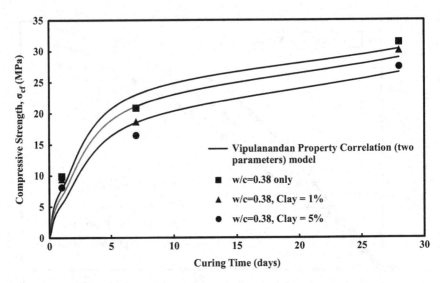

Figure 6.16 Relationship between compressive strength and the curing time for the smart cement modeled with Vipulanandan correlation model.

$$\Delta\rho/\rho_o = (\Delta\rho/\rho_o)_1 - t/(E_1 + F_1 t), \tag{6.4}$$

where

$\Delta\rho/\rho_o$ = Piezoresistivity at failure (%)
$(\Delta\rho/\rho_o)_1$ = Piezoresistivity at failure after 1 day (%)
t = Curing time (day)

Parameters E_1 and F_1 are model material parameters, and parameter E_1 represents the initial rate of change and parameter F_1 determines the ultimate piezoresistivity. The experimental results matched very well, as shown in Figure 6.17, with the proposed model with coefficient of determination (R^2) varied from 0.95 to 0.99. For modified Portland cement only, parameters E_1 and F_1 were found as 0.051 $\Omega m^{-1} day$ and 0.004 Ωm^{-1}. For modified Portland cement with 1% clay, parameters E_1 and F_1 were found as 0.033 $\Omega m^{-1} day$ and 0.003 Ωm^{-1}. For modified Portland cement with 5% clay, parameters E_1 and F_1 were found as 0.551 $\Omega m^{-1} day$ and 0.006 Ωm^{-1}.

6.6.2 Drilling Oil-Based Mud (OBM) Contamination

The test specimens were prepared following API standards. API Class H cement was used with a water-cement ratio of 0.38. For all the samples,

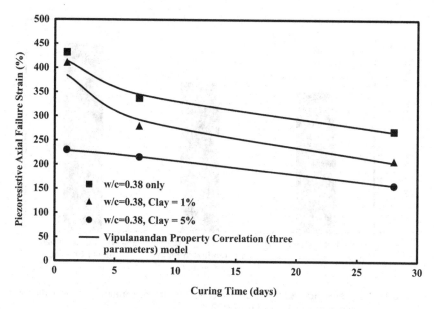

Figure 6.17 Relationship between piezoresistivity at failure and the curing time for the smart cement modeled with Vipulanandan correlation model.

0.075% (by the weight of total, BWOT) of conductive filler (CF) was added to the slurry in order to enhance the piezoresistivity of the cement and to make it more sensing. The smart cement slurry was mixed and then contaminated with up to 3% OBM. Mineral oil was used to prepare the oil-based mud with an oil-to-water ratio of 4:1 with the addition of 1% (by the weight of total, BWOT) of chemical surfactant. After mixing, the slurries were casted into cylindrical molds with a height of 100 mm and a diameter of 50 mm, in which two conductive wires were embedded 50 mm apart from each other in order to monitor the resistivity development of the specimens during the curing time and also to measure the piezoresistivity of the specimens. Specimens were cured up to 28 days under water at room temperature.

(a) Density
The average density of smart cement was 19.4 kN/m³ (16.47 ppg), and the density of the OBM was 8.6 kN.m³ (7.34 ppg). As cement was contaminated with OBM, the density of the cement slurry reduced due to a lower density of OBM. Study showed that 0.1% of OBM contamination reduced the density of the smart cement slurry by 0.5% to 19.3 kN/m³ (16.39 ppg), and 3% of OBM contamination reduced it to 18.3 kN/m³ (15.53 ppg), representing a 5.7% decrease.

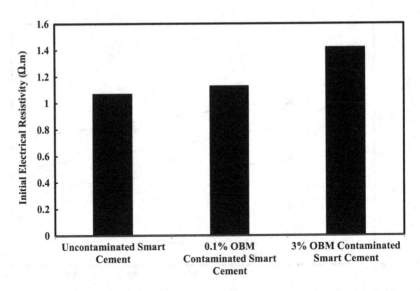

Figure 6.18 Initial electrical resistivity of smart cement slurries contaminated with different percentages of oil-based mud (OBM).

(b) Electrical Resistivity

Initial Resistivity

The initial resistivity of smart cement slurries contaminated with different percentages of OBM was investigated.

- (a) **Smart Cement:** The average initial resistivity of the cement slurry was 1.07 Ω.m.
- (b) **Contaminated Smart Cement:** Contamination of the cement by 0.1% and 3% OBM resulted in an increase in initial resistivity to 1.13 Ω.m and 1.42 Ω.m respectively, as shown in Figure 6.18 and summarised in Table 6.9. Hence, a contamination of 0.1% and 3% of OBM resulted in an increase of 6% and 33% respectively. The changes in density of the smart cement slurry by OBM contamination of 0.1% and 3% were 0.5% and 5.7% respectively.

Hence, comparing the changes in the density of the OBM-contaminated cement slurry and the changes in initial resistivity, it can be concluded that changes in the resistivity of the OBM-contaminated cement slurry were 6 to 12 times more than the changes in density.

The percentage change in resistivity due to OBM contamination was higher than the percentage change in density, an indication of the sensitivity of electrical resistivity.

Table 6.9 Electrical resistivity curing parameters of smart cement slurries contaminated with different percentage of OBM

Smart Cement	ρ_0 (Ω .m)	ρ_{min} (Ω .m)	t_{min} (minute)	ρ_{24} (Ω .m)	$\dfrac{\rho_{24} - \rho_{min}}{\rho_{min}}$ _
Uncontaminated cement	1.07	0.95	85	2.86	201%
0.1% OBM contaminated cement	1.13	0.96	100	2.68	180%
3% OBM contaminated cement	1.42	1.18	120	1.89	84%

One Day Curing

The resistivity decreased initially and reached a minimum resistivity of ρ_{min} at specific time of t_{min}.

(a) **Smart Cement:** The minimum resistivity of smart cement slurry was 0.95 Ω.m, which happened 85 minutes after mixing the sample. The resistivity after 24 hours of curing was 2.86 Ω.m, as summarised in Table 6.9.

(b) **Contaminated Smart Cement:** OBM contamination increased the ρ_{min} of the smart cement slurry by 1% and 24%, from 0.95 Ω.m to 0.96 Ω.m and 1.18 Ω.m respectively, for 0.1% and 3% contamination. OBM contamination also delayed the hydration process owing to the fact that OBM coated the surface of cement particles, which hinders the continuance of hydration. OBM contamination of 0.1% and 3% delayed t_{min} by 15 minutes and 35 minutes respectively, as summarised in Table 6.9.

Twenty-eight Days of Curing

After the setting time, hardened cement has a complete connected network that leads to the forming of a percolated path of C-S-H, causing high resistivity due to its continuous gel micropores. Later, volume fraction of C-S-H will play the main role in changes in the resistivity (Liu et al. 2014).

(a) **Smart Cement:** After 28 days of curing the smart cement under water, the resistivity reached 12.2 Ω.m., more than a 10 times (1,040%) change from the initial resistivity.

(b) **Contaminated Smart Cement:** OBM contamination reduced the development of resistivity during 28 days of curing due to its hindering effect on the hydration process, which caused less production of C-S-H after 28 days. OBM contamination of 0.1% and 3%

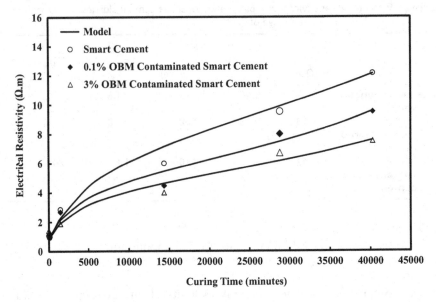

Figure 6.19 Electrical resistivity development of the smart cement contaminated with different percentages of oil-based mud (OBM).

reduced the resistivity of the cement by 22% and 42% to 9.5 Ω.m and 7 Ω.m respectively after 28 days of curing under water, as shown in Figure 6.19.

Electrical Resistivity Index (RI₂₄)

Vipulanandan et al. (2014b, c and 2015b, c, d, e) suggested the resistivity index (RI_{24}) as an indicator of hydration development after 24 hours, which is the maximum percentage change in resistivity after 24 hours with respect to minimum resistivity. This index represents the strength of the hardened cement.

(a) **Smart Cement:** The RI_{24} for the cement was 201%.

(b) **Contaminated Smart Cement:** OBM contamination reduced the development of hydration. OBM contamination of 0.1% and 3% reduced the RI_{24} by 10% and 58% respectively to 180% and 84%, as shown in Figure 6.20.

(c) Compressive Strength

The compressive strength of smart cement was tested after 1 and 28 days of curing under water.

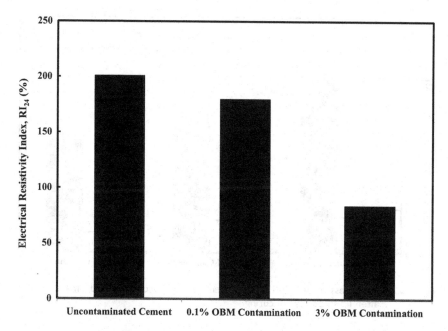

Figure 6.20 Electrical resistivity index of smart cement contaminated with different percentages of OBM.

One Day Curing

(a) **Smart Cement:** The compressive strength of the cement without any contamination or modification was 8.4 MPa (1.22 ksi) after one day of curing, as shown in Figure 6.21.

(b) **Contaminated Smart Cement:** Contamination of the cement with OBM decreased the compressive strength of the smart cement. The compressive strength of the smart cement contaminated with 0.1% and 3% of OBM decreased the compressive strength to 7.6 MPa (1.1 ksi) and 4.6 MPa (0.67 ksi) respectively, a 10% and 45% reduction after one day of curing under water, as shown in Figure 6.21.

Twenty-eight Days of Curing

(a) **Smart Cement:** The compressive strength of the smart cement after 28 days of curing under water was 41 MPa (5.96 ksi), as shown in Figure 6.21.

(b) **Contaminated Smart Cement:** The compressive strength of the smart cement contaminated with 0.1% and 3% of OBM decreased the

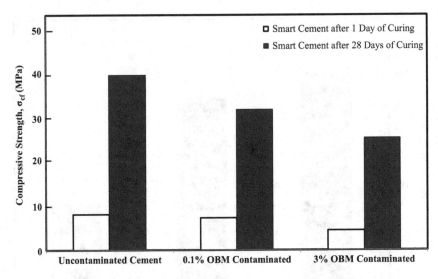

Figure 6.21 Compressive strength development of OBM-contaminated smart cement after 1 and 28 days of curing.

compressive strength by 20% and 36% respectively to 33 MPa (4.78 ksi) and 26,3 MPa (3.81 ksi) after 28 days of curing under water, as shown in Figure 6.21.

(d) Piezoresistivity
The piezoresistive behavior of smart cement was evaluated after 1 and 28 days of curing under water.

One Day Curing

(a) **Smart Cement:** After one day of curing, the piezoresistive axial strain at failure of the smart cement was 375%, as shown in Figure 6.22. The model parameters p_2 and q_2 were 0.13 and 0.55 respectively.

(b) **Contaminated Smart Cement:** Contamination of the smart cement with 3% OBM reduced the piezoresistive axial strain at failure to 320%, about 15% decrease, as shown in Figure 6.23. The p-q model parameters p_2 and q_2 for the 3% OBM-contaminated smart cement were 0.29 and 0.44 respectively.

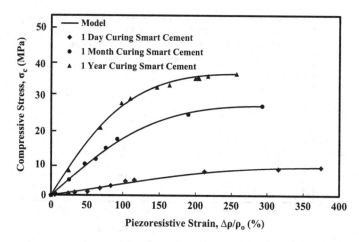

Figure 6.22 Piezoresistivity behavior of the smart cement without OBM contamination.

One Month Curing

(a) **Smart Cement:** After one month of curing, the piezoresistive axial strain at failure of the smart cement was 293%, as shown in Figure 6.22. Parameters p_2 and q for the model were 0.13 and 0.40 respectively.

(b) **Contaminated Smart Cement:** Contamination of the smart cement with 3% of OBM contamination reduced the piezoresistive axial strain at failure of the smart cement to 220%, about a 25% decrease, as shown in Figure 6.23. The model parameters p_2 and q_2 for the 3% OBM-contaminated cement were 0.38 and 0.35 respectively, as summarised in Table 6.10.

One Year Curing

(a) **Smart Cement:** After one year of curing, the piezoresistive axial strain at failure of the smart cement was 257%, as shown in Figure 6.22. The model parameters p_2 and q_2 were 0.48 and 0.42 respectively, as summarised in Table 6.10.

(b) **Contaminated Smart Cement:** Smart cement with 3% of OBM contamination reduced the piezoresistivity of the smart cement to 194%, about a 25% decrease, as shown in Figure 6.23. The model parameters of the p-q model parameters for p_2 and q_2 for the 3% OBM-contaminated cement were 0.52 and 0.32 for p_2 and q_2 respectively, as summarised in Table 6.10.

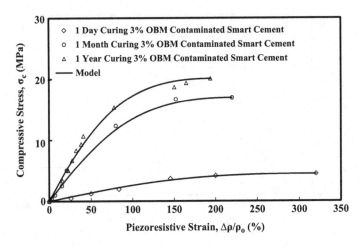

Figure 6.23 Piezoresistivity behavior of the smart cement with 3% OBM contamination up to one year of curing.

Table 6.10 The piezoresistive model parameters for smart cement with and without OBM contamination after one day, one month, and one year of curing

Smart Cement Contamiantion	1 Day Curing				1 Month Curing				1 Year Curing			
	P_2	q_2	R^2	RMSE (MPa)	P_2	q_2	R^2	RMSE (MPa)	P_2	q_2	R^2	RMSE (MPa)
0	0.13	0.55	0.99	0.36	0.22	0.45	0.99	0.56	0.48	0.42	0.99	0.62
3%	0.29	0.44	0.99	0.19	0.38	0.35	0.99	0.57	0.52	0.32	0.99	0.64

6.7 CARBON DIOXIDE (CO₂) CONTAMINATION

The smart cement slurry was prepared with 1% and 3% of dry ice (CO_2) in water. The test specimens were prepared following API standards. API Class H cement was used with a w/c ratio of 0.38. For all the samples, 0.04% (based on weight of cement) of carbon fiber (CF) was added to the slurry in order to enhance the piezoresistivity of the cement and to make it more sensing. After mixing, the slurries were casted into cylindrical molds with a height of 100 mm and a diameter of 50 mm, with two conductive wires embedded 50 mm apart vertically to monitor the resistivity development of the specimens during the curing time. After one day, all the specimens were demolded and were cured for 28 days under water.

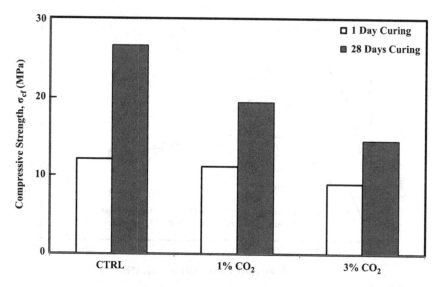

Figure 6.24 Compressive strength development of CO_2-contaminated smart cement after 1 and 28 days of curing under water.

(a) Compressive Strength

The compressive behavior of smart cement was tested after 1 and 28 days of curing under water at room temperature, as shown in Figure 6.24.

One Day Curing

(a) **Smart Cement:** The compressive strength of the smart cement was 12.5 MPa (1.81 ksi) after one day of curing.

(b) **CO_2 Contaminated Smart Cement:** CO_2 contamination decreased the compressive strength of the smart cement. The compressive strength of the smart cement contaminated with 1% and 3% of $CO2$ decreased to 11.5 MPa (1.67 ksi) and 9.2 MPa (1.34 ksi) respectively, a 8% and 26% reduction after one day of curing.

Twenty-eight Days of Curing

(a) **Smart Cement:** The compressive strength of smart cement after 28 days of curing under water was 27.5 MPa (3.98 ksi).

(b) **CO_2-Contaminated Smart Cement:** The compressive strength of smart cement contaminated with 1% and 3% of CO_2 decreased by 27% and 45% respectively to 20.0 MPa (2.90 ksi) and 15.0 MPa (2.17 ksi) after 28 days of curing.

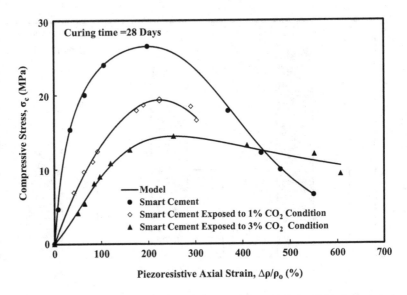

Figure 6.25 Piezoresistivity of CO_2-contaminated smart cement after 28 days of curing.

(b) Piezoresistivity

The stress-piezoresistive strain behavior of smart cement was evaluated after 28 days of curing under water using the Vipulanandan p-q piezoresistive strain softening model (Eqn. 5.6), as shown in Figure 6.25.

Twenty-eight Days Curing

(a) **Smart Cement:** After 28 days of curing, the piezoresistivity of smart cement was 199%. The model parameters p_2 and q_2 were 0.45 and 0.15 respectively, as summarised in Table 6.11.

(b) **CO2-Contaminated Smart Cement:** CO_2 contamination increased the piezoresistive behavior of smart cement, as shown in Figure 6.25. The piezoresistivity of smart cement contaminated with 1% CO_2 was 224% at the failure, a 13% increase. The p-q model parameters p_2 and q_2 for the 1% CO2-contaminated smart cement were 0.35 and 0.47 respectively. The piezoresistive strain at failure of smart cement contaminated with 3% CO_2 was 254% at the failure, a 28% increase. The p-q model parameters p_2 and q_2 for the 3% CO2-contaminated smart cement were 0.15 and 0.76 respectively, as summarised in Table 6.11.

Table 6.11 Piezoresistive model parameters for smart cement contaminated with CO_2 after 28 days of curing

Materials	28 Days Curing				Compressive Strength (MPa)	Piezoresistivity at Failure (%)
	P_2	q_2	R^2	RMSE (MPa)		
Smart Cement	0.45	0.15	0.99	0.49	27.5	199
1% CO2 + Smart Cement	0.35	0.47	0.98	0.68	20.0	224
3% CO2 + Smart Cement	0.15	0.76	0.99	0.52	15.0	254

6.8 SUMMARY

The main focus was to experimentally verify the chemo-thermo-piezoresistive behavior of smart cement. In order to evaluate the chemical (chemo) sensitivity, smart cements with various inorganic additives (sodium meta silicate, metakaolin, fly ash) and inorganic and organic contaminants (clay, oil-based mud, CO_2-carbon dioxide) were tested. Also, the effect of temperature and the curing environments (oven and saturated sand) on smart cement behavior was investigated with and without the sodium meta silicate additive. Compression, direct tension, splitting tension, and bending tests were performed to evaluate the piezoresistivity of the modified smart cements. Based on the experimental study and analytical modeling, the following conclusions are advanced:

1 The additives (inorganic) and contaminants (organic and inorganic), including CO_2 (organic), used in this study changed the density and initial resistivity of smart cement. All the changes in the resistivity have been quantified. The smart cement with the additives and contaminants was a highly sensing chemo-piezoresistive cement.

2 the effect of temperature on smart cement with and without sodium silicate was investigated, and the smart cement was thermo-piezoresistive.

3 The monitoring parameter electrical resistivity was highly sensitive to the type and amount of additive and contaminants compared to other parameters such as density and strength. Also, resistivity can be monitored in the field during the entire service life of smart cement.

4 The effects of adding sodium meta silicate, metakaolin, and fly ash on the mechanical properties and the piezoresistive behavior of smart cement have tested and quantified. Also, the tensile properties with 10% metakaolin and 50% fly ash have been quantified.

5 The effects of contaminants such as clay (inorganic), oil-based mud, and CO_2 solution (organic) on the mechanical properties and piezoresistive behavior of smart cement have tested and quantified.

6 The Vipulanandan p-q stress-strain model and piezoresistive model predicted the experimental results very well based on the root-mean-square error (RMSE) and coefficient of determination (R^2).

7 The relationship between the changes in the compressive strengths of smart cement and the curing time have been modeled with the Vipulanandan property correlation model, and the experimental values matched very well with the model predictions based on coefficient of determination (R^2) and the root-mean-square error (RMSE). The relationship between the piezoresistivity at failure and curing time was also modeled with the Vipulanandan property correlation model, and the predictions agreed very well with the experimental results.

Chapter 7

Smart Cement with Nanoparticles

7.1 BACKGROUND

Nanoparticles are produced using biological, chemical, and physical processes (Li et al. 2003; Harendra et al. 2008; Liu et al. 2017). Past studies have shown that adding nanoparticles has a strong influence on the mechanical and electrical properties of cementitious materials, including all types of cements and cement concretes (Mondal et al. 2008; Choolaei et al. 2012; Zuo et al. 2014; Barbhuiya et al. 2014; Hou et al. 2015; Chitra et al. 2016; Compendex et al. 2016; Quercia et al. 2016; Ubertini et al. 2016; Zhang et al. 2017; Mohammed 2018; Vipulanandan et al. 2015c, e, 2016e, and 2020a). Because nanoparticles have a relatively large surface area, there will be higher chemical reactivity, and the dispersed nanoparticles will fill the voids between the hydrating cement particles, resulting in denser cement. Also well-dispersed nanoparticles act as crystallization centers based on the chemical composition, accelerating the cement hydration, and also producing an additional quantity of calcium silicate hydrate (C–S–H), and promote crystallization of small-sized calcium hydroxide ($Ca(OH)_2$) crystals. There is very limited information in the literature on XRD analyses of reactive cement materials (Vipulanandan et al. 1993, 2016f).

Adding nanoparticles will change hydrating cement microstructures and also the piezoresistivity structure of smart cement, as shown in Figure 5.9. Hence, it is important to investigate the effects of adding nanoparticles on the behavior of smart cement.

7.2 NANOSIO$_2$

One of the most useful nanomaterials that has received considerable attention is nanosilica (Choolaei et al. 2012; Singh et al. 2013). Based on improved compressive strength, it is clear that NanoSiO$_2$ behaves not only as a filler to improve the cement microstructure but also to promote pozzolanic reactions. Therefore, it is of interest to add NanoSiO$_2$ particles to cement mixtures to enhance the performance of cement and concrete. Also, adding $NanoSiO_2$ decreased the amount of lubricating water available in the mixture (Choolaei et al. 2012). Moreover, the effects of $NanoSiO_2$ on the performance of cement under various types of applications have been documented in the literature by several researchers in the past decade (Singh et al. 2013).

Nano-scaled silica particles have a filler effect by filling up the voids between cement grains. With the right composition, higher packing density results in lower water demand of the mixture and it also contributes to strength enhancement due to the reduced capillary porosity. The overall objective of this study was to investigate the effect of up to 1% of $NanoSiO_2$ on smart cement behavior. Smart cement samples were prepared with Class H cement and a w/c ratio of 0.38.

Silicon Dioxide Nanoparticle ($NanoSiO_2$)

Silicon dioxide nanopowder ($NanoSiO_2$) with a grain size of 12 nm and a specific surface area of 175 to 225 m^2/g (from supplier datasheet) was selected for this study.

Results and Discussion

7.2.1 XRD Analysis

In this study commercially available Class H oil-well cement was used.

(a) Cement

The XRD pattern of Class H cement is shown in Figure 7.1 (a). The major constituents in this cement included tricalcium silicate (Ca_3SiO_5), dicalcium silicate (Ca_2SiO_4), calcium aluminoferrite (Ca_2FeAlO_5), calcium sulfate ($CaSO_4$), quartz (SiO_2), and magnesium sulfate ($MgSO_4$).

(b) Cement with 1% NanoSiO2

The addition of 1% $NanoSiO_2$ modified the cement composition after seven days of curing with the formation of magnesium silicate sulfate (Mg_5 $(SiO_4)_2SO_4$: Ma) (2θ peaks at 51.58° and 56.50°) and also adding more quartz (SiO_2) (2θ peaks at 23.55°, 36.90° and 62.50°: Q) based on XRD intensity, as shown in Figure 7.1 (b). Adding $NanoSiO_2$ to the cement resulted in minerology changes, and it is important to investigate these changes in relation to smart cement behavior.

7.2.2 TGA analysis

The thermogravimetric analysis (TGA) and differential thermogravimetric (DTG) results were obtained for oil-well cement, silica nanoparticle ($NanoSiO_2$), and cement modified with 1% $NanoSiO_2$ after seven days

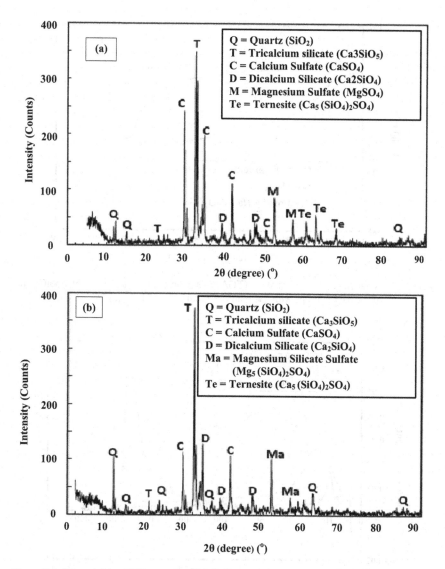

Figure 7.1 (a) and (b) XRD patterns (a) Oil-well cement, (b) Oil-well cement with 1% NanoSiO$_2$.

of curing, as summarised in Table 7.1 and also shown in Figure 7.2. The heating rate used in these tests was 10°C/min, which also shows weight loss at 110°C (standard temperature for determining free water), as shown in Figure 7.2. Four characteristic endothermic effects were observed, as summarised in Table 7.1. The initial effect, in the temperature range from

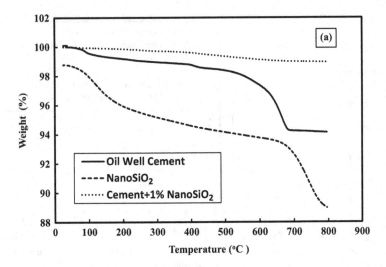

Figure 7.2a Thermogravimetric analysis (TGA) result on the smart cement, NanoSiO₂ and the smart cement with 1% NanoSiO₂ cured for 7 days (a) TGA, (b) DTG.

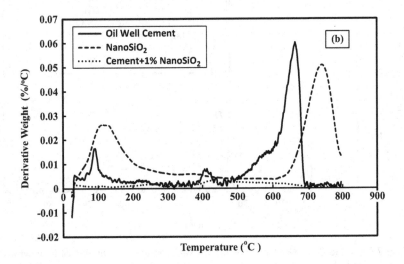

Figure 7.2b

25°C to 120°C, was due to the evaporation of surface-adsorbed water because samples adsorbed water during its curing at room temperature. The second endothermic effect, which was in the temperature range from 120°C to 400°C, is attributed to the dehydration of C-S-H and calcium aluminoferrite.

Table 7.1 TGA results on the NanoSiO$_2$ and the oil-well cement with and without 1% NanoSiO$_2$

Temperature Range	Weight Loss (%)				
Type of Sample	25–120°C	120–400°C	400–600°C	600–800°C	Total (%)
NanoSiO$_2$	1.33	3.0	0.85	5.13	10.31
Smart cement powder	0.56	0.66	1.44	3.44	6.10
Smart cement + 1%NanoSiO$_2$	0.07	0.33	0.49	0.16	1.05

(a) NanoSiO$_2$

The total weight loss for cement between 25°C and 120°C for the NanoSiO$_2$ was 1.33%, as summarised in Table 7.1. When the temperature changed from 120°C to 400°C, the weight loss of the NanoSiO$_2$ increased to 3.0%, as summarised in Table 7.1. For the temperature range between 400°C and 600°C, the weight loss for the NanoSiO$_2$ was 0.85%, as summarised in Table 7.1. When the temperature changed from 600°C to 800°C, the weight loss for the NanoSiO$_2$ was 5.13%, as summarised in Table 7.1. The total weight loss for the NanoSiO$_2$ was 10.31%.

(b) Smart Cement Powder

The total weight loss for cement powder between 25°C and 120°C was 0.56%, as summarised in Table 7.1. When the temperature changed from 120°C to 400°C, the weight loss of the cement increased to 0.66%. For the temperature range between 400°C and 600°C, the total weight loss for the cement was 1.44%. When the temperature changed from 600°C to 800°C, the weight loss for the cement increased to 3.44%, as summarised in Table 7.1. The total weight loss for the cement was 6.10%.

(c) Smart Cement and 1% NanoSiO2

The weight loss for cement modified with 1% NanoSiO$_2$ between 25°C and 120°C was 0.07%, as summarised in Table 7.1. The NanoSiO$_2$ addition strongly modified the free water dehydration (25°C–120°C). When the temperature changed from 120°C to 400°C, the weight loss was 0.33%. The weight loss in the temperature between 400°C and 600°C indicates the decomposition of Ca(OH)$_2$ formed during hydration. For the temperature range between 400°C and 600°C, the weight loss was only 0.49%. When the temperature changed from 600°C to 800°C, the weight loss was 0.16%, a 95% reduction compared to cement in the temperature range of 600°C to 800°C, as summarised in Table 7.1. Finally, an endothermic reaction at 800°C indicates the decarbonation of calcium carbonate in the hydrated cement compound. The total weight loss was only 1.05%, very small in the cement modified with 1% NanoSiO$_2$, which is indicative of the NanoSiO$_2$ interaction with the cement paste.

7.2.3 Density and Initial Resistivity

(a) Smart Cement
The average density of the smart cement with less than 0.1% carbon fiber was 1.975 g/cc (16.47 ppg). The initial electrical resistivity (ρ_o) of the smart cement mix with a w/c ratio of 0.38 was 1.06 Ωm.

(b) Smart Cement with 0.5% NanoSiO$_2$
The average density of the smart cement with 0.5% NanoSiO$_2$ was 1.977 g/cc (16.49 ppg), a 0.1% increase in density. The initial electrical resistivity (ρ_o) of the smart cement with 0.5% NanoSiO$_2$ was 1.20 Ωm, a 13% increase in the electrical resistivity, as shown in Figure 7.3. The resistivity was more sensitive to the NanoSiO$_2$ addition than the density.

(c) Smart Cement with 1% NanoSiO$_2$
The average density of the smart cement with 1% NanoSiO$_2$ was 1.982 g/cc (16.53 ppg), about a 0.4% increase in density. The initial electrical resistivity (ρ_o) of the smart cement with 1.0% NanoSiO$_2$ was 1.39 Ωm, a 31% increase in the electrical resistivity, as shown in Figure 7.3. The resistivity was more sensitive to the 1% NanoSiO$_2$ addition than the density.

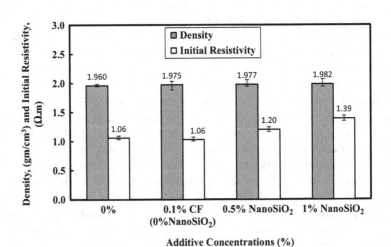

Figure 7.3 Effect of 1% NanoSiO$_2$ on the density and initial resistivity of the smart cement.

7.2.4 Curing

A change in the electrical resistivity with time and the minimum resistivity quantify the formation of solid hydration products, which leads to a decrease in porosity, which influences the cement's strength development.

(a) Smart Cement

The minimum resistivity (ρ_{min}) of the smart cement was 0.85 Ωm, and the time to reach the minimum resistivity (t_{min}) was 99 minutes, as summarised in Table 7.2. The resistivity after 24 hours was 3.90 Ωm, representing a change of about 268% in 24 hours, as shown in Figure 7.4 (a). The resistivity index (RI_{24hr}) for the smart cement was 364%, representing the maximum resistivity change in 24 hours, as summarised in Table 7.2. The resistivity

Table 7.2 Bulk resistivity parameters for the smart cement with NanoSiO$_2$

NanoSiO$_2$ (%)	Initial resistivity, ρ_o (Ω-m)	ρ_{min} (Ω-m)	t_{min} (min)	ρ_{24hr} (Ω-m)	$\rho_{7\,days}$ (Ω-m)	$RI_{24\,hr}$ (%)	$RI_{7\,days}$ (%)
0	1.06 ± 0.02	0.85 ± 0.02	99 ± 5	3.90 ± 0.04	7.7 ± 0.5	364	806
0.5	1.20 ± 0.03	0.95 ± 0.05	110 ± 7	4.16 ± 0.01	8.0 ± 0.4	338	708
1	1.39 ± 0.03	1.11 ± 0.03	122 ± 4	4.76 ± 0.03	8.7 ± 0.8	328	684

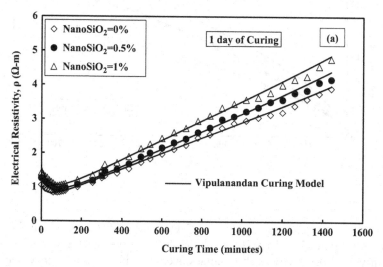

Figure 7.4a Bulk electrical resistivity development of the smart cement with various amounts of NanoSiO$_2$ (a) 1 day, (b) 7 days, (c) 28 days.

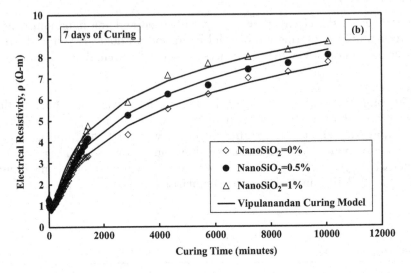

Figure 7.4b

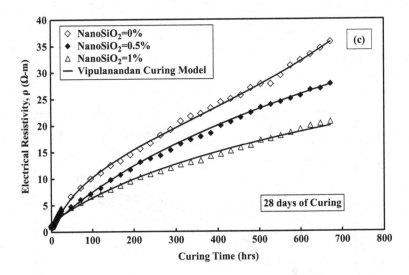

Figure 7.4c

after seven days and 28 days were 7.7 Ωm and 35.7 Ωm respectively, also shown in Figure 7.4 (b) and (c). These observed trends clearly indicate the sensitivity of resistivity to the changes occurring in the curing of cement, as summarised in Table 7.2.

(b) Smart Cement with 0.5% NanoSiO₂

The minimum resistivity (ρ_{min}) of the smart cement with 0.5% NanoSiO$_2$ was 0.95 Ω-m, which was 12% higher compared to the smart cement minimum electrical resistivity. The time to reach the minimum resistivity (t_{min}) was 110 minutes, as summarised in Table 7.2, and it was 11% higher than the smart cement. The resistivity after 24 hours was 4.16 Ωm, representing a change of about 289% in 24 hours, as shown in Figure 7.4 (a). The resistivity index (RI_{24hr}) for the smart cement with 0.5% NanoSiO$_2$ was 338%, as summarised in Table 7.2. The resistivity after seven days and 28 days were 8.0 Ωm and 27.8 Ωm respectively, also as shown in Figure 7.4 (b) and (c). Change in RI_{24hr} decreased with the 0.5% NanoSiO$_2$ content. These observed trends clearly indicate the sensitivity of resistivity to the changes occurring in the curing of cement, as summarised in Table 7.2.

(c) Smart Cement with 1% NanoSiO₂

The minimum resistivity (ρ_{min}) of the smart cement with 1% of NanoSiO$_2$ 1.11 Ω-m was about 30% higher compared to the smart cement minimum electrical resistivity. The time to reach the minimum resistivity (t_{min}) was 122 minutes, as summarised in Table 7.2, which was about 23% higher than the smart cement. Resistivity after 24 hours was 4.76 Ωm, representing a change of about 242% in 24 hours, as shown in Figure 7.4 (a). Resistivity after seven days and 28 days were 8.7 Ωm and 20.6 Ωm respectively, and also as shown in Figure 7.4 (b) and (c). The resistivity index (RI_{24hr}) for the smart cement with 1% of NanoSiO$_2$ was 328%, as summarised in Table 7.2. Change in RI_{24hr} decreased with 1% NanoSiO$_2$ content. These observed trends clearly indicate the sensitivity of resistivity to the changes occurring in the curing of cement, as summarised in Table 7.2.

The Vipulanandan curing model (Eqn. 4.15) was used to predict resistivity changes with the curing time for one day, seven days, and 28 days of curing, as shown in Figure 7.4. The model predicted the experimental results very well. The model parameters p_1, q1 and the ratio q_1/p_1 were all sensitive to the amount of NanoSiO$_2$ added to the cement. Also, the parameters t_{min} and ρ_{min} can be used as quality control indices and were related to the NanoSiO$_2$ content as follows:

$$t_{min} = 23 * \left(NanoSiO_2 (\%)\right) + 99 \qquad R^2 = 0.99 \qquad (7.1)$$

$$\rho_{min} = 0.85 + 0.26 * \left(NanoSiO_2 (\%)\right) \qquad R^2 = 0.98 \qquad (7.2)$$

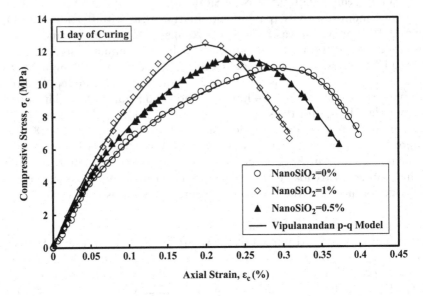

Figure 7.5 Compressive stress-strain relationship for the smart cement with NanoSiO$_2$ after one day of curing.

Hence, the electrical resistivity parameters were linearly related to the NanoSiO$_2$ content.

7.2.5 Compressive Stress-Strain Behavior

Based on the experimental results, the Vipulanandan p-q stress-strain model (Eqn. (5.6)) was used to predict the compressive stress-strain relationship for the smart cement with and without NanoSiO$_2$, and the relationship is as follows:

$$\sigma_c = \frac{\dfrac{\epsilon_c}{\varepsilon_{cf}} * \sigma_{cf}}{q_o + \left(1 - p_o - q_o\right)\dfrac{\epsilon_c}{\varepsilon_{cf}} + p\left(\dfrac{\epsilon_c}{\varepsilon_{cf}}\right)^{\frac{(p_o + q_o)}{p_o}}} \tag{7.3}$$

where σ_c is compressive stress and ε_c compressive strain; and σ_{cf} and ε_{cf} are compressive strength and corresponding strain, as summarised in Table 7.3. The model parameters p_o, and q_o are also summarised in Table 7.3. Model parameters p_o and q_o increased with curing time based on the NanoSiO$_2$ content. The compressive stress-strain relationships for the cement with and without nanoSiO$_2$ are shown in Figure 7.5.

Table 7.3 Compressive stress-strain model parameters for the NanoSiO$_2$-modified smart cement after one day of curing

NanoSiO$_2$ (%)	Curing Time (day)	σ_{cf} (MPa)	εc_f (%)	Ei (MPa)	p_o	q_o	RMSE (MPa)	R^2
0	1	10.9 ± 2	0.30 ± 0.02	10,100	0.034 ± 0.03	0.36 ± 0.02	0.128	0.99
0.5	1	11.6 ± 3	0.25 ± 0.02	10,800	0.061 ± 0.03	0.43 ± 0.02	0.095	0.99
1	1	12.4 ± 2	0.22 ± 0.02	11,000	0.094 ± 0.02	0.53 ± 0.03	0.244	0.99

7.2.5.1 Curing Time

One Day

(a) Smart Cement

The compressive strengths (σ_{cf}) of the smart cement after one day of curing were 10.9 MPa, as shown in Figure 7.5. The failure strain was 0.30% and the initial modulus was 10,100 MPa, as summarised in Table 7.3. The model parameters q_o and p_o were 0.36 and 0.034 respectively. The parameter q_o represents nonlinearity up to peak stress, and the smart cement had the highest nonlinearity, and the lowest value of parameter q_o, as shown in Figure 7.5. The coefficient of determination (R^2) was 0.99 and the RMSE was 0.128 MPa, as summarised in Table 7.3.

(b) Smart Cement with 0.5% NanoSiO$_2$

The compressive strengths (σ_{cf}) of the smart cement with 0.5% NanoSiO$_2$ after one day of curing were 11.6 MPa, 6.4% higher than the smart cement without NanoSiO$_2$, as shown in Figure 7.5. The failure strain was 0.25% and the initial modulus was 10,800 MPa, as summarised in Table 7.3. The addition of 0.5% NanoSiO$_2$ to the smart cement reduced the failure strain and increased the modulus compared to the smart cement without NanoSiO$_2$. The model parameters q_o and p_o were 0.43 and 0.061 respectively. The parameter q_o represents nonlinearity up to peak stress, and was higher than the smart cement without NanoSiO$_2$, as shown in Figure 7.5. The coefficient of determination (R^2) was 0.99 and the RMSE was 0.095 MPa, as summarised in Table 7.3.

(c) Smart Cement with 1% NanoSiO$_2$

The compressive strengths (σ_{cf}) of the smart cement with 1% NanoSiO$_2$ after one day of curing were 12.4 MPa, 13.8% higher than the smart cement without NanoSiO$_2$, as shown in Figure 7.5. The failure strain was 0.22%

and the initial modulus was 11,000 MPa, as summarised in Table 7.3. The addition of 1% NanoSiO$_2$ to the smart cement reduced the failure strain and increased the modulus compared to the smart cement without NanoSiO$_2$. The model parameters q$_o$ and p$_o$ were 0.53 and 0.094 respectively. The parameter q$_o$ represents the nonlinearity up to peak stress, and was higher than the smart cement without NanoSiO$_2$, as shown in Figure 7.5. The coefficient of determination (R^2) was 0.99 and the RMSE was 0.244 MPa, as summarised in Table 7.3.

28 days

(a) Smart Cement
The compressive strengths (σ_{cf}) of the smart cement after 28 days of curing were 19.3 MPa, as shown in Figure 7.6. The failure strain was 0.22% and the initial modulus was 12,500 MPa, as summarised in Table 7.4. The model parameters q$_o$ and p$_o$ were 0.70 and 0.160 respectively. The parameter q$_o$ represents nonlinearity up to peak stress, and the smart cement had the highest linearity, the highest value of parameter q$_o$, as shown in Figure 7.6. The coefficient of determination (R^2) was 0.99 and the RMSE was 0.110 MPa, as summarised in Table 7.4.

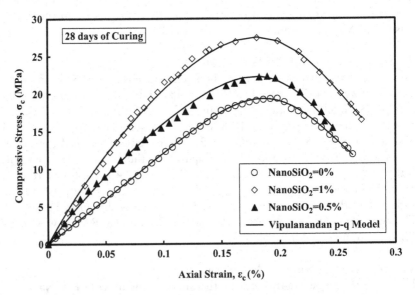

Figure 7.6 Compressive stress-strain relationship for the smart cement with NanoSiO$_2$ after 28 days of curing.

Table 7.4 Compressive stress-strain model parameters for the NanoSiO$_2$-modified smart cement after 28 days of curing

NanoSiO$_2$ (%)	Curing Time (day)	σ_{cf} (MPa)	εc_f (%)	Ei (MPa)	p_o	q_o	RMSE (MPa)	R^2
0	28	19.3 ± 1	0.22 ± 0.02	12,500	0.160 ± 0.01	0.70 ± 0.03	0.110	0.99
0.5	28	22.3 ± 2	0.20 ± 0.02	18,600	0.092 ± 0.02	0.60 ± 0.02	0.085	0.99
1.0	28	27.5 ± 3	0.19 ± 0.02	25,400	0.100 ± 0.01	0.57 ± 0.01	0.150	0.99

(b) Smart Cement with 0.5% NanoSiO$_2$

The compressive strengths (σ_{cf}) of the smart cement with 0.5% NanoSiO$_2$ after 28 days of curing were 22.3 MPa, 15.5% higher than the smart cement without NanoSiO$_2$, as shown in Figure 7.6. The failure strain was 0.20% and the initial modulus was 18,600 MPa, as summarised in Table 7.4. Addition of 0.5% NanoSiO$_2$ to the smart cement reduced the failure strain and increased the modulus compared to the smart cement without NanoSiO$_2$. The model parameters q_o and p_o were 0.60 and 0.092 respectively. The parameter q_o represents nonlinearity up to peak stress, and was lower than the smart cement without NanoSiO$_2$, as shown in Figure 7.6. The coefficient of determination (R^2) was 0.99 and the RMSE was 0.085 MPa, as summarised in Table 7.4.

(c) Smart Cement with 1% NanoSiO$_2$

The compressive strengths (σ_{cf}) of the smart cement with 1% NanoSiO$_2$ after 28 days of curing were 27.5 MPa, 42.5% higher than the smart cement without NanoSiO$_2$, as shown in Figure 7.6. The failure strain was 0.19% and the initial modulus was 25,400 MPa. as summarised in Table 7.4. The addition of 1% NanoSiO$_2$ to the smart cement reduced the failure strain and increased the modulus compared to the smart cement without NanoSiO$_2$. The model parameters q_o and p_o were 0.57 and 0.100 respectively. The parameter q_o represents nonlinearity up to peak stress, and was lower than the smart cement without NanoSiO$_2$, as shown in Figure 7.5. The coefficient of determination (R^2) was 0.99 and the RMSE was 0.150 MPa, as summarised in Table 7.4.

7.2.6 Compressive Piezoresistivity

Based on experimental results, the Vipulanandan p-q piezoresistivity model was developed to predict the change in the electrical resistivity of the smart cement with applied compressive stress for one day, seven days, and 28 days of curing. The Vipulanandan Piezoresistive model is defined as follows:

$$\sigma_c = \frac{\dfrac{x}{x_f} * \sigma_{cf}}{q_2 + \left(1 - p_2 - q_2\right)\dfrac{x}{x_f} + p_2 \left(\dfrac{x}{xf}\right)^{\left(\frac{p_2}{p_2 - q_2}\right)}} \tag{7.4}$$

(MPa);

where σ_c is the stress (MPa); σc_f is the compressive strength at failure

$x = \left(\dfrac{\Delta\rho}{\rho_o}\right)*100$: Percentage piezoresistive axial strain due to applied stress;

$x_f = \left(\dfrac{\Delta\rho}{\rho_o}\right)_f *100$: Percentage piezoresistive axial strain at failure;

$\Delta\rho$: change in electrical resistivity; ρ_o : initial electrical resistivity ($\sigma = 0$ MPa); and p_2 and q_2 are the piezoresistive model parameters.

7.2.6.1 Curing Time

One Day

(a) Smart Cement

The piezoresisitive axial strain of the smart cement at failure $\left(\dfrac{\Delta\rho}{\rho_o}\right)_f$ after one day of curing was 520%, as summarised in Table 7.5 and shown in Figure 7.7. The compressive axial failure strain was 0.30%, so the piezoresistive axial strain was increased by 1,733 times (173,300%), making the smart cement

Table 7.5 Piezoresistive axial strain at failure and strength of the smart cement without and with NanoSiO$_2$ after one day of curing

Material	NanoSiO$_2$ (%)	Curing Time (day)	$\left(\dfrac{\Delta\rho}{\rho_o}\right)_f$ (%)	σc_f (MPa)	p_2	q_2	RMSE (MPa)	R^2
Smart Cement	0%	1	520 ± 10	10.9 ± 3	0.16 ± 0.02	0.29 ± 0.02	0.054	0.99
Smart Cement	0.5%	1	495 ± 8	11.6 ± 5	0.17 ± 0.06	0.19 ± 0.02	0.019	0.99
Smart Cement	1%	1	428 ± 10	12.4 ± 2	0.19 ± 0.02	0.30± 0.04	0.010	0.99

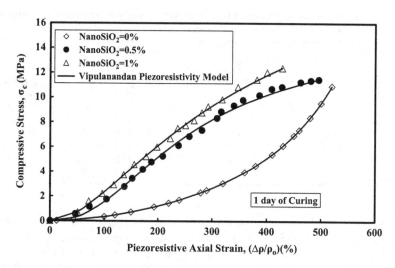

Figure 7.7 Piezoresistive behavior of the smart cement with various amounts of NanoSiO$_2$ after 1 day of curing.

highly sensing. The model parameters q$_2$ and p$_2$ were 0.29 and 0.16 respectively. The coefficient of determination (R^2) was 0.99 and the RMSE was 0.054 MPa, as summarised in Table 7.5.

(b) Smart Cement with 0.5% NanoSiO$_2$

The piezoresisitive axial strain of the smart cement with 0.5% NanoSiO$_2$ at failure $\left(\dfrac{\Delta\rho}{\rho_o}\right)_f$ after one day of curing was 495%, as summarised in Table 7.5 and shown in Figure 7.7. The addition of 0.5% NanoSiO$_2$ reduced the piezoresistive axial strain of the smart cement by 4.8%. The compressive axial failure strain was 0.25%, so the piezoresistive axial strain was increased by 1,980 times (198,000%), making the smart cement highly sensing. The model parameters q$_2$ and p$_2$ were 0.19 and 0.17 respectively. The coefficient of determination (R^2) was 0.99 and the RMSE was 0.019 MPa, as summarised in Table 7.5.

(c) Smart Cement with 1% NanoSiO$_2$

The piezoresisitive axial strain of the smart cement with 1% NanoSiO$_2$ at failure $\left(\dfrac{\Delta\rho}{\rho_o}\right)_f$ after one day of curing was 428%, as summarised in Table 7.5 and shown in Figure 7.7. The addition of 1% NanoSiO$_2$ reduced the piezoresistive axial strain of the smart cement by 17.7%. The

Table 7.6 Piezoresistive axial strain at failure and strength of the smart cement without and with NanoSiO$_2$ after seven days of curing

Material	NanoSiO$_2$ (%)	Curing Time (day)	$\left(\dfrac{\Delta\rho}{\rho_o}\right)_f$ (%)	σc_f (MPa)	p_2	q_2	RMSE (MPa)	R^2
Smart Cement	0%	7	460 ± 14	16.0 ± 3	0.11 ± 0.04	0.04 ± 0.05	0.017	0.99
	0.5%	7	409 ± 9	18.1 ± 2	0.10 ± 0.02	0.18 ± 0.03	0.003	0.99
	1%	7	365 ± 12	21.2 ± 4	0.18 ± 0.03	0.33 ± 0.02	0.018	0.99

compressive axial failure strain was 0.22%, so the piezoresistive axial strain was increased by 1,945 times (194,500%), making the smart cement highly sensing. The model parameters q_2 and p_2 were 0.30 and 0.19 respectively. The coefficient of determination (R^2) was 0.99 and the RMSE was 0.010 MPa, as summarised in Table 7.5.

Seven days

(a) **Smart Cement**

The piezoresisitive axial strain of the smart cement at failure $\left(\dfrac{\Delta\rho}{\rho_o}\right)_f$ after seven days of curing was 460%, as summarised in Table 7.6 and shown in Figure 7.8. The compressive piezoresistive axial strain at failure reduced by 11.5% after seven days of curing. The model parameters q_2 and p_2 were 0.04 and 0.11 respectively. The coefficient of determination (R^2) was 0.99 and the RMSE was 0.017 MPa, as summarised in Table 7.6.

(b) **Smart Cement with 0.5% NanoSiO$_2$**

The piezoresisitive axial strain of the smart cement with 0.5% NanoSiO$_2$ at failure $\left(\dfrac{\Delta\rho}{\rho_o}\right)_f$ after seven days of curing was 409%, as summarised in Table 7.6 and shown in Figure 7.8. The compressive piezoresistive axial strain at failure reduced by 17.3% after seven days of curing and the percentage reduction was higher than the smart cement without the NanoSiO$_2$. The model parameters q_2 and p_2 were 0.18 and 0.10 respectively. The coefficient of determination (R^2) was 0.99 and the root-mean-square error (RMSE) was 0.003 MPa, as summarised in Table 7.6.

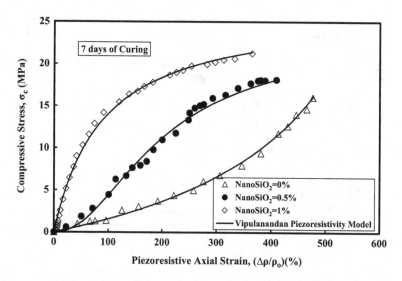

Figure 7.8 Piezoresistive behavior of the smart cement with various amounts of NanoSiO$_2$ after 7 days of curing.

(c) Smart Cement with 1% NanoSiO$_2$

The piezoresisitive axial strain of the smart cement with 1% NanoSiO$_2$ at failure $\left(\dfrac{\Delta\rho}{\rho_o}\right)_f$ after seven days of curing was 365%, as summarised in Table 7.6 and shown in Figure 7.8. The compressive piezoresistive axial strain at failure reduced by 14.7% after seven days of curing, and the percentage reduction was higher than the smart cement without the NanoSiO$_2$. The model parameters q_2 and p_2 were 0.33 and 0.18 respectively. The coefficient of determination (R^2) was 0.99 and the root-mean-square error (RMSE) was 0.018 MPa, as summarised in Table 7.6.

Twenty-eight Days:

(a) Smart Cement

The piezoresisitive axial strain of the smart cement at failure $\left(\dfrac{\Delta\rho}{\rho_o}\right)_f$ after 28 days of curing was 400%, as summarised in Table 7.7 and shown in Figure 7.9. The compressive piezoresistive axial strain at failure reduced by 23% after 28 days of curing. The compressive axial failure strain was

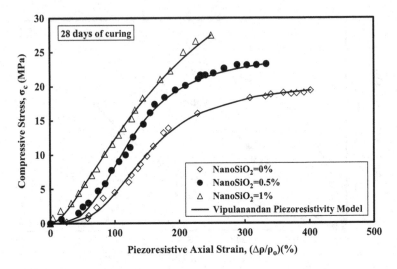

Figure 7.9 Piezoresistive behavior of the smart cement with various amounts of NanoSiO₂ after 28 days of curing.

Table 7.7 Piezoresistive axial strain at failure and strength of the smart cement without and with NanoSiO₂ after 28 days of curing

Material	NanoSiO$_2$ (%)	Curing Time (day)	$\left(\dfrac{\Delta\rho}{\rho_o}\right)_f$ (%)	σc_f (MPa)	p_2	q_2	RMSE (MPa)	R^2
Smart Cement	0%	28	400 ± 10	19.3 ± 2	0.03 ± 0.02	0.05± 0.03	0.022	0.99
	0.5%	28	334 ± 13	22.3 ± 3	0.38 ± 0.05	0.05± 0.02	0.021	0.99
	1.0%	28	250 ± 8	27.5 ± 3	0.14 ± 0.02	0.35± 0.03	0.016	0.99

0.22%, so the piezoresistive axial strain was increased by 1,818 times (181,800%), making the smart cement highly sensing. The model parameters q_2 and p_2 were 0.05 and 0.03 respectively. The coefficient of determination (R^2) was 0.99 and the root-mean-square error (RMSE) was 0.022 MPa, as summarised in Table 7.7.

(b) Smart Cement with 0.5% NanoSiO₂
The piezoresisitive axial strain of the smart cement with 0.5% NanoSiO₂ at failure $\left(\dfrac{\Delta\rho}{\rho_o}\right)_f$ after 28 days of curing was 334%, as summarised in Table 7.7 and shown in Figure 7.9. The compressive piezoresistive axial strain at failure reduced by 32.5% after 28 days of curing, and the percentage reduction was

higher than the smart cement without the $NanoSiO_2$. The addition of 0.5% $NanoSiO_2$ reduced the piezoresistive axial strain of the smart cement by 16.5%. The compressive axial failure strain was 0.20%, so the piezoresistive axial strain was increased by 1,670 times (167,000%), making the smart cement highly sensing. The model parameters q_2 and p_2 were 0.05 and 0.38 respectively. The coefficient of determination (R^2) was 0.99 and the root-mean-square error (RMSE) was 0.021 MPa, as summarised in Table 7.7.

(c) Smart Cement with 1% $NanoSiO_2$

The piezoresisitive axial strain of the smart cement with 1% $NanoSiO_2$ at failure $\left(\dfrac{\Delta\rho}{\rho_o}\right)_f$ after 28 days of curing was 250%, as summarised in Table 7.7 and shown in Figure 7.9. The compressive piezoresistive axial strain at failure reduced by 41.6% after 28 days of curing, and the percentage reduction was higher than the smart cement without $NanoSiO_2$. The addition of 1% $NanoSiO_2$ reduced the piezoresistive axial strain of the smart cement by 37.5%. The compressive axial failure strain was 0.19%, so the piezoresistive axial strain was increased by 1,316 times (131,600%), making the smart cement highly sensing. The model parameters q_2 and p_2 were 0.35 and 0.14 respectively. The coefficient of determination (R^2) was 0.99 and the root-mean-square error (RMSE) was 0.016 MPa, as summarised in Table 7.7.

7.2.7 Nonlinear Model (NLM)

The electrical resistivity (ρ) of the smart cement slurry modified with silica nanoparticle ($NanoSiO_2$) was influenced by the composition of the cement and the curing time (t (day)). It is proposed to relate the model parameters to the independent variables (curing time and $NanoSiO_2$ content) using a nonlinear power relationship (Vipulanandan et al. 2015 a). Hence the effects of curing time (t) and $NanoSiO_2$ content on the model parameters were determined using the nonlinear model (NLM) (Vipulanandan et al. 2015a) as follows:

$$\text{Model parameters} = a * (t)^b + c * (t)^d * (NanoSiO_2(\%))^e \qquad (7.5)$$

The NLM parameters (a, b, c, d, e) were obtained from multiple regression analyses using the least-square method.

(a) Piezoresistive Axial Strain at Failure $\left(\dfrac{\Delta\rho}{\rho_o}\right)_f$

Based on the nonlinear model parameter a (Eqn. 7.5), curing time had the second-highest effect on the parameter $\left(\dfrac{\Delta\rho}{\rho_o}\right)_f$ compared to σ_f, q_2, and p_2.

Based on the NLM parameter c, the addition of NanoSiO$_2$ had the highest effect on the parameter $\left(\dfrac{\Delta\rho}{\rho_o}\right)_f$ compared to σ_f, q_2, and p_2, as summarised in Table 7.8.

(b) Compressive Strength (σ_{cf})

Based on the nonlinear model parameter a (Eqn. 7.5), curing time had the highest effect on increasing the parameter σc_f compared to $\left(\dfrac{\Delta\rho}{\rho_o}\right)_f$, q_2 and p_2. Based on the NLM parameter c, the addition of NanoSiO$_2$ had the highest effect on increasing the parameter σ_f compared to $\left(\dfrac{\Delta\rho}{\rho_o}\right)_f$, q_2 and p_2, as summarised in Table 7.8.

(c) Parameter p$_2$

Based on the nonlinear model parameter a (Eqn. 7.5), curing time had the lowest effect on increasing the parameter p_2 compared to $\left(\dfrac{\Delta\rho}{\rho_o}\right)_f$, σ_f and q_2. Based on the NLM parameter c, the addition of NanoSiO$_2$ had the lowest effect on decreasing the parameter q_2 compared to $\left(\dfrac{\Delta\rho}{\rho_o}\right)_f$, σ_f and q_2, as summarised in Table 7.8.

(d) Parameter q$_2$

Based on the nonlinear model parameter a (Eqn. 7.5), curing time had the second-lowest effect on increasing the parameter q_2 compared to $\left(\dfrac{\Delta\rho}{\rho_o}\right)_f$, σ_f and p_2. Based on the NLM parameter c, the addition of NanoSiO$_2$ had the

Table 7.8 Nonlinear model parameters for the NanoSiO$_2$-modified smart cement

Model Parameters	a	b	c	d	e	No. of Data	RMSE	R^2
$\left(\dfrac{\Delta\rho}{\rho_o}\right)_f$ (%)	3.0	-0.45	447.8	-0.10	-0.02	8	33.75	0.83
σ_f (MPa)	10.8	0.18	2.35	0.36	1.13	8	0.57	0.98
p_2	0.16	-0.32	0.04	0.4	2.52	8	0.01	0.86
q_2	0.23	-0.42	0.10	0.32	3.76	8	0.05	0.86

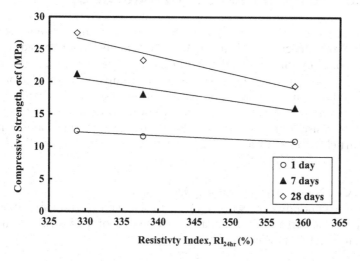

Figure 7.10 Relationship between resistivity index ($RI_{24\,hr}$) and compressive strength of the smart cement modified with NanoSiO₂.

second-lowest effect on increasing the parameter q_2 compared to $\left(\dfrac{\Delta\rho}{\rho_o}\right)_f$, σ_f and p_2, as summarised in Table 7.8.

7.2.8 Compressive Strength-Resistivity Index Relationship

During the entire cement hydration process, both the electrical resistivity and the compressive strength of the cement increased gradually with the curing time. For cement pastes with various NanoSiO₂ content, the change in resistivity varied during the hardening. Cement paste without NanoSiO₂ had the highest electrical resistivity change (RI_{24hr}), as summarised in Table 7.2. The linear relationships between (RI_{24hr}) and the one-day, seven-day, and 28-day compressive strengths (MPa), as shown in Figure 7.10, are as follows:

$$\sigma_{1day} = -7 \times RI_{24hr} + 4066 \qquad R^2 = 0.96 \qquad (7.6)$$

$$\sigma_{7days} = -0.16 \times RI_{24hr} + 73 \qquad R^2 - 0.90 \qquad (7.7)$$

$$\sigma_{28days} = -0.05 \times RI_{24hr} + 28 \qquad R^2 = 0.93 \qquad (7.8)$$

7.3 COMPARING NANO FE$_2$O$_3$ TO NANO AL$_2$O$_3$

There are many types of inorganic (metallic-based) and organic (carbon-based) nanoparticles used to modify cement behavior based on the applications. In this study, the performance of the smart cement modified with 1% nano Fe$_2$O$_3$ (ferric oxide) is compared with 1% of nano Al$_2$O$_3$ (aluminum oxide). The properties of interest are the changes at the microstructural level (TGA analyses) to curing, compressive stress-strain, and piezoresistive relationships. The smart cement with and without nanoparticles and with a water-to-cement ratio of 0.38 up to 28 days of curing at room condition were investigated.

Iron Oxide Nanoparticle (NanoFe$_2$O$_3$)
Iron oxide nanopowder (NanoFe$_2$O$_3$) with the average grain size of 30 nm, specific surface area of 38 m^2/g, and bulk density of 0.25 g/cm^3 (from supplier datasheet) was selected for this study.

Aluminum Oxide Nanoparticle (NanoAl$_2$O$_3$)
Aluminum oxide nanopowder (NanoAl$_2$O$_3$) with average grain size of 50 nm, specific surface area of 35 m^2/g, and bulk density of 0.20 g/cm^3 (from supplier datasheet) was selected for this study.

Results and Discussion

7.3.1 XRD Analyses

(a) **NanoFe$_2$O$_3$**
The NanoFe$_2$O$_3$ had both ferric oxide (Fe$_2$O$_3$) at 2θ peaks 25.48°, 34.92°, 42.20°, 50.40°, 57.92°, 64.72°, and 72.16°. XRD analyses also identified ferrous oxide (FeO) at 2θ peaks 53.26°, as shown in Figure 7.11.

(b) **NanoAl$_2$O$_3$**
The NanoAl$_2$O$_3$ had aluminum oxide (Al$_2$O$_3$) at 2θ peaks at 26.23°, 36.92°, 52.40°, and 72.16°, as shown in Figure 7.12. Also there was silica dioxide (SiO$_2$) at 2θ peak 33.26°. Additionally, there was sodium oxide (Na$_2$O) at 2θ peak 78.16°, as shown in Figure 7.12.

7.3.2 TGA Analyses

The thermogravimetric analysis (TGA) results were obtained for the smart cement and the smart cement modified with 1% NanoFe$_2$O$_3$ and 1% NanoAl$_2$O$_3$ after seven days of curing using a w/c ratio of 0.38. The heating rate used in these tests was 10°C/min and the samples were heated up to 800°C. Four characteristic endothermic effects were observed, as shown in

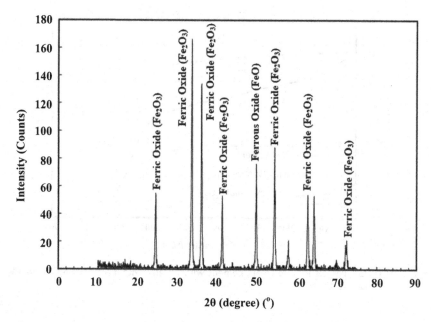

Figure 7.11 XRD analyses of NanoFe$_2$O$_3$ (Ferric Oxide).

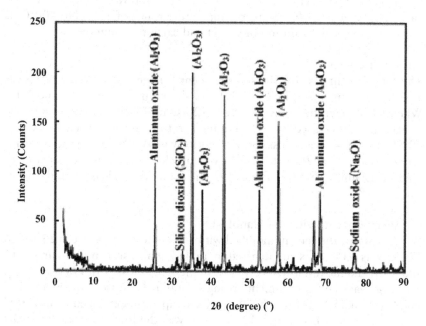

Figure 7.12 XRD analyses of NanoAl$_2$O$_3$.

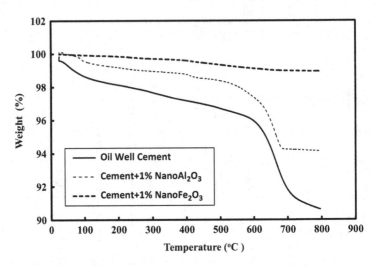

Figure 7.13 TGA results for the smart cement without and with 1% NanoFe$_2$O$_3$ and 1% NanoAl$_2$O$_3$.

Figure 7.13 and Figure 7.14. The initial effect, in the temperature range from 25°C to 120°C was due to evaporation of free and surface adsorbed water by the cement and nanoparticles at room temperature curing. The second endothermic effect, which was in temperature range of 120°C to 400°C, is attributed to the dehydration of C-S-H and calcium aluminoferrite.

(a) **Smart Cement**
The total weight loss for the smart cement between 25°C and 120°C was 1.33%, as summarised in Table 7.9 and shown in Figures 7.13 and 7.14. When the temperature changed from 120°C to 400°C, weight loss of the smart cement increased to 3.0%. For temperature range between 400°C and 600°C, the total weight loss for the cement was 0.85%. When the temperature changed from 600°C to 800°C, weight loss for the cement increased to 5.13%, as summarised in Table 7.9. The total weight loss for the cement was 10.31%.

(b) **Smart Cement with 1% NanoFe$_2$O$_3$**
Weight loss for the cement modified with 1% NanoFe$_2$O$_3$ between 25°C and 120°C was 0.08%, as summarised in Table 7.9 and shown in Figures 7.13 and 7.14, and the result was very close to the weight loss with 1% NanoSiO$_2$. The 1% NanoFe$_2$O$_3$ treatment strongly modified the free water dehydration between 25°C and 120°C. When the temperature changed from 120°C to 400°C, weight loss was 0.32%, which was the lowest for all the smart

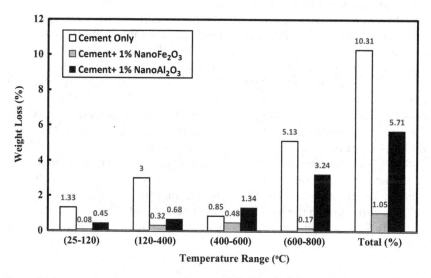

Figure 7.14 Comparing the weight lost during the TGA for the smart cement with and without 1% NanoFe$_2$O$_3$ and 1% NanoAl$_2$O$_3$.

Table 7.9 TGA results for the smart cement without and with 1% NanoFe$_2$O$_3$ and 1% NanoAl$_2$O$_3$

Materials	25–120°C	120–400°C	400–600°C	600–800°C	Total (%)
Smart Cement only	1.33	3.0	0.85	5.13	10.31
Smart Cement+ 1% NanoFe$_2$O$_3$	0.08	0.32	0.48	0.17	1.05
Smart Cement+ 1% NanoAl$_2$O$_3$	0.45	0.68	1.34	3.24	5.71

cement with nanoparticles investigated. Weight loss in the temperature between 400°C and 600°C indicates the decomposition of Ca(OH)$_2$ formed during hydration. For temperature range between 400°C and 600°C, weight loss was only 0.48%. When the temperature changed from 600°C to 800°C, weight loss was 0.17%, a 97% reduction compared to the cement in the temperature range of 600°C to 800°C, as summarised in Table 7.9. Finally, an endothermic reaction at 800°C indicates the decarbonation of calcium carbonate in the hydrated cement compound. The total weight loss was only 1.05%, very small with the cement modified with 1% NanoFe$_2$O$_3$, which is indicative of NanoFe$_2$O$_3$ interaction with the cement paste. The results were very close to the total weight loss with 1% NanoSiO$_2$.

(c) Smart Cement and 1% NanoAl$_2$O$_3$

Weight loss for cement between 25°C and 120°C for cement modified with 1% NanoAl$_2$O$_3$ was 0.45%, as summarised in Table 7.9 and shown in Figures 7.13 and 7.14, much higher than with 1% NanoFe$_2$O$_3$ and 1% NanoSiO$_2$, shown in Table 7.1. The 1% of NanoAl$_2$O$_3$ treatment reduced the free water dehydration (25°C–120°C). When the temperature changed from 120°C to 400°C, weight loss was 0.68%. Weight loss in the temperature between 400°C and 600°C indicates the decomposition of Ca(OH)$_2$ formed during hydration. For temperature range between 400°C and 600°C, weight loss was 1.34%. When the temperature changed from 600°C to 800°C, weight loss was 3.24%, a 37% reduction compared to the cement in the temperature range of 600°C to 800°C, as summarised in Table 7.9. Finally, an endothermic reaction at 800°C indicates the decarbonation of calcium carbonate in the hydrated cement compound. The total weight loss was only 5.71%, higher than the cement modified with 1% NanoFe$_2$O$_3$ and 1% NanoSiO$_2$.

7.3.3 Density and Initial Resistivity

(a) Smart Cement with 1% NanoFe$_2$O$_3$

The average density of the smart cement with 1% NanoFe$_2$O$_3$ was 1.986 g/cc (16.49 ppg), a 0.6% increase in density compared to the smart cement density of 1.975 g/cc. The initial electrical resistivity (ρ_o) of the smart cement with 1% NanoFe$_2$O$_3$ was 0.87 Ωm, a 17.9% decrease in electrical resistivity compared to the smart cement, as shown in Figure 7.15. The resistivity was more sensitive to the 1% NanoFe$_2$O$_3$ addition than the density.

(b) Smart Cement with 1% NanoAl$_2$O$_3$

The average density of the smart cement with 1% NanoAl$_2$O$_3$ was 1.988 g/cc (16.53 ppg), about a 0.6% increase in density compared to the smart cement density of 1.975 g/cc. The initial electrical resistivity (ρ_o) of the smart cement with 1.0% NanoAl$_2$O$_3$ was 1.34 Ωm, a 26% increase in electrical resistivity, as shown in Figure 7.15. The resistivity was more sensitive to the 1% NanoAl$_2$O$_3$ addition than the density.

7.3.4 Curing

The change in resistivity with time is shown in Figure 7.16. There are several parameters that can be used in monitoring the curing (hardening process) of the cement. The parameters are initial resistivity (ρ_o), minimum resistivity (ρ_{min}), time to reach the minimum resistivity (t_{min}), and resistivity after 24 hours of curing (ρ_{24}).

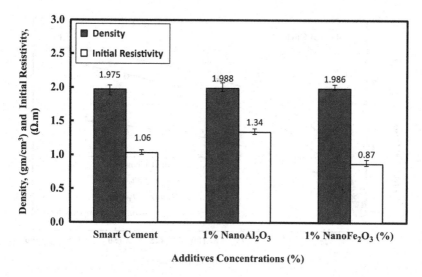

Figure 7.15 Effects of 1% NanoFe$_2$O$_3$ and 1% NanoAl$_2$O$_3$ on the density and initial resistivity of smart cement.

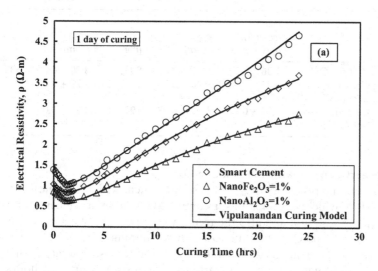

Figure 7.16a Curing of the smart cement with 1% NanoFe$_2$O$_3$ and 1% NanoAl$_2$O$_3$ (a) 1 day (b) 7 days, (c) 28 days.

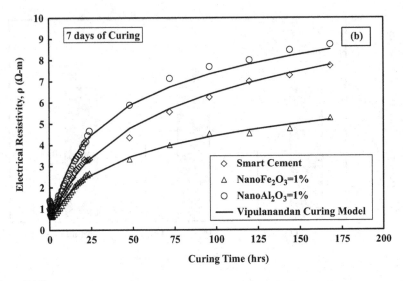

Figure 7.16b

Table 7.10 Bulk resistivity parameters for the smart cement with nanomaterial content

Additive	Initial resistivity, ρ_o (Ω-m)	ρ_{min} (Ω-m)	t_{min} (min)	ρ_{24hr} (Ω-m)	RI_{24} (%)
Smart Cement Only	1.06 ± 0.02	0.85 ± 0.02	99 ± 5	3.70 ± 0.04	335
Smart Cement+ 1% NanoFe$_2$O$_3$	0.87 ± 0.03	0.70 ± 0.05	144 ± 5	2.75 ± 0.01	293
Smart Cement+ 1% of NanoAl$_2$O$_3$	1.34 ± 0.03	1.02 ± 0.03	192 ± 3	4.70 ± 0.03	361

(a) Smart Cement with 1% NanoFe$_2$O$_3$

The minimum resistivity (ρ_{min}) of the smart cement with 1% NanoFe$_2$O$_3$ was 0.70 Ω-m, which was 34% lower compared to the smart cement minimum electrical resistivity, and also the trend was opposite to what was observed with 1% NanoSiO$_2$. The time to reach the minimum resistivity (t_{min}) was 144 minutes, as summarised in Table 7.10, and it was 45% higher than the smart cement without any nanoparticles. The resistivity after 24 hours was 2.75 Ωm, representing a change of about a 216% increase in 24 hours, as shown in Figure 7.16 (a), and the percentage change was lower compared to the 1% NanoSiO$_2$. The resistivity index (RI_{24hr}) for the smart cement with 1% NanoFe$_2$O$_3$ was 293%, lower than 1% NanoSiO$_2$. The resistivity after seven days and 28 days were 5.26 Ωm and 18.2 Ωm respectively, as

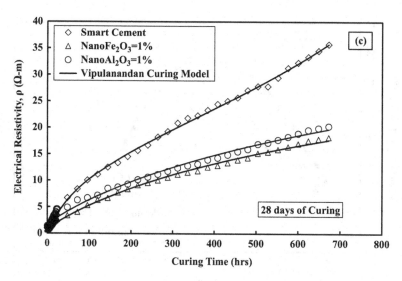

Figure 7.16c

shown in Figures 7.16 (b) and (c). The RI_{24hr} for the smart cement with 1% $NanoFe_2O_3$ was lower than the smart cement with 1.0% $NanoSiO_2$. These observed trends clearly indicate the sensitivity of resistivity to the changes occurring in the curing of the smart cement with different nanoparticles, as summarised in Tables 7.2 and 7.10.

(b) Smart Cement with 1% $NanoAl_2O_3$

The minimum resistivity (ρ_{min}) of the smart cement with 1.0% $NanoAl_2O_3$ was 1.02 Ω-m, which was 20% higher compared to the smart cement minimum electrical resistivity, and also the trend was opposite to what was observed with 1% $NanoFe_2O_3$. The time to reach the minimum resistivity (t_{min}) was 192 minutes, as summarised in Table 7.10, and it was 94% higher than the smart cement without any nanoparticles. The resistivity after 24 hours was 4.70 Ωm, representing a change of about a 251% increase in 24 hours, as shown in Figure 7.16 (a), and the percentage change was high compared to the 1% $NanoFe_2O_3$. The resistivity index (RI_{24hr}) for the smart cement with 1% $NanoAl_2O_3$ was 361%, higher than 1% $NanoFe_2O_3$ and 1% $NanoSiO_2$. The resistivity after seven days and 28 days were 8.73 Ωm and 20.26 Ωm respectively, as shown in Figures 7.16 (b) and (c). The RI_{24hr} for the smart cement with 1% $NanoAl_2O_3$ was higher than the smart cement with 1% $NanoFe_2O_3$ and 1.0% $NanoSiO_2$. These observed trends clearly indicate the sensitivity of resistivity to the changes occurring in the curing of the smart cement with different nanoparticles, as summarised in Tables 7.2 and 7.10.

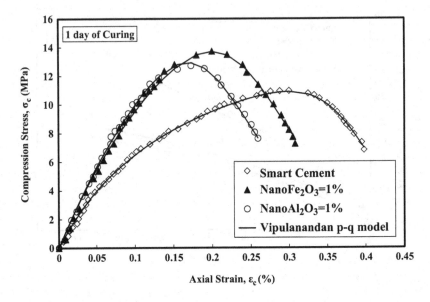

Figure 7.17 Compressive stress-strain relationship of the smart cement with and without 1% NanoFe$_2$O$_3$ and 1% NanoAl$_2$O$_3$ after 1 day of curing.

7.3.5 Compressive Stress-Strain behavior

Based on the experimental results, the Vipulanandan p-q stress-strain model was used to predict the compressive stress-strain relationship (Eqn. 7.3) for the smart cement with and without 1% NanoFe$_2$O$_3$ and 1% NanoAl$_2$O$_3$.

7.3.5.1 Curing Time

One day

(a) Smart Cement with 1% NanoFe$_2$O$_3$

The compressive strengths (σ_{cf}) of the smart cement with 1% NanoFe$_2$O$_3$ after one day of curing were 13.7 MPa, 25.7% higher than the smart cement without NanoFe$_2$O$_3$, as summarised in Table 7.11 and also shown in Figure 7.17. The failure strain was 0.20% and the initial modulus was 13,260 MPa, as summarised in Table 7.11. The addition of 1% NanoFe$_2$O$_3$ to the smart cement reduced the failure strain and increased the modulus compared to the smart cement without NanoFe$_2$O$_3$. The model parameters q_o and p_o were 0.54 and 0.095 respectively. The parameter q_o represents non-linearity up to peak stress, and was higher than the smart cement without NanoFe$_2$O$_3$, as shown in Figure 7.17. The coefficient of determination (R^2) was 0.99 and the RMSE was 0.36 MPa, as summarised in Table 7.11.

Table 7.11 Compressive strength, failure strain, initial modulus, and model parameters for the smart cement without and with 1% NanoFe$_2$O$_3$ and 1% NanoAl$_2$O$_3$ after one day of curing

Material	Curing Time (day)	σ_f (MPa)	ε_f (%)	Ei (MPa)	p_o	q_o	RMSE (MPa)	R^2
Smart Cement	1	10.9±2	0.30 ± 0.02	10,100± 21	0.034± 0.01	0.36± 0.02	0.128	0.99
Smart Cement+1% NanoFe$_2$O$_3$	1	13.7±1.5	0.20 ± 0.01	13,260± 28	0.095± 0.001	0.54± 0.03	0.36	0.99
Smart Cement+1% NanoAl$_2$O$_3$	1	12.7±2	0.18 ± 0.02	12,390± 10	0.10± 0.01	0.53± 0.01	0.24	0.99

(b) Smart Cement with 1% NanoAl$_2$O$_3$

The compressive strengths (σ_{cf}) of the smart cement with 1% NanoAl$_2$O$_3$ after one day of curing were 12.7 MPa, 16.5% higher than the smart cement without NanoAl$_2$O$_3$, as summarised in Table 7.11 and also shown in Figure 7.17. Also the strength was lower compared to the smart cement with 1% NanoFe$_2$O$_3$. The failure strain was 0.18% and the initial modulus was 12,390 MPa, as summarised in Table 7.11. The addition of 1% NanoAl$_2$O$_3$ to the smart cement reduced the failure strain and increased the modulus compared to the smart cement without NanoAl$_2$O$_3$. The model parameters q_o and p_o were 0.53 and 0.10 respectively, as summarised in Table 7.11. The parameter q_o represents the nonlinearity up to peak stress, and was higher than the smart cement without NanoAl$_2$O$_3$, as shown in Figure 7.17. The coefficient of determination (R^2) was 0.99 and the RMSE was 0.24 MPa, as summarised in Table 7.11.

Twenty-eight Days

(a) Smart Cement with 1% NanoFe$_2$O$_3$

The compressive strengths (σ_{cf}) of the smart cement with 1% NanoFe$_2$O$_3$ after 28 days of curing were 27 MPa, 25.7% higher than the smart cement without NanoFe$_2$O$_3$, as summarised in Table 7.12 and also shown in Figure 7.18. The failure strain was 0.20% and the initial modulus was 24,790 MPa, as summarised in Table 7.12. The addition of 1% NanoFe$_2$O$_3$ to the smart cement reduced the failure strain and increased the modulus compared to the smart cement without NanoFe$_2$O$_3$. The model parameters q_o and p_o were 0.57 and 0.097 respectively. The parameter q_o represents nonlinearity up to peak stress, and was lower than the smart cement without NanoFe$_2$O$_3$, as shown in Figure 7.18. The coefficient of determination (R^2) was 0.99 and the RMSE was 0.03 MPa, as summarised in Table 7.12.

Table 7.12 Compressive strength, failure strain, initial modulus, and model parameters for the smart cement without and with 1% NanoFe$_2$O$_3$ and 1% NanoAl$_2$O$_3$ after 28 days of curing

Material	Curing Time (day)	σ_f (MPa)	ε_f (%)	Ei (MPa)	p_o	q_o	RMSE (MPa)	R^2
Smart Cement	28	19.3± 1.5	0.22± 0.02	12,500± 20	0.160± 0.03	0.70± 0.04	0.11	0.99
Smart Cement+1% NanoFe$_2$O$_3$	28	27± 3.2	0.20± 0.01	24,790± 38	0.097± 0.002	0.57± 0.03	0.03	0.99
Smart Cement+1% NanoAl$_2$O$_3$	28	27.5± 3.5	0.17± 0.03	21,500± 20	0.1± 0.01	0.57± 0.01	0.15	0.99

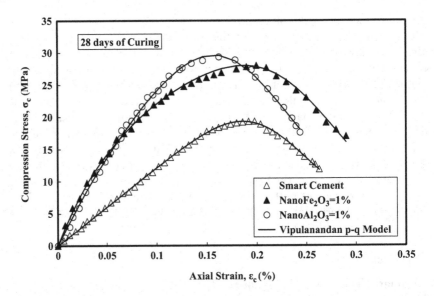

Figure 7.18 Compressive stress-strain relationship of the smart cement with without 1% NanoFe$_2$O$_3$ and 1% NanoAl$_2$O$_3$ after 28 days of curing.

(b) Smart Cement with 1% NanoAl$_2$O$_3$

The compressive strengths (σ_{cf}) of the smart cement with 1% NanoAl$_2$O$_3$ after 28 days of curing were 27.5 MPa, 42.5% higher than the smart cement without NanoAl$_2$O$_3$, as summarised in Table 7.12 and also shown in Figure 7.18. Also, the strength was comparable to the smart cement with 1% NanoFe$_2$O$_3$. The failure strain was 0.17% and the initial modulus was 21,500 MPa, as summarised in Table 7.12. The addition of 1% NanoAl$_2$O$_3$ to the smart cement reduced the failure strain and increased the modulus compared to the smart cement without $NanoAl_2O_3$. The model parameters q_o and p_o were 0.57 and 0.10 respectively, as summarised in Table 7.12. The parameter q_o represents nonlinearity up to peak stress, and was higher than the smart cement without $NanoAl_2O_3$, as shown in Figure 7.18. The coefficient of determination (R^2) was 0.99 and the RMSE was 0.15 MPa, as summarised in Table 7.12.

7.3.6 Compressive Piezoresistivity

The Vipulanandan p-q piezoresistivity model (Eqn. 7.4) was used to predict the stress-piezoresistive axial strain behavior of the smart cement with and without 1% $NanoFe_2O_3$ and 1% $NanoAl_2O_3$ for one day, seven days, and 28 days of curing.

7.3.6.1 Curing Time

One Day

(a) **Smart Cement with 1% $NanoFe_2O_3$**
The piezoresisitive axial strain of the smart cement with 1% $NanoFe_2O_3$ at failure $\left(\dfrac{\Delta\rho}{\rho_o}\right)_f$ after one day of curing was 700%, as summarised in Table 7.13 and shown in Figure 7.19. The addition of 1% $NanoFe_2O_3$ increased the 520% piezoresistive axial strain of the smart cement by 34.6%. The addition of 1% $NanoSiO_2$ reduced the piezoresistive axial failure strain to 428%. The compressive axial failure strain was 0.20%, so the piezoresistive axial strain was increased by 3,500 times (350,000%), making the smart cement highly sensing. The model parameters q_2 and p_2 were 2.70 and 0.0 respectively. The coefficient of determination (R^2) was 0.99 and the RMSE was 0.031 MPa, as summarised in Table 7.13.

Table 7.13 Compressive piezoresistive properties and model parameters after one day of curing

Material	Additives (%)	Curing Time (day)	$(\Delta\rho/\rho_o)_f$ (%)	σ_f (MPa)	q_2	p_2	RMSE (MPa)	R^2	Fig. No.
Smart Cement	0	1	520 ± 10	10.9±3	0.29± 0.02	0.16± 0.06	0.05	0.99	8 (a)
Smart Cement+	1% $NanoFe_2O_3$	1	700 ± 20	13.7 ± 1.5	2.70 ± 0.04	0.0	0.031	0.99	8 (a)
	1% $NanoAl_2O_3$	1	420 ± 10	12.7 ± 2	0.19 ± 0.02	0.30 ± 0.03	0.019	0.99	8 (a)

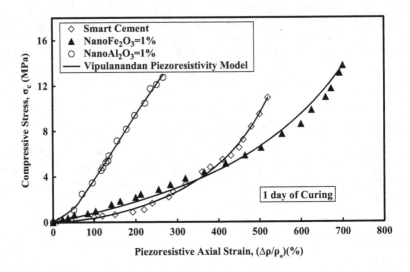

Figure 7.19 Piezoresistive behavior of the smart cement without and with 1% NanoFe$_2$O$_3$ and 1% NanoAl$_2$O$_3$ after 1 day of curing.

(b) Smart Cement with 1% NanoAl$_2$O$_3$

The piezoresisitive axial strain of the smart cement with 1% NanoAl$_2$O$_3$ at failure $\left(\dfrac{\Delta\rho}{\rho_o}\right)_f$ after one day of curing was 420%, as summarised in Table 7.13 and shown in Figure 7.19. The addition of 1% NanoAl$_2$O$_3$ reduced the 520% piezoresistive axial strain of the smart cement by 19.2%. The addition of 1% NanoSiO$_2$ also reduced the piezoresistive axial failure strain to 428%. The compressive axial failure strain was 0.19%, so the piezoresistive axial strain was increased by 2,211 times (221,100%), making the smart cement highly sensing. The model parameters q_2 and p_2 were 0.19 and 0.30 respectively. The coefficient of determination (R^2) was 0.99 and the RMSE was 0.019 MPa, as summarised in Table 7.13.

Seven Days

(a) Smart Cement with 1% NanoFe$_2$O$_3$

The piezoresisitive axial strain of the smart cement with 1% NanoFe$_2$O$_3$ at failure $\left(\dfrac{\Delta\rho}{\rho_o}\right)_f$ after seven days of curing was 560%, as summarised in Table 7.14 and shown in Figure 7.20. The addition of 1% NanoFe$_2$O$_3$ increased the 460% piezoresistive axial strain of the smart cement by 21.7%. The addition of 1% NanoSiO$_2$ reduced the piezoresistive axial failure strain to 365%. The compressive axial failure strain was 0.23%, so the piezoresistive axial strain was increased by 2,435 times (243,500%)

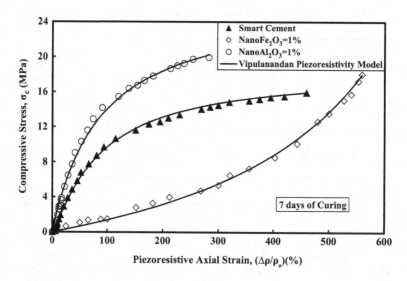

Figure 7.20 Piezoresistive behavior of the smart cement without and with 1% NanoFe$_2$O$_3$ and 1% NanoAl$_2$O$_3$ after 7 days of curing.

making the smart cement highly sensing. The model parameters q$_2$ and p$_2$ were 2.60 and 0.10 respectively. The coefficient of determination (R^2) was 0.99 and the RMSE was 0.02 MPa, as summarised in Table 7.14.

(b) Smart Cement with 1% NanoAl$_2$O$_3$

The piezoresisitive axial strain of the smart cement with 1% NanoAl$_2$O$_3$ at failure $\left(\dfrac{\Delta\rho}{\rho_o}\right)_f$ after seven days of curing was 356%, as summarised in Table 7.14 and shown in Figure 7.20. The addition of 1% NanoAl$_2$O$_3$ reduced the 460% piezoresistive axial strain of the smart cement by 22.6%. The addition of 1% NanoSiO$_2$ also reduced the piezoresistive axial failure strain to 365%. The compressive axial failure strain was 0.22%, so the piezoresistive axial strain was increased by 1,618 times (161,800%), making the smart cement highly sensing. The model parameters q$_2$ and p$_2$ were 0.15 and 0.22 respectively. The coefficient of determination (R^2) was 0.99 and the RMSE was 0.02 MPa, as summarised in Table 7.14.

Twenty-eight Days

(a) Smart Cement with 1% NanoFe$_2$O$_3$

The piezoresisitive axial strain of the smart cement with 1% NanoFe$_2$O$_3$ at failure $\left(\dfrac{\Delta\rho}{\rho_o}\right)_f$ after 28 days of curing was 409%, as summarised in Table 7.15

Table 7.14 Compressive piezoresistive properties and model parameters after seven
days of curing

Material	Additives (%)	Curing Time (day)	$(\Delta\rho/\rho o)_f$ (%)	σ_f (MPa)	q_2	p_2	RMSE (MPa)	R^2
Smart Cement	0%	7	460±14	16±1	0.11± 0.02	0.04± 0.03	0.017	0.99
	1% NanoFe$_2$O$_3$	7	560 ± 15	18.0 ± 2.5	2.60 ± 0.03	0.10 ± 0.02	0.02	0.99
	1% NanoAl$_2$O$_3$	7	356± 11	22±3	0.15± 0.03	0.22± 0.03	0.02	0.99

Table 7.15 Compressive piezoresistive properties and model parameters after 28 days
of curing

Material	Additives (%)	Curing Time (day)	$(\Delta\rho/\rho o)_f$ (%)	σ_f (MPa)	q_2	p_2	RMSE (MPa)	R^2
Smart Cement	0	28	400±10	19.4± 2	0.05± 0.03	0.03± 0.02	0.022	0.99
	1% NanoFe$_2$O$_3$	28	409 ± 10	27.0 ± 3.2	0.21± 0.02	0.13± 0.01	0.02	0.99
	1% NanoAl$_2$O$_3$	28	250± 10	27.5 ± 3.5	0.14± 0.007	0.35± 0.2	0.016	0.99

and shown in Figure 7.21. The addition of 1% NanoFe$_2$O$_3$ increased the
400% piezoresistive axial strain of the smart cement by 2.3%. Addition of
1% NanoSiO$_2$ reduced the piezoresistive axial failure strain to 250%. The
compressive axial failure strain was 0.20%, so the piezoresistive axial strain
was increased by 2,045 times (204,500%), making the smart cement highly
sensing. The model parameters q_2 and p_2 were 0.21and 0.13 respectively.
The coefficient of determination (R^2) was 0.99 and the RMSE was 0.02
MPa, as summarised in Table 7.15.

(b) Smart Cement with 1% NanoAl$_2$O$_3$

The piezoresisitive axial strain of the smart cement with 1% NanoAl$_2$O$_3$ at
failure $\left(\dfrac{\Delta\rho}{\rho_o}\right)_f$ after 28 days of curing was 250%, as summarised in Table 7.15
and shown in Figure 7.21. The addition of 1% NanoAl$_2$O$_3$ reduced the
400% piezoresistive axial strain of the smart cement by 37.5%. The addition of 1% NanoSiO$_2$ also reduced the piezoresistive axial failure strain to
250%. The compressive axial failure strain was 0.17%, so the piezoresistive
axial strain was increased by 1,471 times (147,100%), making the smart

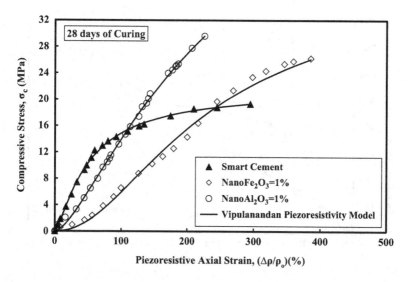

Figure 7.21 Piezoresistive behavior of the smart cement without and with 1% NanoFe$_2$O$_3$ and 1% NanoAl$_2$O$_3$ after 28 days of curing.

cement highly sensing. The model parameters q_2 and p_2 were 0.14 and 0.35 respectively. The coefficient of determination (R^2) was 0.99 and the RMSE was 0.016 MPa, as summarised in Table 7.15.

7.3.7 Gauge Factor

Vipulanandan Gauge Factor Model

Based on the experimental results, the compressive strain at failure did not change much with increasing the curing time for the smart cement without and with 1% of NanoFe$_2$O$_3$ and 1% of NanoAl$_2$O$_3$ contents, while the percentage of the piezoresistive axial strain $\left(\dfrac{\Delta\rho}{\rho_o}\right)$ changed by 200% to 700% based on the curing time, as shown in Figure 7.22. The piezoresistive axial strain at failure $\left(\dfrac{\Delta\rho}{\rho_o}\right)_f$ for the smart cement increased and decreased with the addition of 1% of NanoFe$_2$O$_3$ and 1% of NanoAl$_2$O$_3$ respectively.

The relationship between the strain with the change in electrical resistivity (gauge factor) for the smart cement after one day and 28 days of curing was investigated and modeled using the Vipulanandan gauge factor model as follows:

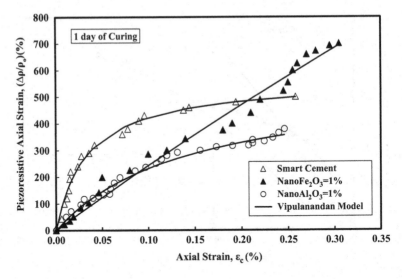

Figure 7.22 Piezoresistive axial strain and compressive strain relationships for the smart cement without and with 1% NanoFe$_2$O$_3$ and 1% NanoAl$_2$O$_3$ after 1 day of curing.

$$\frac{\Delta\rho}{\rho_o} = \frac{\epsilon}{D_2 + E_2\epsilon} \qquad (7.8)$$

where $\left(\dfrac{\Delta\rho}{\rho_o}\right)p$ percentage of piezoresistive axial strain due to the applied stress and ϵ is the compressive strain and D_2 and E_2 are the model parameter. The initial slope of this relationship will be equal to $\dfrac{1}{D_2}$. Also, if the parameter E_2 is equal to zero, the relationship between the piezoresistive axial strain $\left(\dfrac{\Delta\rho}{\rho_o}\right)$ and axial strain (ϵ) will be linear.

7.3.7.1 Curing Time

One Day

(a) Smart Cement
The piezoresistive axial strain and compressive strain relationship was non-linear, and the model parameter E_2 was not zero, as shown in Figure 7.22 and also summarised in Table 7.16. The piezoresistive axial strain at failure

Table 7.16 Gauge factor model parameters for smart cement without and with 1% NanoFe$_2$O$_3$ and 1% NanoAl$_2$O$_3$ after one day of curing

Material	Curing Time (day)	D_2	E_2	RMSE (%)	R^2
Smart Cement	1	6*10^{-5}	0.002	18.1	0.98
Smart Cement+1% NanoFe$_2$O$_3$	1	0.0004	0.0001	32.4	0.98
Smart Cement+1% NanoAl$_2$O$_3$	1	0.0003	0.0002	12.5	0.99

was 520% and the corresponding axial strain was 0.30%, with the lowest secant value 1733, which represents the minimum magnification of strain. The model parameters D_2 and E_2 were 6*10^{-5} and 0.002 respectively. The coefficient of determination (R^2) was 0.98 and the RMSE was 18.1%, as summarised in Table 7.16.

(b) Smart Cement with 1% NanoFe$_2$O$_3$

The piezoresistive axial strain and compressive strain relationship was non-linear and the model parameter E_2 was not zero, as shown in Figure 7.22 and also summarised in Table 7.16. The piezoresistive axial strain at failure was 700% and the corresponding axial strain was 0.20%, with the lowest secant value 3500, and was the highest for the materials tested in this study. The model parameters D_2 and E_2 were 0.0004 and 0.0001 respectively. The coefficient of determination (R^2) was 0.98 and the RMSE was 32.4%, as summarised in Table 7.16.

(c) Smart Cement with 1% NanoAl$_2$O$_3$

The piezoresistive axial strain and compressive strain relationship was non-linear and the model parameter E_2 was not zero, as shown in Figure 7.22 and also summarised in Table 7.16. The piezoresistive axial strain at failure was 420% and the corresponding axial strain was 0.19%, with the lowest secant value 2210, which represents the minimum magnification of strain. The model parameters D_2 and E_2 were 0.0003 and 0.0002 respectively. The coefficient of determination (R^2) was 0.99 and the RMSE was 12.5%, as summarised in Table 7.16.

Twenty-eight Days

(a) Smart Cement

The piezoresistive axial strain and compressive strain relationship was linear and the model parameter E_2 was zero, as shown in Figure 7.23 and also summarised in Table 7.17. The piezoresistive axial strain at failure was 400% and the corresponding axial strain was 0.22%, with the lowest secant value 1818, which represents the minimum magnification of strain. The model parameters D_2 and E_2 were 0.0005 and 0.0 respectively. The

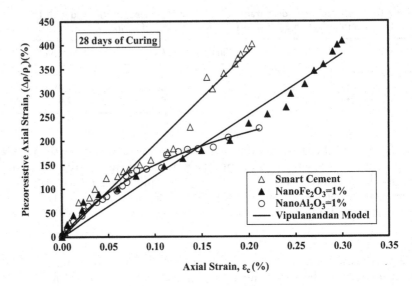

Figure 7.23 Piezoresistive axial strain and compressive strain relationships for the smart cement without and with 1% NanoFe$_2$O$_3$ and 1% NanoAl$_2$O$_3$.

coefficient of determination (R^2) was 0.97 and the RMSE was 22.4%, as summarised in Table 7.17.

(b) Smart Cement with 1% Nano Fe2O3

The piezoresistive axial strain and compressive strain relationship was linear and the model parameter E_2 was not zero, as shown in Figure 7.23 and also summarised in Table 7.17. The piezoresistive axial strain at failure was 409% and the corresponding axial strain was 0.20%, with the lowest secant value 2045, which represents the minimum magnification of strain. The model parameters D_2 and E_2 were 0.0008 and 0.0 respectively. The coefficient of determination (R^2) was 0.98 and the RMSE was 21.7%, as summarised in Table 7.17.

(c) Smart Cement with 1% NanoAl$_2$O$_3$

The piezoresistive axial strain and compressive strain relationship was non-linear and the model parameter E_2 was not zero, as shown in Figure 7.23 and also summarised in Table 7.17. The piezoresistive axial strain at failure was 250% and the corresponding axial strain was 0.17%, with the lowest secant value 1471, which represents the minimum magnification of strain. The model parameters D_2 and E_2 were 0.0004 and 0.002 respectively. The coefficient of determination (R^2) was 0.98 and the RMSE was 6.20%, as summarised in Table 7.17.

Table 7.17 Gauge factor model parameters for the smart cement without and with 1% NanoFe$_2$O$_3$ and 1% NanoAl$_2$O$_3$ after 28 days of curing

Material	Curing Time (day)	D$_2$	E$_2$	RMSE (%)	R^2	Fig. No.
Smart Cement	28	0.0005	0	22.4	0.97	10 (b)
Smart Cement+1% NanoFe$_2$O$_3$	28	0.0008	0	21.7	0.98	10 (b)
Smart Cement+1% NanoAl$_2$O$_3$	28	0.0004	0.002	6.20	0.98	10 (b)

7.4 SUMMARY

Based on the experimental and analytical study of the smart cement modified with nanoparticles (NanoSiO$_2$, NanoFe$_2$O$_3$ and NanoAl$_2$O$_3$) up to 1%, the following conclusions are advanced:

1 Based on the XRD analyses, with the addition of 1% NanoSiO$_2$, changes in cement mineralogy were observed and the new constituents were magnesium silicate sulfate (Mg$_5$ (SiO$_4$)$_2$SO$_4$) (2θ peaks at 51.58° and 56.50°) and quartz (SiO$_2$) (2θ peaks at 23.55°, 36.90° and 62.50°). Hence, some of the changes observed in the modified cement with NanoSiO$_2$ behavior could have been due to the changes in the cement mineralogy. TGA analyses also showed a great reduction in the total weight loss of the cement at 800°C when it was modified with 1% NanoSiO$_2$.

2 The addition of 1% NanoSiO$_2$ increased the compressive strength of the smart cement by 14% and 42% after one day and 28 days of curing respectively. Also, the modulus of elasticity of the smart cement increased with the addition of 1% NanoSiO$_2$. The Vipulanandan p-q stress-strain model predicted the behavior very well.

3 Resistivity was sensitive to the amount of NanoSiO$_2$ used to modify the smart cement. The amount of NanoSiO$_2$ can be detected based on the change in the initial resistivity. An addition of 1% NanoSiO$_2$ increased the initial electrical resistivity (ρ_o) of the smart cement by 35% and also increased the time to reach the minimum resistivity by 31 minutes. Initial electrical resistivity can be used as a good indicator for quality control. The Vipulanandan p-q curing model predicted the behavior very well.

4 The addition of the NanoSiO$_2$ reduced the piezoresistivity (changed the resistivity at peak stress) of the smart cement. The Vipulanandan p-q piezoresisitive model predicted the behavior very well.

5 Linear correlations were found between the resistivity index (RI$_{24}$) and compressive strength at different curing ages. Also, a nonlinear

model (NLM) was used to correlate the model parameters to the curing time and NanoSiO$_2$ contents.

6 Resistivity was sensitive to the addition of 1% of NanoFe$_2$O$_3$ and 1% of NanoAl$_2$O$_3$ used to modify the smart cement. An addition of 1% of NanoFe$_2$O$_3$ and 1% of NanoAl$_2$O$_3$ increased the initial electrical resistivity (ρ_o) of the smart cement by 35% and 40% respectively and also increased the time to reach the minimum resistivity by 45 minutes and 93 minutes respectively. Initial electrical resistivity can be used as a good indicator for quality control. The Vipulanandan p-q curing model predicted the behavior very well.

7 The addition of 1% of NanoFe$_2$O$_3$ and 1% of NanoAl$_2$O$_3$ increased the compressive strength of the smart cement by 17% to 42% based on the curing time. Also, the modulus of elasticity of the smart cement increased with the addition of 1% NanoFe$_2$O$_3$ and 1% of NanoAl$_2$O$_3$. The Vipulanandan p-q stress-strain model predicted the behavior very well.

8 The addition of the 1% of NanoFe$_2$O$_3$ and 1% of NanoAl$_2$O$_3$ reduced the piezoresistivity (changed the resistivity at peak stress) of the smart cement. The Vipulanandan p-q piezoresistive model predicted the behavior very well.

9 The modulus of elasticity of the smart cement increased with the addition of 1% of NanoFe$_2$O$_3$ 1% of NanoAl$_2$O$_3$. The Vipulanandan p-q stress-strain model predicted the behavior very well.

10 Changes in the electrical resistive axial strains were more sensitive than the change in the compressive axial strains and the Vipulanandan gauge factor model was used to predict the relationships between the piezoresistive axial strains and compressive axial strains.

Chapter 8

Gas Leak Detection, Fluid Loss, and Sensing Dynamic Loadings Using Smart Cement

8.1 BACKGROUND

Cement is used in multiple infrastructure constructions and maintenance applications such as storage facilities, water and waste water pipelines and treatment plants, production wells, CO_2 sequestration wells, and foundations, hence making cementitious materials sensing to fluid (gases, liquids) leaks one of the important requirements to minimise losses. Fluid loss from cement during various operations will not only impact the quality of the cementing job but also the hydration of the cement and the hardened cement quality. Also, dynamic loadings occur very frequently due to earthquakes, wind storms, traffic on highways, and fluid leaks through pipelines, resulting in the degradation of critical infrastructures and premature failures that need to be minimised with real-time monitoring.

Gas leaks are a major problem in both operational wells and abandoned wells, as are leaking pipelines around the world. Gas leaks are a problem in carbon dioxide (CO_2) sequestration wells too (Radenti et al. 1972; Nelson 1990; Clavert 1990; Bonett 1996; Eoff et al. 2009; Vipulanandan et al. 2019a, 2020a, b). With the advancements in smart cement, nanotechnology, and polymers, these materials can be used for solving some of the problems encountered in oil and gas wells (Mangadlao et al. 2015, Vipulanandan et al. 2015 and 2017; Houk 2017). Gas and liquid leakage through cement install wells and pipelines may be divided into three stages: initial stage during placement (slurry), intermediate stage (during hardening), and long-term stage (gas in service).

At the initial stage of well installation, gas migration occurs due to incorrect cement densities, contamination on account of mud and filter cake, premature gelation leading to loss of hydrostatic pressure control, and high shrinkage of the cement. Excessive fluid loss creates increased voids in cement slurry for gas to enter, and high stresses around the cement sheath surrounding the casing, leading to the formation of micro annulus (Bonett et al. 1996). Various methods have been suggested to prevent gas migration through cement, such as adding polymers to the cement and also using thixotropic and high-gel-strength cements. In the case of thixotropic cement, transmitted hydrostatic pressure should revert to the interstitial water and its high gel strength offers resistance to gas leaks. There are many types of polymers that are being used to enhance the performance of cement.

Real-time monitoring of gas leaks and dynamic loadings has become a major issue during the entire service of cemented wells (oil, gas, and water) and infrastructures (Zhang et al. 2010; Vipulanandan et al. 2014b, c, d, and 2019a). In this study, smart cement was modified with commercially available styrene butadiene polymer. Styrene butadiene rubber (SBR) was first developed in Germany in 1929 as an alternative to natural rubber and started to be manufactured in the United States in 1942. Styrene butadiene promotes the reaction of calcium aluminate with gypsum, while restraining the formation of C-S-H gel and C_4AH_{13}. (Bonnet et al. 1996). Styrene butadiene polymer enhances bond strength and also prevents the annular gas flow until 175°C (350° F). Different manufacturers recommend different percentages of polymer for controlling fluid loss and gas migration. Unfortunately, the conditions in-situ cannot be monitored with the currently available technologies, and there is a need for developing new real-time monitoring technology.

Also, there is a need for detecting and monitoring dynamic loadings such as impact, cyclic, and earthquake loadings to ensure safety and to evaluate protection systems and the conditions of infrastructures as well.

8.2 FLUIDS FLOW MODELS AND VERIFICATIONS

(a) Darcy Law (1856)

Both compressible and incompressible fluids flow through porous media. Darcy's law was formulated by Henry Darcy in 1856 based on experimental results on the flow of water through a bed of sand representing a porous medium. Water is an incompressible fluid and the sand formation was not reactive with the water. Darcy's law identified the hydraulic gradient as the major factor causing the flow of water through the sand. Darcy's law relates macroscopic pressure gradients to the flow velocity vector in the fluid phase as:

$$v = -\frac{1}{\mu} K \cdot (\nabla P) \tag{8.1}$$

in which v is the Darcy velocity vector (m/s); μ is the fluid viscosity (Pa.s); K is the permeability tensor (m²); P is the macroscopic pressure; and ∇ is the vector operator.

(b) Scheidegger Model (1974)

In the case of compressible gas as pore fluid, Scheidegger (1974) expressed the average gas permeability to be calculated as follows:

$$\frac{Q}{A} = \frac{k_g}{\mu L} \frac{\left(P_{up}\right)^2 - \left(P_{down}\right)^2}{2\,P_{down}} \tag{8.2}$$

where Q is the volume of fluid flow rate measured; A is the cross-sectional area of the sample; μ is viscosity of the pore fluid; L is sample length; k_g is expressed as the average permeability; and $P_{up}\,P_{down}$ are the pore pressure of the upper and lower ends of the specimen, respectively.

(c) Vipulanandan Fluid Flow Model (2017)

The rate of fluid flows through a porous media, defined as connected pores, is influenced by not only the piezometric gradient but also the changes in the nano- and micro structures of the porous media due to biological and chemical reactions between the flowing fluids and the porous media, the diffusion process, and the bio-chemo-thermo-stress changes in the fluids and porous media. The Vipulanandan generalised fluid flow model is as follows:

$$q_i = q_{i0} + \frac{\nabla_j h}{A_{ij} + B_i \nabla_j h} \tag{8.3}$$

and the vector operator ∇ is

$$\nabla = \hat{i}\frac{\partial}{\partial x} + \hat{j}\frac{\partial}{\partial y} + \hat{k}\frac{\partial}{\partial z}$$

and q_i is the specific rate of discharge (average velocity = discharge per unit area) in the direction i (x, y, or z) and h is the piezometric head causing the flow. The q_{io} represents the specific rate of flows in the direction i due to diffusion and non-hydraulic gradient. The material parameter A_{ij} is a second-order tensor and is related to the permeability of the porous media and affected by the flow patterns. Parameter B_i also represents the directional material (porous media and flowing fluid) properties (vector) that will be influenced by the fluid properties, flow patterns, and biological, chemical, temperature, and stress conditions (bio-chemo-thermo-stress).

1-Dimensional (1-D) Flow

The Vipulanandan fluid flow model for the porous media includes the changes in the fluid and also porous media in regard to pressure, temperature, and biochemical reactions and is represented as:

$$Q = Q_0 + \frac{i}{C + Di} A \tag{8.4}$$

where the volume of flow in 1-D is equal to $Q = qi * A$ and $Q_0 = q_{io} * A$, where A is the effective flow area. The variable i is the piezometric gradient. The first derivative of the discharge when $i = 0$ is defined as the initial permeability (K_o) of the porous medium as follows:

$$\frac{1}{A} \frac{dQ}{di}\bigg|_{i=0} = K_0 = \frac{1}{C} \tag{8.5}$$

Substituting K_o into the equation leads to following equation as:

$$q = q_0 + K_0 \frac{i}{1 + Ni} \tag{8.6}$$

where q is the rate of volume discharge per unit area (discharge velocity). Hence, the first derivation of the equation is:

$$\frac{dq}{di} = K_0 \frac{1}{(1 + Ni)^2} \tag{8.7}$$

For the laboratory test, the hydraulic gradient i is defined as:

$$i = \frac{\left(\frac{p_1}{\rho g} + h_1\right) - \left(\frac{p_2}{\rho g} + h_2\right)}{L} = \frac{\Delta P}{\rho g L} \tag{8.8}$$

where L is the length/thickness of the specimen (h_1-h_2) tested. Since (h_1-h_2), the thickness of the specimen in the numerator is very small, it is neglected. Hence, we apply the hydraulic gradient into the flow equation as:

$$Q = Q_0 + A K_0 \frac{\frac{\Delta P}{\rho g L}}{1 + N \frac{\Delta P}{\rho g L}} \tag{8.9}$$

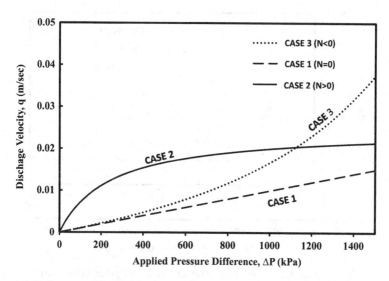

Figure 8.1 Different cases (trends) for the fluid discharge per unit area versus pressure difference based on the Vipulanandan fluid flow model.

Using the new parameters M and N, the Vipulanandan 1D flow model in Eqn. 8.9 is as follows:

$$q = q_0 + \frac{\Delta P}{M + N.\Delta P} \qquad (8.10)$$

where q is the rate of volume discharge per unit area (discharge velocity); ΔP is the pressure difference between the top and bottom of the test specimen since the specimen thickness is relatively small (h_1-h_2); and M (kPa.s/m) and N (s/m) are the model parameters.

Based on this model, when there is fluid flow, always $\frac{dq}{di} > 0$. Also, $\frac{d^2q}{di^2}$ leads to three different possible cases for the fluid flow in the porous media, as shown in Figure 8.1 and Figure 8.2.

CASE 1: Fluid Flow with Constant Permeability (N = 0)

$$\frac{d^2q}{di^2} = 0 \qquad (8.11)$$

which will represent Darcy's law, in which the permeability of the fluid is considered constant, as shown in Figure 8.2.

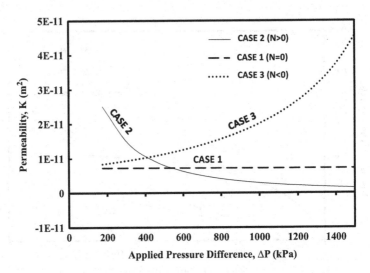

Figure 8.2 Vipulanandan fluid flow model predictions of the permeability variations with the applied pressure (Constant M and Various N).

CASE 2: Fluid Flow with Decreasing Instantaneous (Tangent) Permeability (N > 0)

$$\frac{d^2q}{di^2} < 0, \tag{8.12}$$

which will represent the compressible fluid flow in porous media, where with the increment of pressure gradient, the permeability decreases, as shown in Figure 8.2.

CASE 3: Fluid Flow with Increasing Instantaneous (Tangent) Permeability (N < 0)

$$\frac{d^2q}{di^2} > 0, \tag{8.13}$$

which can be used for fluid flow in porous media, where with the increment of pressure gradient, the permeability increases due the varying viscosity and density of the fluids, fluid-porous media interaction, and also the modification to the porous structure of the media, as shown in Figure 8.2.

8.3 CIGMAT TESTING FACILITY

It is important to develop unique test methods to detect and quantify the liquid and gas flow through different porous media such as sand and cement. In sand there is no chemical reaction between the sand particles and also water, in the test that was performed by Darcy in 1856. In the cement there are continuous hydration reactions, which will result in changes in the porous structure of the cement and also interaction with the flowing fluid. So cement is a reactive porous medium with water compared to sand, which is a non-reactive porous medium. Also, tests have to be developed to real-time monitor the changes in the medium.

A unique high-pressure, high-temperature (HPHT) testing facility has been developed with real-time monitoring probes, as shown in Figure 8.3. In each test chamber the outlet base has two rings to separate the side wall leakage flow and the bulk flow through the testing materials. The testing chambers can be used to test drilling fluids, cements, soils, rocks, and also simulate accelerated corrosion with corroding fluids under high pressure

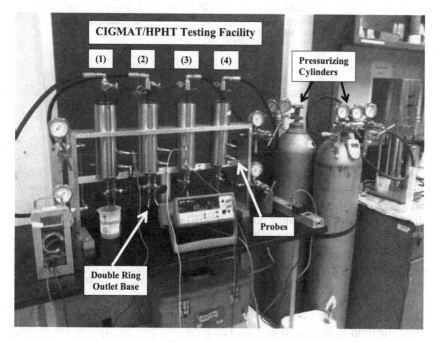

Figure 8.3 CIGMAT high temperature high pressure (HPHT) testing facility with four chambers integrated with real-time monitoring probes and double ring bases.

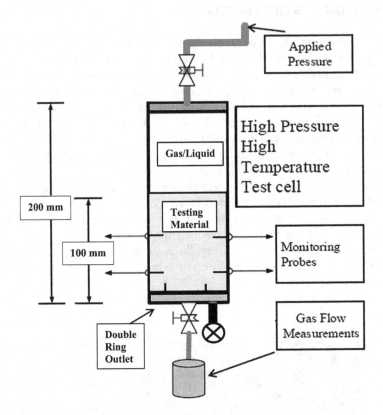

Figure 8.4 Cross-sectional view of the HPHT test chamber.

(14 MPa) and high temperature (500°F). There are many pressure probes to control the applied pressures.

The vertical cross-sectional view of the test chamber is shown in Figure 8.4. The testing material is placed in the chamber and can be tested with liquids and gases. There are also four probes insulated and attached to the chamber, which can be used to monitor continuously the resistance changes vertically, horizontally, and also diagonally using LCR meters. The four probes were tested for air tightness and performed very well. The ring at the bottom separates the area equally and also the fluid flow and fluid losses based on the type of tests. Additionally, the four chambers shown in Figure 8.3 can be independently operated.

In this study both sand (nonreactive) and cement (reactive) were used to verify the fluid flow models and also quantify the permeability of the testing materials with water and gas. Also, the sensitivity of smart cement with gas leak was investigated and quantified. Moreover, the smart cement was modified with polymers and the performances were investigated.

(a) Sand

Experiments were performed using the HPHT chambers filled with sand. Fluid flow tests were performed using nitrogen gas and water representing compressible and incompressible fluids respectively. Fluid was pressurised from the top of the sand with different pressures, and the discharges were measured from the bottom part of the chamber, which was at atmospheric pressure of 101 kPa. During the test the change in electrical resistance was measured continuously using an LCR meter. To minimise the contact resistances, the resistance was measured at 300 kHz using the two-probe method.

(b) Smart Cement

Specimens were prepared using Class H cement with a water-to-cement ratio of 0.38 and 0.04% of carbon fibers (CF) in order to enhance the chemo-thermo-piezoresistivity of the cement to make it more sensing (Figure 5.19). After mixing, the cement slurry was placed into the HTHP chambers and monitored using the probes. After a specified time period of curing, the specimens were first tested with the bottom valves closed (Figure 8.4) to evaluate the piezoresistivity of the smart cement. After that, the specimens were pressurised to 700 kPa (100 psi) for three hours with the bottom valves open for a fluid-loss test using nitrogen gas. For the fluid flow study, both nitrogen gas and water were used as compressible and incompressible fluids. Fluid was pressurised to different pressures, and the discharges were measured from the bottom outlet of the chamber, which was at an atmospheric pressure of 101 kPa. During the test, the changes in resistances were also measured continuously using an LCR meter at 300 kHz with the two-wire method.

8.4 RESULTS AND ANALYSES

8.4.1 In Sand

(a) Water

The water flow through the sand with an initial porosity of 33.6% was investigated by varying the pressure, and the results are shown in Figure 8.5. Experiments showed that with the increase in pressure gradient, the water discharge (average discharge velocity) increased linearly, representing Darcy's law and CASE 1 of the Vipulanandan generalised fluid loss model. For the fluid flow model, the parameter M was 75 kPa.sec/m and the parameter N was zero. As summarised in Table 8.1, the sand porosity reduced to 31.5% with applied pressure of 2,500 kPa.

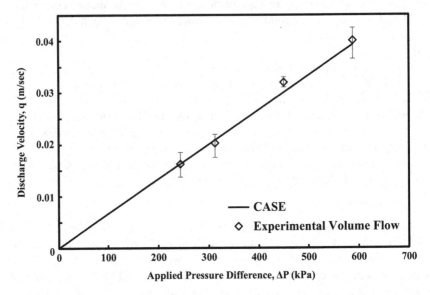

Figure 8.5 Comparison of the experimental results of water flowing through sand with the Vipulanandan fluid flow model (CASE 1).

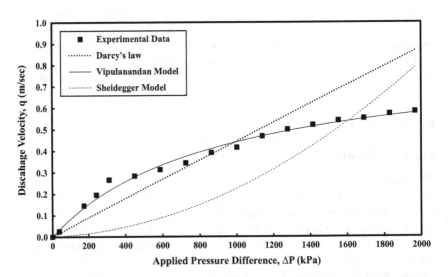

Figure 8.6 Variation of the discharge velocity of gas flowing through sand with the applied pressure difference and predicted with the Vipulanandan fluid flow model (CASE 2).

Table 8.1 Porosity of dry sand with applied pressure

Applied Pressure	Total Weight	Total Volume	Density	Ms	Mw	Vs	Vw	Va	e=Vv/Vs	Porosity
kPa	(g)	(cm³)	(g/cm³)	(g)	(g)	(cm³)	(cm³)	(cm³)	_	%
0	673.1	382.4	1.76	673.1	0.0	254.0	0.00	128.4	0.51	33.57
150	673.1	381.7	1.76	673.1	0.0	254.0	0.00	127.7	0.50	33.46
1,000	673.1	377.1	1.78	673.1	0.0	254.0	0.00	123.1	0.48	32.64
1,500	673.1	373.5	1.80	673.1	0.00	254.0	0.00	119.50	0.47	31.99
2,480	673.1	370.9	1.81	673.1	0.00	254.0	0.00	116.92	0.46	31.52

Ms is the weight of solids; Mw is the weight of water; Vs is the volume of solids; Vw is the volume of water; Va is the volume of air; e is the void ratio.

(b) Gas

Experimental results showed that the discharge velocity was not linear, with increased pressure difference due to changes in both the fluid (gas) and porous media properties with pressure (Figure 8.6). The porosity of the sand in the testing chamber decreased with the increased pressure, which resulted in reduced discharge per unit area (discharge velocity), representing the CASE 2 of the Vipulanandan fluid flow model with the parameter $N > 0$. The Scheidegger model predicted the increasing rate for discharge with a change in pressure gradient. This increasing trend predicted was mainly due to increment in the density and viscosity of nitrogen gas with pressure increase. However, the pressure increment leads to reduction in the porosity of the sand, as summarised in Table 8.1, which reduced the discharge volume and velocity of the nitrogen gas.

The gas permeability of the sand was determined using the models. Darcy's law predicted the gas permeability to be $2.02 \times 10^{-12} m^2$. The Scheidegger model showed reduction in gas permeability from $2.55 \times 10^{-12} m^2$ at pressure difference of $\Delta P = 101.3 kPa$ to $1.22 \times 10^{-13} m^2$ at a pressure difference of $\Delta P = 2070 kPa$. The Vipulanandan model also predicted reduction in the gas permeability from $1.24 \times 10^{-12} m^2$ at a pressure difference of $\Delta P = 101.3 kPa$ to $1.35 \times 10^{-13} m^2$ at a pressure difference of $\Delta P = 2070 kPa$, as shown in Figure 8.7.

8.4.2 Smart Cement

(a) Water

Experiments showed the piezoresistivity of the smart cement after six hours of curing with the outlet valve closed and zero fluid loss, as shown in Figure 8.8. Resistivity increased by about 35% with the 1.4 MPa applied pressure, as shown in Figure 8.8, representing the piezoresistivity of the smart cement. Electrical resistivity of the smart cement was measured during

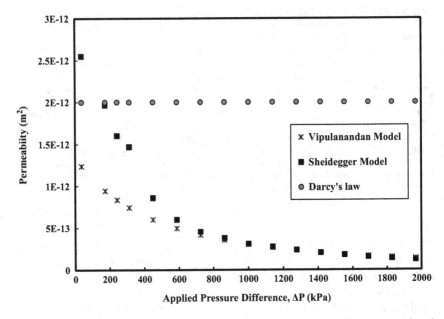

Figure 8.7 Gas permeability in sand based on the Darcy's Law, Scheidegger model and Vipulanandan model.

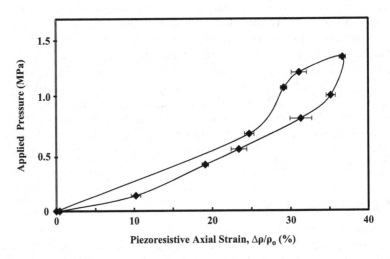

Figure 8.8 Changes in the electrical resistivity of the piezoresisitive smart cement under different pressures.

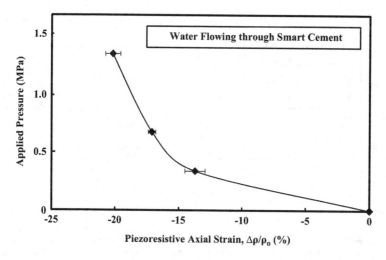

Figure 8.9 Changes in electrical resistivity of the smart cement under water flow with different applied pressures.

the flow of the water through the smart cement, and it decreased the electrical resistivity of the cement. Water flow with a pressure difference of 1.4 MPa reduced the electrical resistivity by 20%, as shown in Figure 8.9, due to the change in the stress within the cement and also the reaction between the cement and the flowing water.

(b) Gas

Experimental results showed that with the increase in the applied pressure gradient, the gas discharge volume per unit area (discharge velocity) increased in a nonlinear manner due to changes in the gas and porous media properties with pressure (Figure 8.10). This trend was predictable based on the Vipulanandan model and CASE 2, as shown in Figure 8.10, and also as compared to the linear prediction of Darcy's law. The Scheidegger model predicted the nonlinear increase in the rate of discharge per unit area with the change in pressure gradient. This predicted increasing trend was due to the model discharge and pressure relationship in Eqn. (8.2). The pressure increase reduced the porosity from 54.5% to 36%, as summarised in Table 8.2, which reduced the rate of volume discharge per unit area (discharge velocity) of the nitrogen gas, as in CASE 2 of the Vipulanandan Fluid Flow model ($N > 0$), as shown in Figure 8.10.

Gas permeability in the smart cement was evaluated using the different models. Darcy's law predicted gas permeability to be $3.0 \times 10^{-13} m^2$. The Scheidegger model showed reduction in gas permeability from $2.82 \times 10^{-13} m^2$ at a pressure difference of $\Delta P = 174 \, kPa$ to $3.88 \times 10^{-14} m^2$ at a pressure difference of $\Delta P = 1553 kPa$. The Vipulanandan model predicted

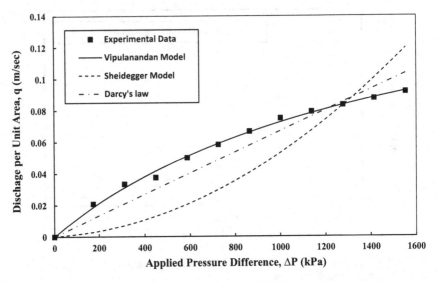

Figure 8.10 Variation of the discharge velocity of nitrogen gas with the applied pressure difference in the smart cement cured for 6 hours.

Table 8.2 Porosity changes in cement slurry under pressure in the HPHT during gas leak

	Total Weight	Total Volume	Density	Mc	Mw	Vc	Vw	Va	Porosity
Smart Cement	(g)	(cm³)	(g/cm³)	(g)	(g)	(cm³)	(cm³)	(cm³)	%
Slurry	987.0	506.7	1.95	715.2	271.8	230.7	271.8	4.2	54.5
Filter Cake	586.0	255.0	2.30	505.7	80.3	163.1	80.3	11.6	36.0

Mc is the weight of cement; **Mw** is the weight of water; **Vs** is the volume of cement; **Vw** is the volume of water; **Va** is the volume of air; **e** is the void ratio.

gas permeability as $1.40 \times 10^{-13} m^2$ at a pressure difference of $\Delta P = 174 \text{kPa}$ and $4.27 \times 10^{-14} m^2$ at a pressure difference of $\Delta P = 1553 \text{kPa}$, as shown in Figure 8.11.

8.5 GAS LEAK DETECTION WITH SMART CEMENT

Electrical resistivity change in the smart cement was measured after six hours of curing under nitrogen gas flow with different applied pressures. After six hours of cement curing and testing under a 2 MPa nitrogen gas flow, the electrical resistivity of the smart cement decreased by 12%, as shown in Figure 8.12. Also, the variation in the electrical resistivity with the gas discharge velocity is shown in Figure 8.13, and this is a clear indication

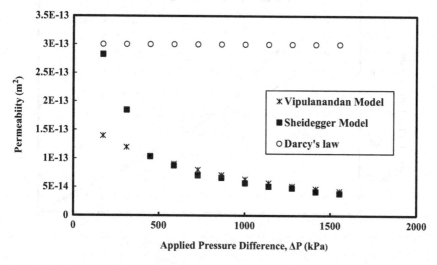

Figure 8.11 Gas permeability through the smart cement based on Darcy's Law, Scheidegger model and Vipulanandan model.

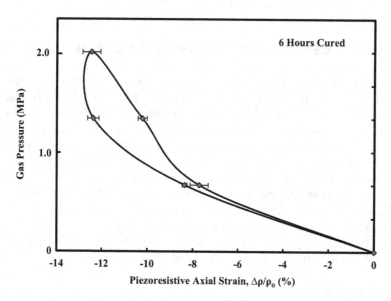

Figure 8.12 Changes in electrical resistivity of the smart cement under nitrogen gas flow at different pressures.

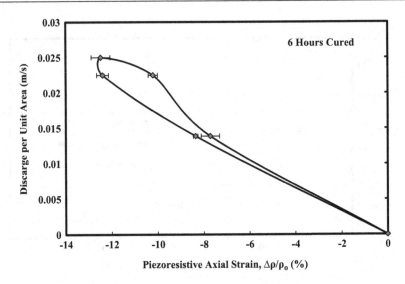

Figure 8.13 Variation in electrical resistivity of the smart cement with nitrogen gas specific discharge rate.

of the sensitivity of the smart cement to detect and quantify the gas leaks by monitoring the changes in the electrical resistivity.

8.6 SMART CEMENT WITH POLYMER TREATMENT

The HPHT device in Figure 8.4 was modified to measure the resistance in real time during the fluid loss for 30 min (API 13A and API 13 B) and in the gas leak study. The HPHT device has an area of 22.58 cm² and can withstand a pressure of around 14 MPa (2,000 psi), and for this study up to 5 MPa (700 psi) was used with the cement height of 100 mm (4 inches). Smart cement with and without SBR polymer modifications was placed in devices used for gas migration tests. Gas migration tests were performed using nitrogen gas. During the entire test, samples cured for 30 minutes and 24 hours were tested, The vertical resistances of the samples were measured using an LCR device. The change in the resistance was used to determine the resistivity (Eqn. (4.10)), which is a material property. Also, a gas flow meter was used to determine gas leak with the applied pressures.

Results and Discussion

8.6.1 Curing of Cement Slurry

Initial resistivity was measured immediately after mixing the smart cement with and without the SBR polymer. Initial resistivity of the smart cement

Table 8.3 Initial resistivity and curing model parameters for the smart cement with and without SBR

SBR Content (%)	Initial Resistivity, ρ_o (Ω-m)	ρ_{min} (Ω-m)	t_{min} (min)	ρ_{24h} (Ω-m)	RI_{24} (%)	p_I	q_I	t_o (min)
0	0.99 ± 0.03	0.97 ± 0.01	58 ± 5.0	3.48 ± 0.02	259	0.65	0.298	142.8
1	1.03 ± 0.02	1.00 ± 0.03	65 ± 4.5	4.07 ± 0.02	307	0.65	0.325	173.4
3	1.11 ± 0.04	1.05 ± 0.04	80 ± 3.5	4.63 ± 0.02	341	0.65	0.337	183.3

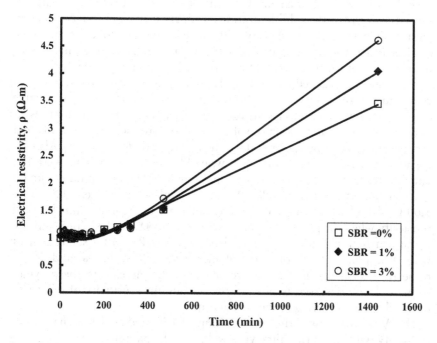

Figure 8.14 Variation of resistivity with curing time for the smart cement with and without styrene butadiene rubber (SBR) polymer.

was 0.99 Ωm (Table 8.3). With the addition of 1% SBR it increased to 1.03 Ωm, a 4% increase, four times the amount of SBR added. With 3% SBR, the initial resistivity increased to 1.11 Ωm, a 12% increase, four times the amount of SBR added. Hence, initial resistivity could be adopted as a quality control measure in the field to determine the amount of SBR added and also the quality of mixing. During the curing process, the resistivity rapidly changed with the time, as shown in Figure 8.14. Hence, there are several parameters that can be used in monitoring the curing (hardening process) of cement. The parameters are initial resistivity (ρ_o), minimum resistivity

(ρ_{min}), time to reach minimum resistivity (t_{min}), resistivity after 24 hours of curing (ρ_{24}), and percentage of maximum change in resistivity, referred to as the resistivity index. $\left[RI_{24hr} = \left(\dfrac{\rho24 - \rho min}{\rho min} \right) * 100 \right]$. The test results from various smart cement compositions are summarised in Table 8.3. The change in electrical resistivity (ρ) during curing for modified smart cement with SBR polymer is shown in Figure 8.14. After initial mixing, the electrical resistivity reduced to a minimum value (ρ_{min}), and then it gradually increased with time. Time to reach minimum resistivity, t_{min}, can be used as an index of the speed of chemical reactions and cement set times. With the formation of resistive solid hydration products that block the conduction path, resistivity increased sharply with curing time. The following increase in electrical resistivity was caused by the formation of large amounts of hydration products in the cement matrix. Finally, a relatively stable increase in the trend was reached by the ions diffusion control of the hydration process, and resistivity increased steadily for up to 24 hours, reaching a value of ρ_{24hr}. Change in the electrical resistivity with respect to minimum resistivity quantifies the formation of solid hydration products, which leads to a decrease in porosity and, hence, the cement's strength development. Also the t_{min} was increased by 12% and 38% when SBR polymer contents were 1% and 3% respectively, as summarised in Table 8.3. The minimum resistivity (ρ_{min}) of smart cement with 0%, 1%, and 3% of polymers were 0.97 Ω-m, 1.00 Ω-m, and 1.05 Ω-m, a 3% and 8% increase in the minimum electrical resistivity when the polymer addition increased to 1% and 3% respectively, as summarised in Table 8.3. The resistivity index (RI_{24hr}) for smart cement with 0%, 1%, and 3% of SBR polymer were 259%, 307%, and 341% respectively. Change in RI_{24hr} increased with the increase in the polymer content. These observed trends clearly indicate the sensitivity of resistivity to the changes occurring in the curing cement.

The Vipulanandan curing model (Eqn. 4.15) predicted the changes in resistivity with the curing time very well, and the parameters p_1 and q_1 are summarised in Table 8.3. In the range of variables investigated, up to 24 hours of curing parameter q_1 and t_0 were influenced by the polymer content, and parameter p1 was not. The parameter t_0 is influenced by the initial resistivity ρ_0. The parameter q1 increased with the addition of polymer and also the ratio q_1/p_1 increased with the addition of polymer.

8.6.2 Compressive Strength

The compressive strength of the smart cement increased with the SBR polymer addition. The smart cement average compressive strength after one day of curing was 9.66 MPa (1,400 psi). With the addition of 1% SBR polymer, the strength increased to 11.30 MPa (1,650 psi), an 18% increase. With the addition of 3% SBR polymer, the one-day compressive strength increased to 12.76 MPa (1,850 psi), a 32% increase in strength.

8.6.3 Piezoresistivity Behavior

The piezoresisitive responses (stress-resistivity change relationship) for the smart cement with and without polymer and cured for one day are shown in Figure 8.15. The piezoresistivity of the smart cement at failure after one day of curing $\left(\dfrac{\Delta\rho}{\rho_o}\right)_f$ was 171%, as summarised in Table 8.4. The addition of 1% and 3% of SBR polymer to the smart cement decreased the electrical resistivity at failure $\left(\dfrac{\Delta\rho}{\rho_o}\right)_f$ to 125% and 104% respectively, as summarised in Table 8.4. But the smart cements with SBR polymer are piezoresistive, and the responses were over 500 times compared to the compressive failure strain of cement of 0.2%.

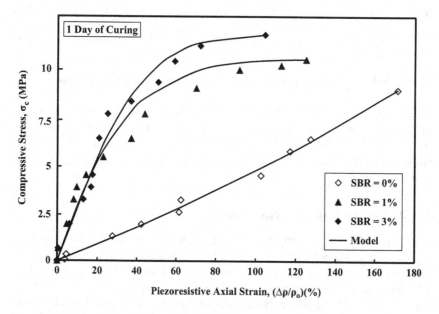

Figure 8.15 Piezoresistive behavior of the smart cement with and without styrene butadiene rubber (SBR) polymer.

Table 8.4 Piezoresistive model parameters for the smart cement with and without SBR polymer

Mix Type	p_2	q_2	Piezoresistive Axial Strain at Peak Stress, (%)	Compressive Strength, (σ_c MPa)	R^2
SBR = 0%	0	1.25	171	9.66	0.99
SBR = 1%	0.12	0.301	125	11.38	0.96
SBR = 3%	0.43	0.467	104	12.76	0.99

Vipulanandan p-q Piezoresistivity Model

The Vipulanandan p-q piezoresistivity model was used to predict the observed trends for the smart cement with and without SBR (Vipulanandan et al. 2014). The Vipulanandan p-q piezoresistive model used was as follows:

$$\sigma = \frac{\dfrac{x}{x_f} * \sigma_f}{q_2 + \left(1 - p_2 - q_2\right)\dfrac{x}{x_f} + p_2 \left(\dfrac{x}{xf}\right)^{\left(\frac{p_2}{p_2 - q_2}\right)}} \qquad (8.14)$$

where σ is the stress (MPa); σ_f: compressive stress at failure (MPa);

$x = \left(\dfrac{\Delta\rho}{\rho_o}\right) * 100$ is percentage of piezoresistive axial strain due to the stress;

$x_f = \left(\dfrac{\Delta\rho}{\rho_o}\right)_f * 100$ is the percentage of piezoresistive axial strain at failure;

$\Delta\rho$: change in electrical resistivity; ρ_o : initial electrical resistivity ($\sigma = 0$ MPa); and p_2 and q_2 are piezoresistive model parameters. Both parameters were sensitive to the polymer content in the smart cement (Table 8.4). As summarised in Table 8.4, model parameter p_2 increased with polymer content while q_2 decreased.

8.6.4 Fluid Loss and Gas Leak

One reason for adding polymer to cement slurry is to reduce the fluid loss. Styrene-butadiene polymer reduced fluid loss when added to the smart cement. Fluid loss observed for the smart cement without and with 3% polymer was 141 mL and 67mL respectively, as shown in Figure 8.16 and Figure 8.17. Hence, adding 3% of SBR polymer reduced fluid loss by 52%. A gas leak was observed after an initial fluid loss 46 mL and 25mL respectively in the smart cement without and with 3% polymer. Therefore the gas leak occurred 33%

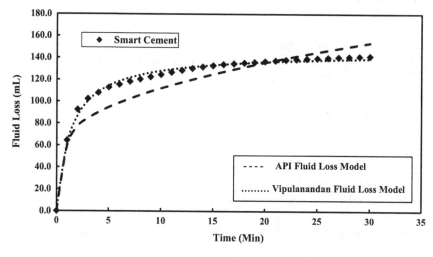

Figure 8.16 Fluid loss from the smart cement after 30 minutes of curing.

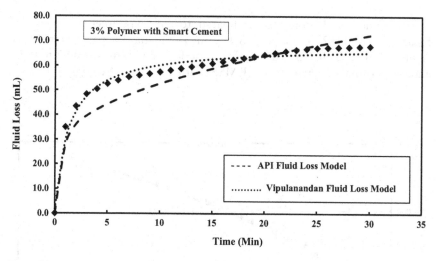

Figure 8.17 Fluid loss from the smart cement with 3% SBR polymer after 30 minutes of curing.

and 37% of the total fluid loss without polymer and with 3% polymer. This is also a clear indication of the importance of fluid loss model predictions.

Reduction in the amount of fluid loss is a clear indicator that styrene butadiene polymer fills the pores of cement slurry and also retains water. When the fluid loss test was done after 24 hours, the observed fluid loss in the smart cement without and with 3% polymer was 26 mL and 15 mL respectively.

(a) Static Model (API Model)
In this model the filter cake properties are assumed to be constant with time and the relationship is as follows:

$$FL_f - FL_o = M_1 * \sqrt{t} \qquad (8.15)$$

where:

FL$_f$: volume of fluid loss (cm^3);
FL$_o$: initial volume of fluid loss (spurt) (cm^3);
t: time; and
M$_1$: model parameter (mL/(min)$^{0.5}$).
From Eqn. (8.15),

$$\text{the maximum fluid loss} (FL_u) \text{ when } t \to \infty \Rightarrow FL_u = \infty \qquad (8.16)$$

API model predictions are compared to the experimental results in Figures 8.16, 8.17, 8.18, and 8.19. Also, the model parameter M is summarised in Table 8.5. The coefficient of determination was 0.97 to 0.99 and the root-mean-square error (RMSE) varied from 0.44 to 0.67 mL.

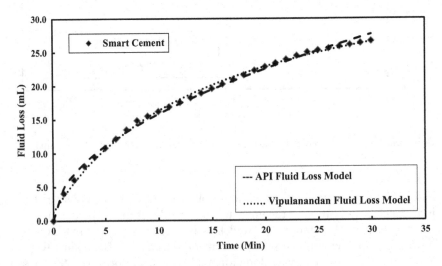

Figure 8.18 Fluid loss from the smart cement after 24 hours of curing.

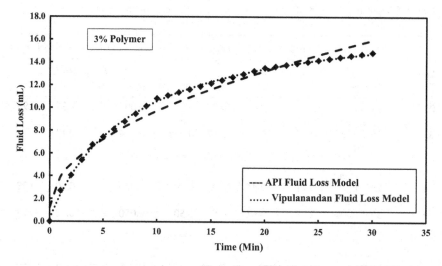

Figure 8.19 Fluid loss from the smart cement with 3% SBR polymer after 24 hours of curing.

(b) Vipulanandan Fluid Loss (FL) Model

Taking into account the changes in the permeability and cake thickness with time, the Vipulanandan fluid loss model was developed, and the relationship is as follows (Vipulanandan et al. 2014e, 2020b):

$$\frac{dFL}{dt} > 0 \quad \text{and} \quad \frac{d^2FL}{d^2t} < 0$$

$$FL(t) - FL_0 = \frac{t}{D + E * t} \tag{8.17}$$

where:

D (min/mL) and E (mL^{-1}): model parameters.
From Eqn. (7), the maximum fluid loss (FL$_u$)

$$t \to \infty \Rightarrow \qquad FL_{max} = FL_0 + \frac{1}{E} \tag{8.18}$$

(c) Test Results and Analyses

The Vipulanandan fluid loss model predictions are compared to the experimental results in Figures 8.16, 8.17, 8.18, and 8.19. Also, the model

Table 8.5 Vipulanandan and API fluid loss model parameters for the smart cement without and with 3% polymer-modified smart cement

Curing Time(hrs)	Cement Type	API Fluid Loss Model			Vipulanandan Fluid Loss Model				
		M_i (mL)	RMSE (mL)	R^2	D min/ mL	E $(mL)^{-1}$	FL_{max} (mL)	RMSE (mL)	R^2
0.5	Smart Cement	5.74	0.65	0.99	0.020	0.007	150	0.35	0.99
0.5	3% Polymer with Smart Cement	2.68	0.67	0.97	0.033	0.014	70	0.12	0.99
24	Smart Cement	5.11	0.44	0.99	0.30	0.025	40	0.40	0.99
24	3% Polymer with Smart Cement	2.68	0.67	0.97	0.50	0.050	20	0.12	0.99

parameters D and E are summarised in Table 8.5. The coefficient of variation was 0.99 and the root-mean-square error (RMSE) varied from 0.12 to 0.40 mL. In all cases, the root-mean-square error (RMSE) for the Vipulanandan model was lower than the API model; hence, it shows a better prediction of the test results.

(d) Gas Leak

The smart cement with and without 3% polymer was tested for gas leak after one hour of curing (slurry) and after 24 hours of curing (solid). The tests were performed to investigate the sensing characteristics of smart cement during gas leak. The maximum pressure gradient of 100 MPa/m (2,100 psi/ft) was used in this study. The pressure gradient was increased, and the discharges were measured using HPHT test chambers (Figure 8.3). The resistance changes were monitored continuously during the test. In Figure 8.20, the discharge velocity (volume discharge per unit area) for smart cement without and with 3% polymer addition is compared for specimens cured for one hour. Polymer addition reduced the discharge velocity of the gas at all the pressure gradients tested. The maximum reduction in discharge velocity was about 18%. The velocity of discharge and pressure gradient relationship was nonlinear, as shown in Figure 8.20. In Figure 8.21, the gas leak velocities with pressure gradients are compared for the smart cement with and without 3% polymer cured for 24 hours, and the relationship was nonlinear. With the 3% polymer addition, the maximum gas velocity reduction was about 33%.

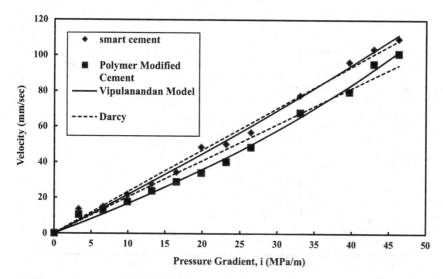

Figure 8.20 Variation of the gas discharge velocity with pressure gradient after one hour of curing.

To quantify the gas leak (average discharge per unit area = V) occurring in the cement at different pressure gradients i and the curing time, the new Vipulanandan Fluid flow model (Eqn. 8.3) is represented as follows:

$$V = \frac{i}{M + Ni}$$

(8.19)

The model parameters M and N are summarised in Table 8.6 with the coefficient of determination and root-mean-square error (RMSE). The Vipulanandan fluid flow model predictions are compared to the experimental results in Figure 8.20 and Figure 8.21, and the predictions were very good based on the root-mean-square error (RMSE).

8.7 SENSING GAS LEAKS

In the smart cement slurry without and with 3% polymer and cured for one hour due to gas leak, the resistivity increased, as shown in Figure 8.22. The resistivity increased by 45% and 32% for the smart cement slurry (SC) and smart cement with 3% polymer (PMSC1). In the case of hardened cement, the trends were opposite to that of slurry cement. In the smart cement cured

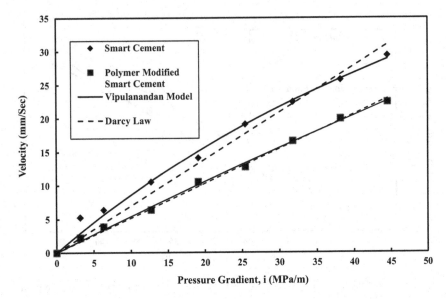

Figure 8.21 Variation of the gas discharge velocity with pressure gradient after 24 hours of curing.

Table 8.6 Vipulanandan gas flow model parameters for the smart cement with and without SBR polymer

	Darcy's Law			Vipulanandan Fluid Flow Model			
	k (m/sec)	R²	RMSE (mm/s)	M MPa.sec/m²	N sec/m	R²	RMSE (mm/s)
1-hour Curing							
	2.34	0.99	3.23	0.47	-0.001	0.99	2.72
	2.04	0.99	4.67	0.63	-0.003	0.99	2.34
24 Hour Curing							
	0.69	0.99	1.56	1.05	0.010	0.99	0.90
	0.51	0.99	0.43	1.81	0.003	0.99	0.38

for 24 hours, with the addition of pressure the resistivity increased by +80% for compressive stress of 5 MPa, as shown in Figure 8.15, but with gas leaks the resistivity decreased by -35%, as shown in Figure 8.23. This is a clear indication of the sensing characteristic of smart cement. In the case of the smart cement with 3% polymer, the maximum resistivity change during the gas leak was -12%. By monitoring the resistivity change, the gas migration in cement can be detected and quantified.

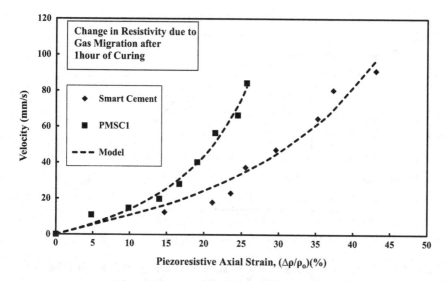

Figure 8.22 Changes in piezoresistive axial strain with the gas leak through the smart cement with and without 3% SBR after one hour of curing.

Table 8.7 Correlation model parameters for discharge velocity and the change in resistivity

Curing Time	A_3 (sec%/mm)	B_3 (sec/ mm)	R^2	RMSE (mm/s)
1 Hour Curing				
Smart Cement (SC)	1.13	-0.01	0.96	5.97
3% PMSC1	0.96	-0.02	0.98	3.39
24 Hour Curing				
Smart Cement (SC)	0.83	0.016	0.99	1.17
3% PMSC1	0.59	0.006	0.99	0.62

The Vipulanandan correlation model was used to relate the discharge velocity (V) to the change in resistivity as follows:

$$V = \frac{x}{A_3 + B_3 * x} \qquad (8.20)$$

where $x = Piezoresistive\ Axial\ Strain\left(\frac{\Delta\rho}{\rho_0}\right)\%$. The model parameters A_3 (sec%/mm) and B_3 (sec/mm) are summarised in Table 8.7.

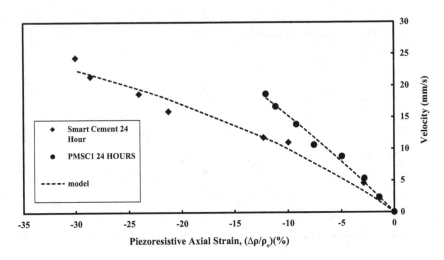

Figure 8.23 Changes in piezoresistive axial strain with the gas leak through the smart cement with and without 3% SBR after 24 hours of curing.

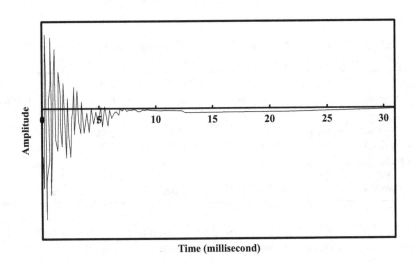

Figure 8.24 Accelerometer (amplitude) with time response for the smart cement along the longitudinal mode.

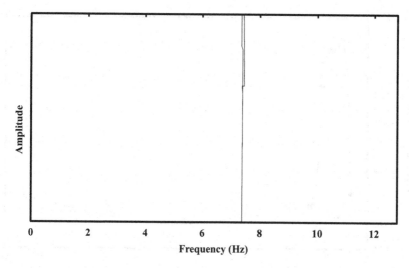

Figure 8.25 Fast Fourier Transformation (FFT) of the Frequency Spectrum for Smart Cement in Longitudinal Mode.

8.8 DYNAMIC LOADINGS

(a) Impact

After 28 days of curing, the smart cement was tested for impact loading responses. An accelerometer was placed on the test specimen and connected to the dynamic analyzer to characterise the impact loading, as shown in Figure 8.24. Based on fast Fourier transform (FFT) analyses, the resonance frequency of the applied impact on the smart cement was 7.3 Hz, as shown in Figure 8.25. At the time of the impact, the electrical resistivity showed response and increased from 15.87 Ω.m to 16.21 Ω.m, a 2% increase, as shown in Figure 8.26.

In order to evaluate the response of the smart cement to dynamic (repeated) impacts, with the frequency of 1 Hz, tests were done and the change in resistance was monitored. The impact caused increment of up to 4% in the electrical resistivity of the smart cement, as shown in Figure 8.27.

(b) Cyclic Loading

The smart cement specimens were first preloaded and then tested for the cyclic loading response. The preloading stress was 3 MPa, and the cyclic stress applied was 1 MPa at different rates of 0.20, 0.40, and 0.80 mm/min up to 60 seconds and then unloaded, as shown in Figure 8.28. The increment in loading increased the electrical resistivity, and the response was sensitive to the rate of loading, as shown in Figure 8.29. With the faster

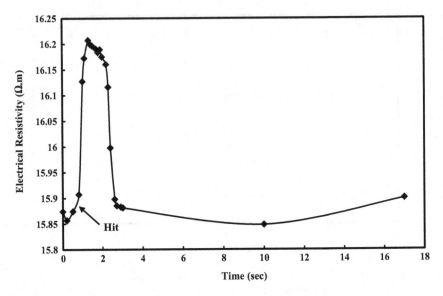

Figure 8.26 Axial electrical resistivity response of the smart cement with the impact load.

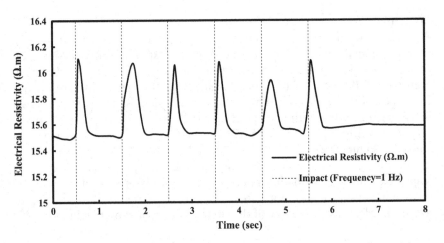

Figure 8.27 Change in the electrical resistivity based on dynamic impacts of 1 Hz frequency.

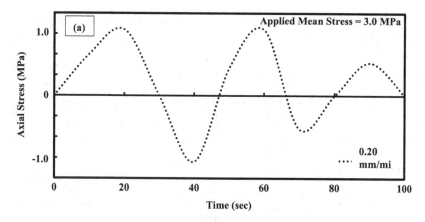

Figure 8.28a Cyclic stress loading patterns for various constant displacement rates (a) 0.20 mm/min., (b) 0.40 mm/min., (c) 0.80 mm/min.

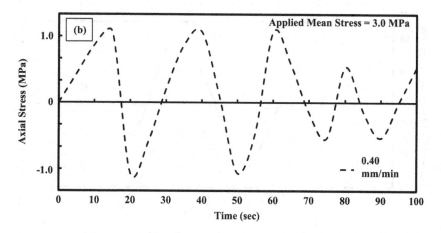

Figure 8.28b

rate of loading, the resistivity change was less compared to the lower rate of loading, and the trend agreed with the time to reach the peak load. With 1 MPa cyclic loading, the change in the piezoresistive axial strain (electrical resistivity) was 19.45%, 14.56%, and 9.9% respectively for 0.20, 0.40, and 0.80 mm/minute displacement rates, as shown in Figure 8.29. The test results showed that the smart cement was sensitive to the rate of cyclic loading.

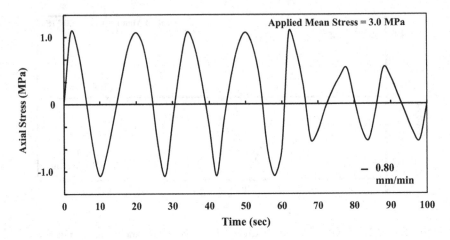

Figure 8.28c

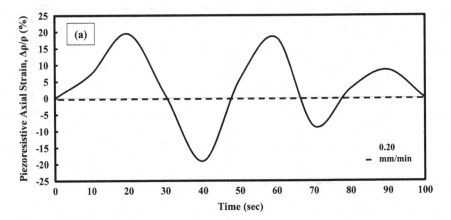

Figure 8.29a Change in the piezoresistive axial strain responses due to various cyclic
loading rates (a) 0.2 mm/min., (b) 0.40 mm/min., (c) 0.80 mm/min.

8.9 SUMMARY

This study was related to the fluid flow model development and verification with the gas and liquid flow through porous media represented
by the sand and smart cement and detecting impact and cyclic dynamic
loadings. Also, the gas leak detection using the smart cement without and
with polymer modification was investigated and following conclusions are
advanced:

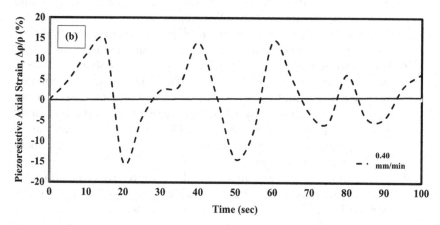

Figure 8.29b

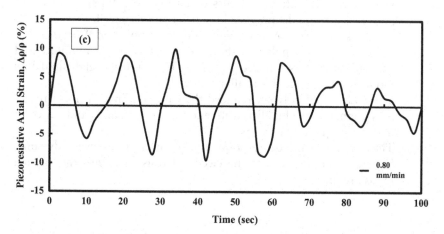

Figure 8.29c

1 The generalised Vipulanandan Fluid Flow model was developed based on theory and verified with gas and liquid flow through porous sand and cement. The three parameters in the model are related to the properties of the porous medium, flowing gas and liquid, and their interaction. Compared to the other models in the literature, the Vipulanandan Fluid Flow model predicted the experimental results with the water and gas very well in the sand and smart cement.

2 Smart cement detected the gas leak. With the application of pressure to the smart cement without any fluid loss, the resistivity increased, but with the gas leak it decreased. For cement cured for one-hour,

application pressure up to 2MPa (300 psi) increased gas flux and reduced the resistivity 15%. With aged smart cement, the trends were similar but the numbers were different. Changes in the resistivity were correlated to the gas leak velocity (discharge per unit area).

3 The addition of 3%styrene betadine rubber (SBR) polymer increased the resistivity of the smart cement. The addition of 1% polymer increased the initial resistivity by 4%, and adding 3% polymer increased the resistivity by 12%. Hence, resistivity could be a good quality control parameter in the field. The Vipulanandan curing model predicted the curing trends very well for the smart cement with polymer modifications.

4 The addition of polymer increased the compressive strength of the smart cement after 24 hours of curing. The compressive strength increased by 32% with 3% addition of polymer.

5 With the addition of polymer, the smart cement was piezoresisitive. The resistivity change at peak stress with 3% polymer addition was over 100% after 24 hours of curing. The Vipulanandan p-q piezoresisitive model predicted the behavior very well.

6 Fluid loss was reduced with the addition of polymer. With the addition of 3% SBR polymer, the fluid loss for the 30-minute cured cement slurry was reduced by 52%. The Vipulanandan fluid loss model predicted the test results very well.

7 The addition of polymer reduced the gas leak. Smart cement resistivity changes were highly sensitive to the gas leak and can be used to detect and quantify gas leaks.

8 The smart cement also responded to the impact and cyclic dynamic loadings. The changes in the electrical resistivity with the applied loads are quantified.

Chapter 9

Laboratory and Field Model Tests

9.1 BACKGROUND

In general, laboratory and field model tests are performed to demonstrate the development of new instruments and also verify the load-bearing capacity of piles to support bridges and structures (Wong et al. 1992; Vipulanandan et al. 2007b, 2009b, 2018e, f). With some of the reported major failures and growing interest in environmental safety, sustainability, and economic concerns in the oil and gas industry, the integrity of the cement sheath supporting and isolating wells is of major importance for both onshore and offshore wells. With multiple variables, it is also important to predict the behavior of smart cement in the field.

Also, with the advancement of various technologies, there is a need to integrate them for more efficient field applications for real-time monitoring, the minimising of failures, and for safety issues. The use of AI in various applications with multiple variables is becoming popular. The cementing of oil wells has been used for over 200 years. Cementing failures during installation and other stages of operations have been clearly identified as some of the safety issues that have resulted in various types of delays in cementing operations and oil production, and also have been the cause of some of the major disasters around the world. For successful oil-well cementing operations, it is essential to monitor them in real time due to the varying environmental and geological conditions regarding depth and also the performance of the cement sheath after hardening during the entire service life.

AI, otherwise known as machine learning or computational intelligence, is the science and engineering aimed at creating intelligent tools, devices, and machines. Its application in solving complex problems and case-based complications in various field applications has become more and more popular and acceptable over time (Opeyemi et al. 2016). AI techniques are developed and deployed worldwide in a myriad of applications as a result of AI's symbolic reasoning, explanation capabilities, potential, and flexibility (Demrican et al. 2011). Most AI techniques or tools have shown tremendous potential for generating accurate analysis and results from large historical databases, the kind of data an individual may find extremely difficult for conventional modeling and analysis processes (Shahab 2000). AI is currently employed in various sections of the oil and gas industry, from the selection of drill bits to well bore risk analysis. In recent years, there has been a drastic increase in the application of AI in the petroleum industry due to the presence of digital data and case studies. AI can provide real-time prediction in the oil and gas industry from selecting, monitoring, diagnosing, predicting, and optimising, thus leading to better production efficiency and profitability. (Opeyemi et al. 2016).

Model tests are important to demonstrate the applications of smart cement in a more realistic scaled-up environment. This involves also mixing of large volumes of smart cement based on the selected applications. In the laboratory model and field model tests, smart cement slurry and hardened cement were subjected to varying stresses and curing environments so that the sensitivity of the smart cement could be verified. Scaled-up tests will also verify the applicability of a real-time monitoring system under different environmental conditions.

9.2 LABORATORY MODEL TESTS

Laboratory models were designed and built at the Center for Innovative Grouting Materials and Technology (CIGMAT) Laboratory at the University of Houston (Texas). Several laboratory models were built to monitor the slurry levels (drilling fluid, spacer fluid, and cement) during the installation and the hardening of the cement. The drilling fluid and spacer fluid were characterised using the impedance-frequency response discussed in Chapter 2. The drilling fluids used in this study were characterised as CASE 2 (Chapter 2), so the electrical resistance was selected to be the real-time monitoring parameter. The measured resistance with time will clearly indicate the level of the pumping slurry, which can be used to determine the depth at which the drilling fluid and cement were located. Several models were tested separately by varying the drilling fluids (water based, oil based) to demonstrate real-time monitoring.

9.2.1 Small Model

The models were built using plexiglass and metal pipes, as shown in Figure 9.1, to simulate the formation and casing. Form the laboratory model test, a 1 m-long casing was instrumented with electrical wires (very flexible) to monitor the resistance change. The distance between the two wires (wire sensor) varied from 100 mm to 150 mm (vertically), and there were seven levels of wire sensors, as shown in Figure 9.1. Different combinations of

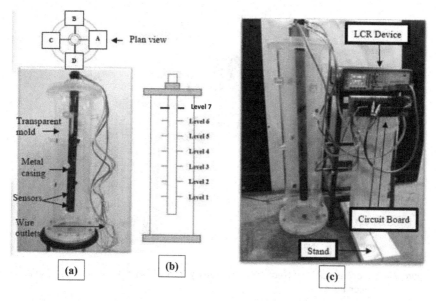

Figure 9.1 Laboratory model (a) actual model, (b) schematic of the model configuration, (c) LCR monitoring setup.

sensors were connected to an LCR meter to measure the resistance between those wire sensors both vertically and horizontally. The horizontal electrical wire sensors were marked A, B, C, and D at every level, as shown in Figure 9.1. The vertical wire leads were marked as numbers 1 through 7 to differentiate the various levels to simulate well depth.

Results and Discussion

Approximately 10 L of drilling mud was placed in the annulus of the well model using the pressure chamber, as shown in Figure 9.2. The electrical resistances of different combinations of wire sensors were measured at each level to determine the effect of the drilling mud as more mud was added to the annulus. After measuring the resistance for all seven levels, the drilling mud was gradually replaced by 10 L of spacer fluid by applying pressure of 35 kPa (5 psi) in the pressure chamber. The same procedure was used for the spacer fluid as the drilling mud, in order to measure the resistance when the spacer fluid reached each level. After measuring the resistance of the spacer fluid for all seven levels, it was replaced by the cement slurry. This too was done by applying pressure of about 35 kPa (5 psi) in the pressure chamber, and the same procedure was followed as before to continuously monitor the electrical resistances between the selected wire leads.

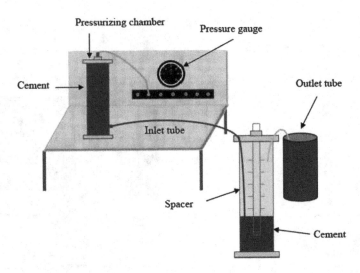

Figure 9.2 Experimental setup of the laboratory model where cement is displacing spacer fluid.

(a) Real-Time Monitoring

Monitoring the drilling fluid: Resistance was measured at different vertical levels in the well model with time, as shown in Figure 9.3. When there was no fluid in the well, the resistance was in the range of 155 to 205 kΩ, which represented the air resistance for the particular distance monitored at a relative humidity of 50% in the laboratory. When the oil-based mud level reached level 1 (visible through the plexiglass chamber), all of the vertical resistances started to decline to the 35 to 60 kΩ range. This sudden change clearly showed that oil-based mud had reached level 1. Similar reductions in electrical resistances were observed with horizontal measurements.

When the oil-based mud reached level 2 (as observed through the plexiglass model), the sensors A1-A2 (resistance in the vertical direction between level 1 and 2) and represented as A2 in Figure 9.3 showed a sudden change in the electrical resistance, and the resistance dropped to 14 kΩ from about 36 kΩ. The change occurred because the oil-based drilling mud, with a resistivity of 110 Ω.m, filled the space between levels 1 and 2. When the mud reached level 7 (as observed through the plexiglass model), the resistance between A1 and A7, which is represented as A7 in Figure 9.3, also dropped. The same pattern was also observed for the other sets of readings. This consistent behavior showed that the level of the drilling fluid can be monitored effectively by measuring the resistance.

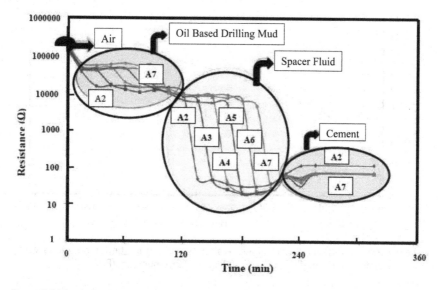

Figure 9.3 Vertical resistance measurements in the laboratory well mode.

Monitoring of spacer fluid: The electrical resistivity of the spacer fluid was 0.80 Ωm. The same procedure was used to monitor the spacer fluid. Figure 9.3 shows the changes in the resistance readings for the spacer fluid. When the spacer fluid filled the annular space, the resistance rapidly declined due to the lower resistivity of the spacer fluid. The resistances measured were in the range of 25 to 35 Ω.

Monitoring of smart cement slurry: The electrical resistivity of the smart cement slurry was 1.0 Ωm. The spacer fluid was displaced by the cement slurry. As shown in Figure 9.3, the vertical resistance readings changed for the cement slurry at various vertical levels. The monitoring resistances increased slightly because the electrical resistivity of the cement slurry was slightly higher than the replaced spacer fluid.

All of the vertical resistances increased to between 55 and 67 Ω when the cement slurry reached level 1. This change clearly showed that the cement slurry had reached level 1. The increase occurred because the resistivity of the cement was higher than that of the spacer fluid. The cement became contaminated with the spacer fluid in the process of displacing it, and during the initial hydration resistivity dropped; therefore, the resistance values dropped to a range of 30 to 50 Ω between level 1 and 7 (A7). With time, the hydration process in the cement caused the resistance to increase because cement resistivity increased with time as the cement hydrated and cured.

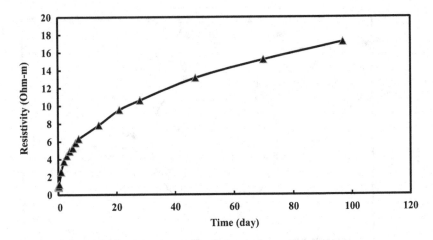

Figure 9.4 Variation of the smart cement resistivity with curing time for samples cured under room conditions (23°C and 50% relative humidity (RH)).

(b) Curing of Cement Sheath

Resistivity of the smart cement: The resistivity of the smart cement slurry with curing time of up to 100 days was measured using smart cement samples 50 mm (2 inch) in diameter and 100 mm (4 inch)-high cylindrical molds. The resistivity increased with curing time under room temperature and humidity, as shown in Figure 9.4.

Predicted (Electrical Resistance Model; ERM) and measured resistance for hardening cement sheath: Using the calibrated parameters K in Eqn. 4.10 (with G = 0) and the resistivity-time relationship as shown in Figure 9.4, the changes in the cement sheath resistance (R) in the laboratory model were predicted. The variations in the predicted resistance values are compared to the actual measured values for different wire sensor combinations, as shown in Figure 9.5, Figure 9.6, and Figure 9.7.

Vertical resistance: The vertical measured resistances are compared to the ERM model prediction to verify the model predictions.

Wire setup-A: For the wire setup-A, the wire combination A1-A2, 100 mm apart, showed that the predicted values were lower than the measured values for up to 14 days of curing, but after that time the measured resistance values were in the range of the predicted resistance (Figure 9.5). This may be because in the small model the cement was hydrating under pressure and temperature, whereas the resistivity used (Figure 9.4) to predict the resistance was cured under room conditions under atmospheric pressure.

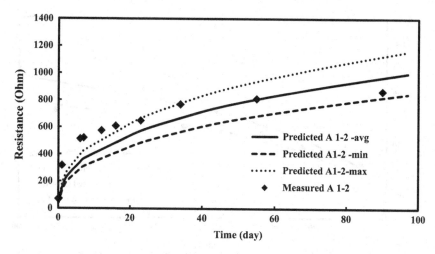

Figure 9.5 Comparing the predicted and measured resistance for wire setup-A for wire combination A1-A2. 100 mm apart.

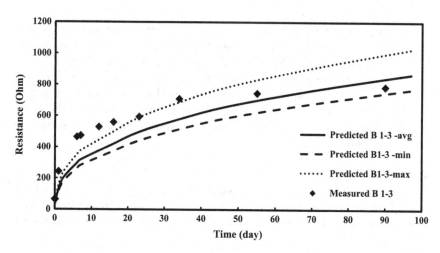

Figure 9.6 Comparing the predicted and measured resistance for wire setup-B for wire combination B1-B3, 200 mm apart.

Wire setup-B: For the wire setup-B, the wire combination B1 and B3, 200 mm apart, showed that the predicted values were lower for about 30 days of curing, as shown in Figure 9.6. The initial hydration of the small model was different from the samples curing in the mold under room condition. The measured values matched very well with the predicted values after one month since the curing temperatures were very similar.

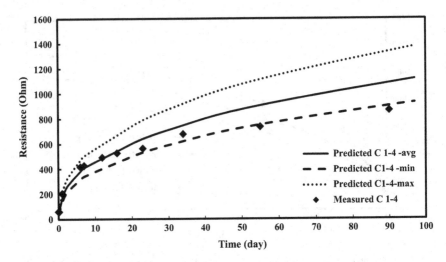

Figure 9.7 Comparing the predicted and measured resistance for wire setup-C for wire combination C1-C4, 300 mm apart.

> *Wire setup-C:* For wire setup-C. the wire combination C1 and C4, 300 mm apart, showed that all the data points are within the range of the predicted values, as shown in Figure 9.7, and somewhat different for setup-A (100 mm apart) and setup-B (200 mm apart). It also confirms that 300 mm apart curing is almost under room condition of curing similar to the smart cement samples cured under room condition, as shown in Figure 9.4.

(c) Pressure Test

To simulate a pressure test, air pressure (P_i) was applied inside the casing to load the cement sheath, and the electrical resistances (R in Ohms) were measured between Level 1–2, Level 2–3, Level 3–4, and Level 4–5, as shown in Figure 9.8.

> **Step 1: Initial Condition** (No pressure, Pi = 0): The variation of the initial resistance measured is shown in Figure 9.9. The initial resistance was higher at the bottom (level 1–2) due to the weight applied by the cement sheath and was lower at the top level (level 4–5). The electrical resistance increased with depth and varied from 400 to 800 ohms. This is partly due to the piezoresistive property of the smart cement, in which the electrical resistance will be higher because of the increase in pressure with depth due to the weight of the cement above.

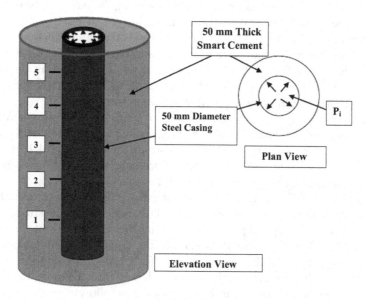

Figure 9.8 Configuration of the pressure applied inside the steel casing.

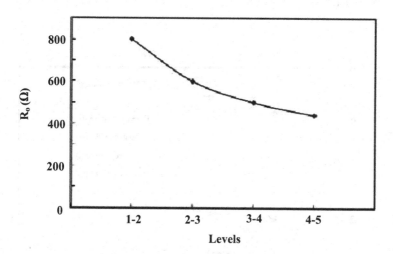

Figure 9.9 Variation of the initial resistance with depth after 100 days of curing.

<u>**Step 2: Pi = 420 kPa**</u>: Internal pressure of 420 kPa (60 psi) was applied and the resistance changes were measured. The change in resistance was normalised with initial resistance $\Delta R/R(\%)$, equivalent to the piezoresistive axial strain $\Delta\rho/\rho$ (%) (Eqn. (4.13)), as shown in Figure 9.10. The resistivity change in the smart cement

due to the applied pressure was about 0.5%–0.6%, indicating the piezoresistivity of the smart cement.

Step 3: Pi = 700 kPa: Pressure of 700 kPa (100 psi) was applied and the resistance changes were measured and reported in the form of ΔR/R (%) in Figure 9.10. The piezoresistive axial strain in the smart cement due to the applied pressure of 700 kPa was about 2.5%, indicating the piezoresistivity of the smart cement.

Step 4: Pi = 980 kPa: Internal pressure of 980 kPa (140 psi) was applied and the resistance changes were measured and reported in the form of ΔR/R in Figure 9.10. The resistivity change (Eqn. (4.13)) in the smart cement due to the applied pressure was about 6.5% to 7%, indicating the piezoresistivity of the smart cement.

Piezoresistive Modeling: The stress at every point can be separated into mean stress and deviatoric stress. The change in the deviatoric stress due to an applied pressure (Pi) along the vertical axis of the casing (z-axis) is represented as ΔS_{zz}. Using equilibrium and stress analyses, it can be shown that S_{zz} is directly proportional to the applied internal pressure, Pi (Eqn. 9.1). Hence, the change in deviatoric stress (shear stress) can be represented as follows:

$$\Delta S_{zz} = f(P_i) \tag{9.1}$$

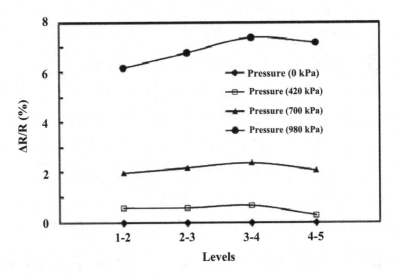

Figure 9.10 Changes in resistance with applied pressure in the smart cement supporting the casing after 100 days of curing.

The variation of internal applied pressure in the casing with the resistivity of smart cement ($\Delta\rho_z/\rho_z$) is shown in Figure 9.11, and the response of the smart cement was nonlinear. This is another clear indication of the 3D sensing capabilities of the smart cement.

(d) Vipulanadan p-q Piezoresistive Model

The nonlinear p-q piezoresistive model developed by Vipulanandan et al. (2014b, c) was used to predict $\Delta\rho_z/\rho_z$ variation with the applied pressure inside the casing. The relationship was represented as follows:

$$P_i = \frac{\left(\dfrac{\Delta\rho_z}{\rho_z}\right)}{q_2 + (1 - q_2 - p_2)\left(\dfrac{\Delta\rho_z}{\rho_z}\right) + p_2 \left(\dfrac{\Delta\rho_z}{\rho_z}\right)^{\frac{p_2 + q_2}{p}}} \tag{9.2}$$

The model parameters p_2 and q_2 were 1.2 and 3.0 respectively. Hence, it is possible to predict the pressure in the casing and also the stress in the cement sheath by measuring the change in the vertical resistivity in the smart cement and also the 3D sensing of the smart cement.

9.2.2 Large Model Test

Large models were designed and built at the Center for Innovative Grouting Materials and Technology (CIGMAT) laboratory, University of Houston

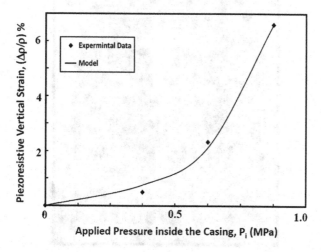

Figure 9.11 Model prediction of the vertical resistivity strain with applied pressure for the smart cement after 100 days of curing.

(Texas), as shown in Figure 9.12. The models were built to monitor the slurry levels during installation and the behavior of hardening smart cement. The main focus was to demonstrate the effectiveness of the instrumentation and monitoring of the 50 mm (2 inch)-thick cement sheath supporting the 100 mm (4 inch)-diameter steel casing. The observed resistance with time clearly indicated the level of slurry and determined the depth at which the smart cement was located. Two models were built separately and tested separately to demonstrate the real-time monitoring concept and do long-term monitoring of the hardened cement.

(a) Materials and Method

Two large models were built, based on the small model test experience and by placing the steel casing in a larger water-tight PVC mold (200 mm in diameter) and cementing the casing, as shown in the Figure 9.12. The height of the model was about 2.4 m (8 feet) more than double the height of the small model. As mentioned, the main focus was to demonstrate the effectiveness of the instrumentation and monitoring of the 50 mm (2 inch)-thick cement sheath supporting the 100 mm (4 inch)-diameter steel casing. The outside of the casing was instrumented with electrical wires to monitor the resistance change in the smart cement. The distances between two probes were varied, and there were 13 vertical levels (which will be referred to

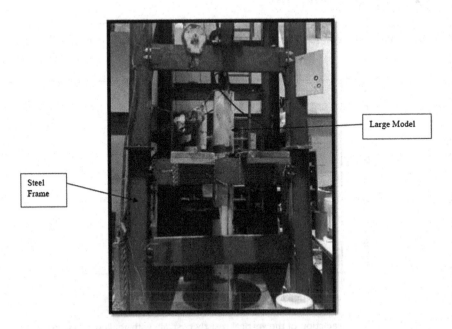

Figure 9.12 Large laboratory model testing facility.

numbers) of probes. Also, there were four horizontal probes (A, B, C, D), similar to the small model (Figure 9.1) at each level, resulting in a total of 52 probes and 2,652 combinations of two-probe measurements. Different combinations of the probes were connected to a 300 kHz LCR meter to measure resistance between those probes, to monitor the behavior of the cement. Also, thermocouples were installed with the instrumentation to monitor and to verify the changes in the temperature in the cement.

(b) Installation

Vertical Resistance

Wire Setup A

Changes in the vertical electrical resistances with the rising cement slurry are shown in Figure 9.13. It can be observed from Figure 9.13a that the resistance between wire combinations A1-A3 initially was about 12.5 kΩ, which rapidly decreased about 17 Ω when the cement slurry reached level 3. With the rising cement level, the resistance A1-A3 remained almost unchanged. Similarly, for wire combinations A1-A5, A1-A7, A1-A9, A1-A11, and A1-A13 the resistance varied from 10.8 kΩ to 13.1 kΩ before the two levels of wires became submerged in the cement slurry and the resistance dropped sharply to about 19 Ω, as shown in Figure 9.13. The sharp reduction in the resistance was an indicator of the cement slurry reaching the wire level.

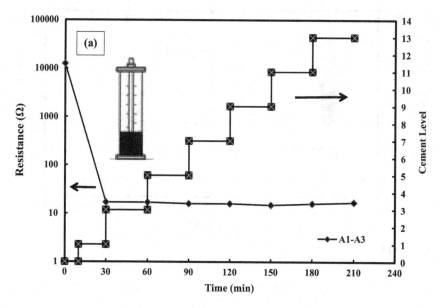

Figure 9.13a Changes in the vertical resistance along probe setup-A with the rising cement slurry in the large model (a) A1-A3 probe, (b) A1-A9 probe, (c) A1-A13 probe.

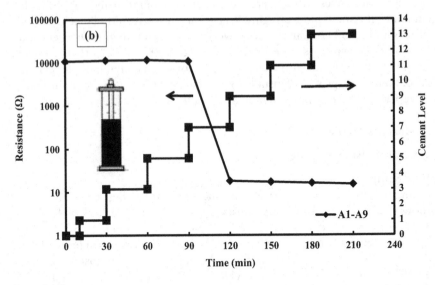

Figure 9.13b

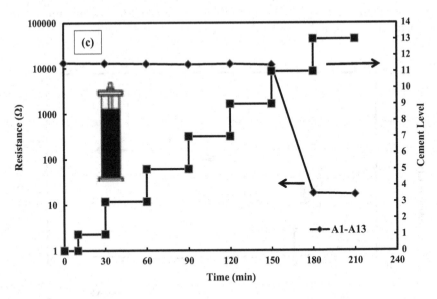

Figure 9.13c

(c) Cement Curing

Vertical Resistance

Wire Setup-A

For the vertical wire setup-A, the wire combination A1-A5 (600 mm, 24-inch spacing Figure 9.14) showed that the predicted values using the room condition curing resistivity-curing time relationship were comparable to the measured resistance over a period of over 250 days. This agreement in the predicted trend ensures the hardening of the smart cement.

The wire combination A1-A7, 0.9 m (36-inch spacing) and A1-A9, 1.2 m (48-inch) spacing showed similar trends where the predicted values matched very well with the measured values up to 250 days of curing, as shown in Figure 9.15 and Figure 9.16 respectively. This agreement in the predicted trend ensures the hardening of the smart cement.

The measured resistances for wire combination A1-A11 (60-inch spacing) and A1-A13, 1.8 m (72-inch) spacing showed similar trends where the predicted values matched very well with the measured values up to 250 days of curing, as shown in Figure 9.17 and Figure 9.18 respectively. As mentioned, the curing conditions in Figure 9.4 will influence the predictions and ensure the hardening of the smart cement.

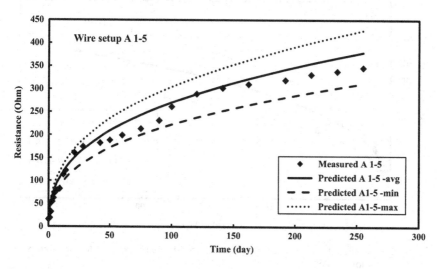

Figure 9.14 Predicted and measured resistance for the wire setup A for wire combination A1-A5.

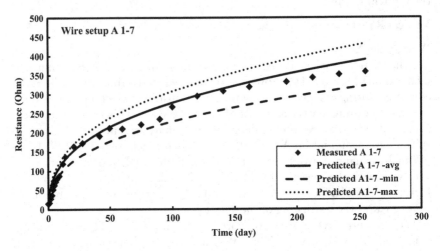

Figure 9.15 Predicted and measured resistance for the wire setup A for wire combination A1-A7.

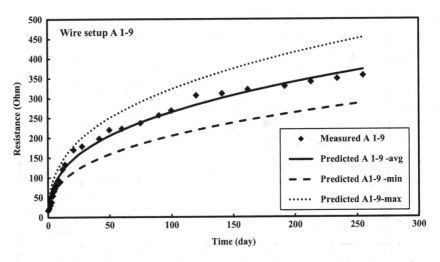

Figure 9.16 Predicted and measured resistance for the wire setup A for wire combination A1-A9.

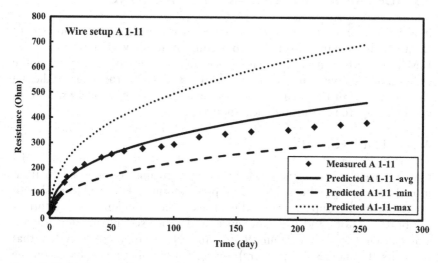

Figure 9.17 Predicted and measured resistance for the wire setup A for wire combination A1-A11.

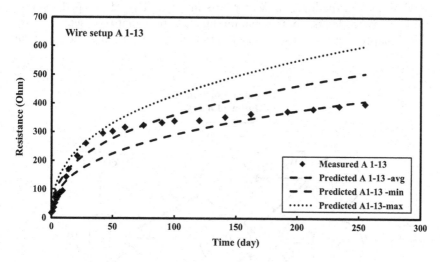

Figure 9.18 Predicted and measured resistance for the wire setup-A for wire combination A1-A13.

9.3 NUMERICAL MODEL (MACHINE LEARNING)

It is important to quantify the changes in the resistivity of the cement sheath with time and also take into account multiple variables in the field. Developing AI models representing the machine learning to predict the changes in resistivity and adopting it as the machine learning for predicting changes is one of the focuses of this study. Also, the AI model predictions will be compared to the analytical models.

Artificial Intelligent (AI)

An artificial neural network (ANN) is a computational numerical model that is based on, at some level, brain-like learning as opposed to traditional computing, which is based on programming. The model consists of interconnected groups of artificial neurons, which simulate the structure of the brain to store and use experience, and processes information using a connectionist approach (Figure 9.19). An ANN is an adaptive system that trains itself (changes its structure) during the learning phase based on the information flowing through the network.

Researchers who have studied neural networks aimed to model the fundamental cell of the living brain: the neuron. The recognised U.S. pioneers who first introduced the concept of an ANN were neurophysiologist Warren McCulloch and the logician Walter Pitts in 1943. They developed a simple model of variable resistors and summing amplifiers that represent the variable synaptic connections or weights that link neurons together, and the

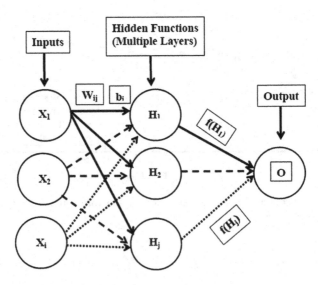

Figure 9.19 Artificial intelligence (AI) with integrated artificial neural networks (ANN).

operation of the neuron body, respectively. The popularity of an ANN increased in 1962 with the introduction of 'perceptron' by Frank Rosenblatt, who used the term to refer to a system that recognised images using the McCulloch and Pitts model (Alexander 1990).

AI can be defined as a collection of new analytical algorithms and tools that attempt to imitate and predict situations. These AI techniques exhibit an ability to learn and deal with new situations. ANNs, evolutionary programming, and fuzzy logic are among the paradigms that are classified as AI (Sadiq et al. 2000). The main root principles of AI include reasoning, knowledge, planning, learning, communication, perception, and the ability to manipulate objects (Bhattacharyya 2011).

Artificial Neural Networks (ANN)

Neural network research was first published by McCulloch and Pitts in 1943 (Hubbert et al. 1957). The ANN is a numerical model that mimics the functional aspects of the neural network in the human brain system. It consists of many artificial neurons interconnected where each of them gives a single output (O) induced from all inputs (X_i) (Hammoudi et al. 2019). The predictive capability of ANNs comes from the ability to learn and adapt to new situations in which additional data becomes available. In an ANN, a training set comprising input and output data is entered and the neural network algorithms attempt to map the process by which inputs become outputs (Sadiq et al. 2000; Saridemir 2009). An ANN is a multilayer perceptron (MLP) including three layers (Figure 9.19). The first layer (input layer) consists of neurons representing the independent variables (inputs X_i), the second one is the hidden layer (Hi, f(H$_i$)), and the last one is the ANN responses (output layer, representing AI). The number of neurons required in the hidden layer is determined in a way to minimise both prediction error and number of neurons.

The general forms of the equations are as follows:

$$Hj = \sum WijXi + bj \tag{9.3}$$

where Xi represents the inputs (Figure 9.19, neurons) and subscript index i represents the inputs (1 to n). The Wij is the weighing matrix for each input term Xi connecting it to the hidden term Hj, The bj is the bias input function.

Using sigmoid as the transfer function, f(Hj) is represented as

$$f\left(H_j\right) = 1 + 1/(1 + e^{-H_j}) \tag{9.4}$$

Two accurate neural network algorithms are back propagation neural networks (BPNNs) and generalised regression neural networks (GRNNs). The following are the summaries on these algorithms.

Back Propagation Neural Networks (BPNNs)

BPNNs are the most widely used type of ANNs (Sadiq et al. 2000). BPNNs consist of an input layer that is propagated through the network with a set of weights to have a predicted output. A BPNN is set with an objective to adjust the set of weights so that the difference between output prediction and required output is reduced.

Generalised Regression Neural Networks (GRNNs)

The generalised regression neural network (GRNN) is a feedforward neural network based on nonlinear regression theory consisting of three or more layers: the input layer (one layer), the pattern layer with the summation layer (one or more layers), and the output layer. While the neurons in the first three layers are fully connected, each output neuron is connected only to some processing units in the summation layer. The individual pattern units compute their activation using a radial basis function (bj), which is typically the Gaussian kernel function (Sadiq et al. 2000). The training of the GRNN is quite different from the training used for the BPNN. It is completed after presentation of each input-output vector pair from the training set to the GRNN input layer only once. What this means is that both the centers of the radial basis functions of the pattern units and the weights in connections of the pattern units and the processing units in the summation layer are assigned simultaneously.

Development of Neural Network and Design

The data used in this study was obtained from a set of laboratory and field scale studies on smart cement. The obtained data comprises electrical resistance, piezoresistive axial strain data from the laboratory for one year and field for a period of 4.5 years. Database preparation for the training of a neural network is very important in neural network modeling. The suggested neural network model does not consider weather and temperature factors. Neural network architecture with different hidden layers was used to predict laboratory and field measurements. Attempts were made to use hidden layers from one to four to obtain a good fit to the data. BPNN architecture with four hidden layers exhibited better correlation to laboratory and field measurements. The coefficient of determination (R^2) and root-mean-square error (RMSE) were used to evaluate the statistical significance of the ANN models. The data for this study was collected from the laboratory tests and field test, and a total of over 3,000 data were used.

9.4 FIELD MODEL TEST

After reviewing a few potential test sites based on availability and accessibility, the Energy Research Park (ERP) at the University of Houston (Texas) was selected to install the field well. Many factors including geology, swelling and soft clays, changing surrounding conditions (weather, active zone in the ground), environmental regulations, and accessibility to the site for long-term monitoring had to be considered in selecting the test site since the focus of the study was to demonstrate the sensitivity of the smart cemented field well. The selected site had swelling clays with fluctuating moisture conditions (active zone), which represented the nearly the worst conditions that could be encountered when installing oil wells. The top 6.1 m (20 feet) of the soil was swelling clay soil with liquid limit of over 50%. Based on ASTM D 2487 classification, this soil was characterised as CH soil. The active zone in the Houston area is about 5 m (15 feet), indicating relatively large moisture fluctuation in the soil causing it to swell and shrink (Sivruban et al. 2007; Tand et al. 2011; Vipulanandan et al. 2020c). The water table was 6.5 meters (20 feet) below ground, and soil below the water table was also clay with less potential for swelling, and the liquid limit was below 40%. Based on the Unified Soil Classification System (ASTM D 2487), this soil was characterised as CL soil.

9.4.1 Instrumentation

It has been shown that two probes with AC current can be used to determine electrical resistance changes in smart cement and drilling fluid (Vipulanandan 2015a–d). It was also important to use other standard tools for measuring changes in the cement sheath and comparing them to resistance changes. For practical reasons no instrument was placed on the casing and a totally independent system was developed to be placed in the cement sheath. The probes were placed at various vertical depths at 15 levels, starting from level 1 one at the bottom of the casing to level 15 at the top of the casing. Also, eight probes (coded as A, B, C, D, E, F, G, and H) were placed horizontally at each level, for a total of 120 probes amounting to 7,140 two-probe combinations. The probe coding is similar to the small models. Also, nine stain gauges and nine thermocouples were included in the instrumentation (Figure 9.20).

(a) Installation of the Field Well

A commercial company familiar with the drilling and cementing wells in an urban setting was selected to install the field well. A very large drilling truck with a 356 mm (14 in)-diameter drill was used to drill the hole and place the 244 mm ($9^5/_8$ in)-diameter standard steel casing. The total length of the casing was 12.8 m (42 feet), and needed pieces (including well head and needed connections to lift the casing) were welded together to make a single unit. An initial 4.6 m (15 feet) was drilled without any drilling fluid. Polymer-based drilling fluid was used to drill the rest of the borehole. After completing the drilling, the casing and the instrumentation units were centered and lowered into the borehole. The resistance between the selected probes, temperature, and stains (strain gauges) were measured.

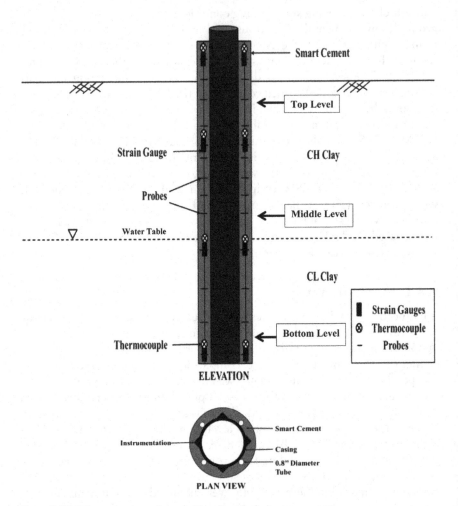

Figure 9.20 Schematic view of the field well with the instrumentation.

(b) Pressure Test

To simulate a pressure test to verify the piezoresistive response of the smart cement, air pressure (P_i) was applied inside the 20 mm (0.8 in)-diameter tube to the entire depth of 12.8 m (42 feet), as shown in Figure 9.20. Initially, the electrical resistances (R_o in Ohms) were measured along the entire depth before applying the pressure. This test was done regularly to demonstrate the sensitivity of the smart cement to the applied small pressures of up to 0. 55 MPa (80 psi).

(c) Materials and Methods

In this study, polymer drilling fluid and smart cement were used. The type of drilling fluid was selected based on the soil formation at the location.

Polymer Drilling Fluid

Polymer-based drilling fluids are used to drill through reactive geological formation. Since in this study the drilling was to be done through swelling soft montmorillonite clay, polymer drilling fluid was used. It is less reactive with clay formations and also controls the fluid loss into the formations. The density of the polymer drilling fluid was 1.04 g/cc (8.7 ppg), and the electrical resistivity was in the range of 2 Ω.m to 3 Ω.m.

Smart Cement

Cement slurry was prepared using a water-to-cement ratio of about 0.6, making the mixing and pumping easier in the field. The cement was modified with an addition of 0.075% carbon fibers by the weight of the cement. The initial resistivity of the cement slurry was in the range of 1.20 to 1.24 Ω.m. A total of 42 samples were collected for characterising the smart cement behavior. A commercially available conductivity probe and digital resistivity meters were used in determining the resistivity of the cement (Vipulanandan et al. 2014b, c, d).

Resistivity of Smart Cement

A LCR meter was used to measure impedance (resistance, capacitance, and inductance) in the frequency range of 20 Hz to 300 kHz. Based on the impedance (z)–frequency (f) response, it was determined that the smart cement was a resistive material (Vipulanandan et al. 2013). Hence, the resistance measured at 300 kHz using the two-probe method was correlated to the resistivity (measured using the digital resistivity device) to determine the K factor (Eqn. 4.10) for a time period of an initial five hours of curing. This K factor was used to determine the resistivity of the cement sheath with the curing time.

Piezoresistivity Test

In this study a compression test was used to characterise the piezoresisitivity of the smart cement. Piezoresistivity describes the change in the electrical resistivity of a material under stress. Since oil-well cement serves as a pressure-bearing part of oil and gas wells in real applications, the piezoresistivity of smart cement (stress–resistivity relationship) with different w/c ratios was investigated under compressive loading at different curing times. During the compression test, electrical resistance was measured in the direction of the applied stress. To eliminate the polarisation effect, AC resistance measurements were made using an LCR meter at a frequency of 300 kHz (Vipulanandan et al. 2013a).

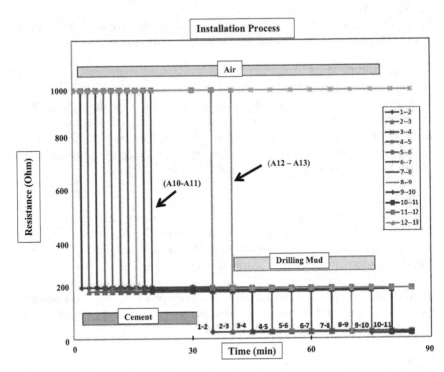

Figure 9.21 Vertical resistance changes for drilling fluid and cement slurry reaching various levels.

9.4.2 Monitoring

(a) Installation

During the initial 6.1 m (20 feet) of drilling, no drilling fluid was used. In order to stabilise the borehole polymer, drilling fluid was used to drill the rest of the hole. The total length of the borehole was about 12 m (39 feet). The steel casing with external instrumentation was lowered into the bore-hole, and the changes in the resistance started to be monitored. The vertical resistance between the adjacent probes was of the order of 1,000 Ω, as shown in Figure 9.21. The resistance between probes A1-A2 reduced to about 200 Ω when probe 2 reached the drilling fluid. Similarly, the resistance between other adjacent probes reduced when the probes were submerged into the drilling fluid. In 20 minutes, the probe A10-A11 reduced to 200 Ω, indicating the rate of the lowering of the casing about 12 m (39 feet). This sudden change in the resistance clearly showed the level of the casing that was lowered and submerged in the drilling fluid.

The cementing was started after 30 minutes. The resistance of probe A1-A2 reduced to about 20 Ω after 35 minutes, indicating that the cement had reached vertical level 2. The rising of the cement lowered the resistances, as shown in Figure 9.21. In about 80 minutes, the cement reached vertical level 11 and the resistance dropped 20 Ω (A10-A11). The electrical resistance changes observed during the placement of the drilling fluid and cement were very similar to the laboratory model test (Vipulanandan et al. 2015c). Cement was displacing the drilling fluid at the top of the borehole, and the vertical resistance (A12-A13) dropped from 1,000Ω to 200Ω, indicating that the drilling fluid had reached level 13 after 40 minutes of the operation.

(b) Cement Curing
First Day
Typical changes measured in the strain gauge, thermocouple, and resistance probe during the first day of curing of the cement in the borehole are shown in Figure 9.22. The thermocouple shows the increase in the temperature due to the hydration of the cement. The cement initial resistance was 24 Ω and reduced to about 20 Ω, and then increased to about 50 Ω in 24 hours, a 150% change, as shown in Figure 9.22. The change in the strain gauge resistance of 120 Ω was very small. Hence, change in the electrical resistance was the largest of the parameters that were being monitored during the hydration of the cement.

(c) Piezoresistivity Verification
Collected cement samples were cured under different conditions and were tested under compressive loading after 45 days to determine the piezoresistivity. The bottom- and middle-level samples were cured under water and the top-level sample was cured under room condition (23°C and relative humidity of 50%). For samples cured under water, the change in the resistivity at peak stress varied from 70% to 160% based on the different batches of mixing of the smart cement, as shown in Figure 9.23. The compressive failure strain for the smart cement was 0.2% at peak compressive stress (Vipulanandan et al. 2015b), and hence the resistivity change was 350 (35,000%) to 800 times (80,000%) more sensitive.

Based on experimental results, the Vipulanandan p-q model was used to predict the change in the electrical resistivity of the cement with applied compressive stress after 45 days of curing. The model is defined in Eqn. (5.16) as follows:

$$\frac{\sigma}{\sigma_f} = \left[\frac{\dfrac{x}{x_f}}{q_2 + (1 - p_2 - q_2)\dfrac{x}{x_f} + p_2\left(\dfrac{x}{xf}\right)^{\left(\frac{p_2+q_2}{p_2}\right)}} \right] \tag{9.5}$$

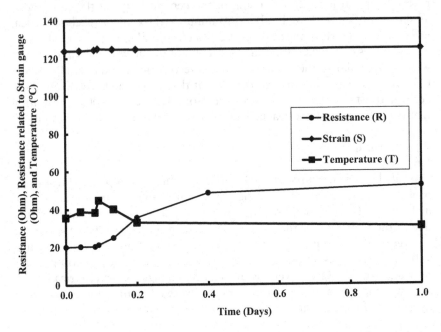

Figure 9.22 Variation in the strain gauge, temperature and smart cement resistance during the first day of cement curing in the bottom of the borehole.

Table 9.1 Piezoresistive model parameters for the field samples

Borehole Level	p_2	q_2
Bottom level	0.011	0.55
Upper level	0.012	0.46
Top level	0.005	0.34

where σ is the stress (MPa); σ_f: compressive stress at failure (MPa); $x = \left(\dfrac{\Delta\rho}{\rho_o}\right)*100$: Percentage of change in electrical resistivity due to the stress; $x_f = \left(\dfrac{\Delta\rho}{\rho_o}\right)_f *100$: Percentage of change in electrical resistivity at failure; $\Delta\rho$ is the change in electrical resistivity; and ρ_o is the initial electrical resistivity ($\sigma=0$ MPa). The model parameters q_2 and p_2 are summarised in Table 9.1. The coefficient of determinations (R^2) varied from 0.98 to 0.99.

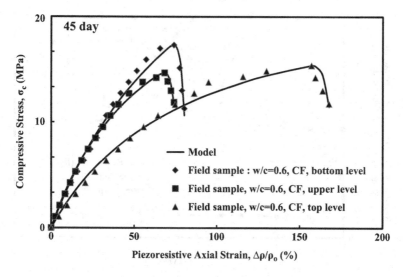

Figure 9.23 Piezoresistive behavior of the smart cement at various depths in the field well.

(d) Monitoring of Resistance, Strain, and Temperature

The smart cement was mixed in the field and used for cementing the field well. It is important to identify the measurable parameters in the cement sheath and also determine the changes with time and depth. Fiber optics is used for monitoring and it depends on the changes in the strain in the cement sheath. The strain in the cement will be influenced by the cement curing, stress, and temperature in the cement sheath. Over the past 4.5 years (over 1,600 days), thousands of data have been collected on monitoring parameters. It is important to quantify the changes in measuring parameters with important variables such as depth. In order to investigate the changes in depth, top-level (CH soil), middle-level (above the water table, CH soil), and the bottom-level (below the water table, CL soil) samples were selected for investigation.

Top Level

Resistance (R): The top level was about 0.3 m (1 ft.) below the ground surface. The initial resistivity of the smart cement measured using the two probes was 1.03 Ω.m, which is comparable to the laboratory-mixed cement of 1.05 Ω.m. The resistance in the top level changed from 22 Ω to 221 Ω, about a 9.05 times (905%) change in the resistance, as shown in Figure 9.24. The changes in the cement sheath resistance were not uniform but overall showed continuous increase. The rapid increase in the cement resistance

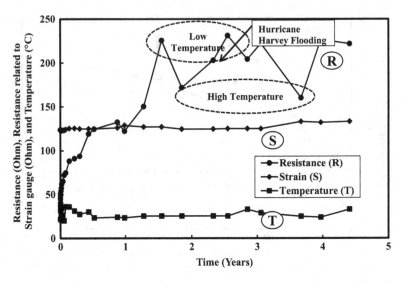

Figure 9.24 Electrical resistance, strain and temperature variation in top level during the 4.5 years.

was due to the lowering of the environmental temperature and the loss of moisture in the cement. The rapid decrease in the cement resistance was due to the increase in the environmental temperature and the saturation of the cement due to flooding.

Temperature (T): The temperature continuously fluctuated with time with no clear trend. Over the 4.5 years, the minimum and maximum measured temperature in the cement sheath was 20.1°C (68°F) and 36.2°C, (97.2°F), with a maximum change of 42.8%, as shown in Figure 9.24. The average temperature at the top level was about (25.4 °C), (77.7°F), a 14% decrease from the initial temperature of 32.4°C (90.3°F), which would have been influenced by cement hydration.

Strain (S): The strain gauge resistance increased from 123 Ω to 133 Ω during the period of 4.5 years with some fluctuations. The change in strain gauge resistance was about 8.1%. The tensile strain at the top level was about $3.3xE^{-6}$.

Based on the measured monitoring parameters in the cement sheath, the electrical resistance showed the largest change compared to the changes in temperature and strain. Hence, it is important to develop models to predict this change with time for monitoring the well.

Middle Level

Resistance (R): The middle level was about 4.6 m (15 ft) below ground level and above the water table. The initial resistivity of the smart cement measured using the two probes was 1.24 Ω.m higher than the top level of 1.03 Ω.m and the laboratory mixed cement of 1.05 Ω.m. The resistance in the top level changed from 26.5 Ω to 182.9 Ω, about a 5.90 times (590%) change in the resistance, as shown in Figure 9.25. The changes in the cement sheath resistance were not uniform but overall showed continuous increase. The rapid increase in the cement resistance was due to the lowering of the environmental temperature and loss of moisture in the cement. The rapid decrease in the cement resistance was due to an increase in the environmental temperature and saturation of the cement due to the rising of the water table because of flooding.

Temperature (T): The temperature continuously fluctuated with time with no clear trend. Over the 4.5 years, the minimum and maximum measured temperature in the cement sheath was 21.6°C (70.9°F) and 34.7°C (95.5°F), with a maximum change of 34.7%, as shown in Figure 9.25. The average temperature at the middle level was about 26 °C (78.8°F), a 18% decrease from the initial temperature of 35.8°C (96.4°F), which would have been influenced by cement hydration.

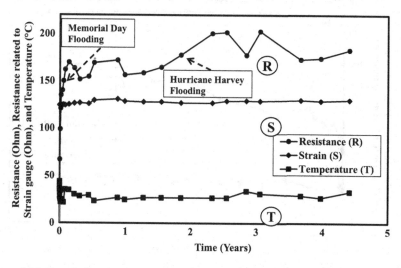

Figure 9.25 Electrical resistance, strain and temperature variation in middle level during the 4.5 years.

Strain (S): The strain gauge resistance increased from 124 Ω to 132 Ω during the period of 4.5 years with some fluctuations. The change in the strain gauge resistance was about 6.5%. The tensile strain in the middle level was $3.65\text{x}E^{-6}$.

Based on the measured monitoring parameters in the cement sheath, the electrical resistance showed the largest change compared to the changes in temperature and strain. Hence, it is important to develop models to predict this change with time for monitoring the well.

Bottom Level

Resistance (R): The bottom level was at 10.9 m (36 ft.) below the ground and was under the water table. The initial resistivity of the smart cement measured using the two probes was 1.32 Ω.m higher than the top level of 1.03 Ω.m and the laboratory-mixed cement of 1.05 Ω.m. The resistance in the bottom level changed from 28.2 Ω to 104.9 Ω, about a 2.72 times (272%) change in the resistance, as shown in Figure 9.26. The changes in the cement sheath resistances were uniform and overall showed a continuous increase. The minor fluctuations are due to changes in the water table level due to flooding.

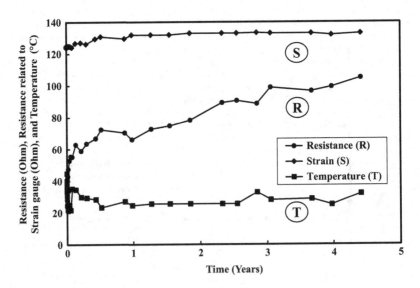

Figure 9.26 Electrical resistance, strain and temperature variation in bottom level during the 4.5 years.

Temperature (T): The temperature fluctuated with time but was much less than the middle and top levels. Over the 4.5 years, the minimum and maximum measured temperature in the cement sheath was 21.7°C (71.1°F) and 33°C (91.4°F), a maximum change of 28.6% (Figure 9.26). The average temperature at the bottom level was about 25°C (77°F), a 15.8% decrease from the initial temperature of 33°C (91.4°F), which would have been influenced by cement hydration.

Strain (S): The strain gauge resistance increased from 124 Ω to 133 Ω during the period of 4.5 years with some fluctuations. The change in strain gauge resistance was about 8.6%. The tensile strain at the bottom level was $4.8 \times E^{-6}$.

Based on the measured monitoring parameters in the cement sheath, the electrical resistance showed the largest change compared to the changes in temperature and strain. Hence, it is important to develop models to predict this change with time for monitoring the well.

Comparing Resistance Changes

From the measurements made at all levels, clearly the electrical resistance change was the highest. Hence, it is of interest compare the changes and trends in the electrical resistance with the depth. The electrical resistance change was not uniform in the top and middle levels in the field well. The electrical resistance changed by 905% in the top level close to the surface. The top level also showed the largest fluctuation in the resistance changes based on weather patterns. Both the environmental temperature and rainfall influenced the fluctuation in the resistance at the top level, as shown in Figure 9.27. The electrical resistance changed by 590% in the middle level (4.6 m (15 feet below the ground)), with much less fluctuation compared to the top level. The electrical resistance change at the bottom level, below the water table, was 272%, as shown in Figure 9.27. Also, the difference in the electrical resistance changes was due to the difference in cement curing conditions of the field well. The top level was exposed to outside temperature and had air curing, while the middle level was under moisture curing and the bottom level was cured under water.

Prediction of Electrical Resistivity of Smart Cement

Top Level

The value of the initial resistivity of the smart cement was 1.03 Ω.m. immediately after mixing. The electrical resistivity of the smart cement was 10.4 Ω.m. after 4.5 years of curing, as shown in Figure 9.28. The time for minimum resistivity was 195 minutes after mixing. As summarised in Table 9.2, based on the preliminary analyses, an AI model with four layers of ANNs was selected to predict the resistivity change with time. Over 30 data were

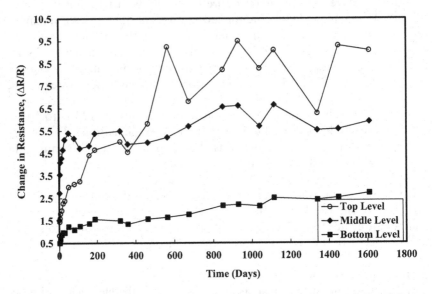

Figure 9.27 Electrical resistance data for top, middle and bottom levels in field well during the 4.5 years.

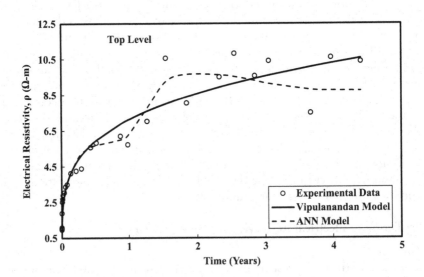

Figure 9.28 Comparing the prediction of the electrical resistivity at the top level using the AI model and Vipulanandan curing model up to 4.5 years.

Table 9.2 Electrical resistivity model parameters for the smart cement in the field for 4.5 years

Level	ρ_o (Ω.m.)	t_{min} (min)	t_o (min)	p_1	q_1
			Vipulanandan Curing Model		
Top	1.03	195	220	1.3	0.261
Middle	1.24	195	615	286	1.05
Bottom	1.32	288	165	3.68	0.253

Table 9.3 Correlation parameters for the ANN model and the Vipulanandan curing model for the smart cement in field after 4.5 years

| Level | ANN Model | | Curing Model | |
	R^2	RMSE	R^2	RMSE
Top	0.91	1.02	0.97	0.86
Middle	0.61	1.73	0.91	1.44
Bottom	0.86	0.43	0.95	0.25

used perform the BPNN and also predict the experimental trend. Curing model parameters p_1 and q_1 were 1.3 and 0.261 respectively after 4.5 years of curing (Table 9.2). The other curing model parameters are summarised in Table 9.2. The value of the root-mean-square error (RMSE) for the curing model was 0.86 Ω.m, while it was 1.02 Ω.m for the AI model. The value of R^2 for the curing model was 0.97, while it was 0.91 for the AI model, as summarised in Table 9.3. Thus, the Vipulanandan curing model had comparatively better prediction for the long term compared to the AI model.

Middle Level
The value of the initial resistivity of smart cement was 1.24 Ω.m. immediately after mixing. The electrical resistivity of the smart cement was 8.5 Ω.m. after 4.5 years of curing, as shown in Figure 9.29. The time for minimum resistivity was 195 minutes after mixing, as summarised in Table 9.2. Based on the preliminary analyses, an AI model with four layers of ANNs was selected to predict the resistivity change with time. Over 30 data were used to perform the BPNN and also predict the experimental trend. Curing model parameters p_1 and q_1 were 2.86 and 1.05 respectively after 4.5 years of curing, as summarised in Table 9.2. The other curing model parameters are summarised in Table 9.2. The value of the root-mean-square error (RMSE) for the electrical resistivity model was 1.44 Ω.m, while it was 1.73 Ω.m for the AI model. The value of R^2 for the electrical resistivity model was 0.91, while it was 0.61 for the AI model, as summarised in Table 9.3. Thus,

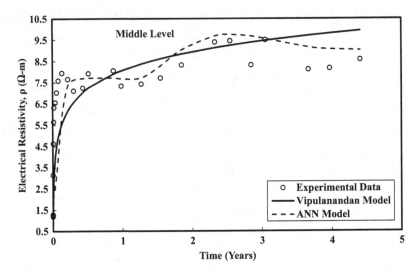

Figure 9.29 Comparing the prediction of electrical resistivity at the middle level using the AI model and Vipulanandan curing model up to 4.5 years.

the Vipulanandan curing model had comparatively better prediction for the long term compared to the AI model.

Bottom Level

The value of initial resistivity of the smart cement was 1.32 Ω.m immediately after mixing. The electrical resistivity of the smart cement was 4.91 Ω.m. after 4.5 years of curing (Figure 9.30). The time for minimum resistivity was 288 minutes after mixing (Table 9.2). Based on the preliminary analyses, the AI model with four layers of ANNs was selected to predict the resistivity change with time. Over 30 data were used to perform the BPNN and also predict the experimental trend. Curing model parameters p_1 and q_1 were 3.68 and 1.05 respectively after 4.5 years of curing (Table 9.2). The other curing model parameters are summarised in Table 9.2. The value of the root-mean-square error (RMSE) for electrical resistivity model was 0.25 Ω.m, while it was 0.43 Ω.m for the AI model. The value of R^2 for the electrical resistivity model was 0.95, while it was 0.86 for the AI model, as summarised in Table 9.3. Thus, the Vipulanandan curing model had comparatively better prediction for the long term compared to the AI model.

Piezoresistivity Prediction

It is important to demonstrate the piezoresistivity of smart cement in the field. Also, it is important to show the sensitivity of smart cement for small pressure changes. Hence, the test was performed at 0.07 MPa (10

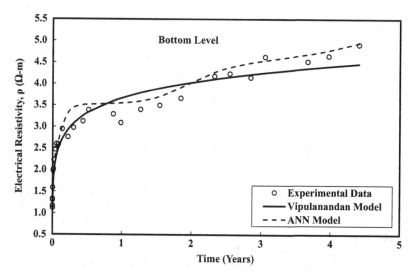

Figure 9.30 Comparing the prediction of the electrical resistivity at the bottom level using the AI model and Vipulanandan curing model up to 4.5 years.

Table 9.4 Correlation parameters for the ANN model and the piezoresistivity model for the smart cement in the field after 4.5 years

Pressure Test on Smart Cement			
ANN Model		Piezoresistivity Model	
R^2	RMSE(Ωm)	R^2	RMSE (Ωm)
0.99	0.015	0.99	0.02

psi) increments up to 0.55 MPa (80 psi). The maximum value of vertical piezoresistive strain (3D) for smart cement after 4.5 years of curing was 13.5% at an applied pressure of 0.55 MPa, as shown in Figure 9.31. This is a clear demonstration of the 3D sensitivity of the smart cement. Also, by measuring the piezoresistive strain in the smart cement, it is possible to predict the pressure in the casing using the Vipulanandan p-q piezoresistive model (Eqn. 9.2). The value of model parameters p_2 and q_2 for the piezoresistivity model are 0.025 and 0.417, and the root-mean-square error (RMSE) of 0.02 Ωm with R^2 of 0.99. The AI model had a root-mean-square error (RMSE) of 0.015 Ωm value compared to the piezoresistivity model root-mean-square error (RMSE) of 0.02 Ωm with a coefficient of determination of 0.99, as summarised in Table 9.4. Hence, both models predicted the piezoresistive behavior of the smart cement.

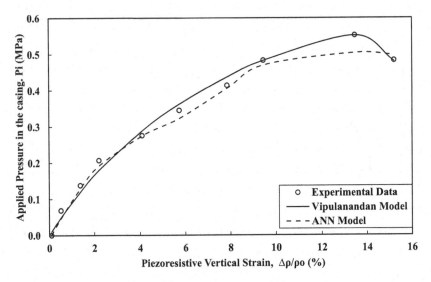

Figure 9.31 Vertical piezoresistive strain–pressure relationship (3D) for the smart cement in the field after 4.5 years of curing.

9.5 SUMMARY

In order to evaluate the performance of smart cement with a real-time monitoring system, both laboratory and field model tests were performed. The changes in the resistances with depth were monitored using alternative current (AC) at 300 kHz and two probes in the laboratory models for about 250 days and the field model study for 4.5 years. Also, the AI models were used with the Vipulanandan models to predict the changing properties of the smart cement in the field well and the following conclusions are advanced.

1 Based on the laboratory model data and the field model data, electrical resistivity showed the largest variation compared to strain and temperature changes. Hence, electrical resistivity was selected as the monitoring parameter for the smart cement.

2 The two-probe method was effective in measuring the bulk resistance of the drilling mud, spacer fluid, and smart cement slurries. Based on the changes in resistance measurements, it was possible to identify the fluid rise in the laboratory and field well borehole.

3 Using the laboratory models and field model, it was possible to demonstrate the adaptability of the real-time monitoring system at different scales of the well bore with drilling mud, spacer fluid, and smart cement slurries.

4 Based on the concept developed in this study, it is possible to use the K parameter to predict changes in the resistance of hardening smart cement.

5 Using the Vipulanandan p-q model, the change in the vertical piezoresistive axial strain in the smart cement was related to the applied pressure in the casing. The smart cement 3D sensor was very sensitive to the applied pressure in the casing.

6 AI models representing machine learning with one, two, three, and four layers of artificial neural networks (ANNs) were evaluated using the laboratory and field data with the statistical parameter coefficient of determination (R^2) and the root-mean-square error (RMSE). Based on the type of available data, both generalised regression neural networks (GRNN) and back propagation neural network (BPNN) were used to train the AI models.

7 The AI model predicted the long-term smart cement curing with the resistivity parameter well and was comparable to the Vipulanandan curing model.

Chapter 10

Smart Cement Grouts

10.1 BACKGROUND

Cement grouts and polymer grouts are used in a number of applications including construction and the maintenance and repair of bridges, highways, airport runways, buildings, tunnels, pipelines, piles, and wells (oil, gas, and water) (Dharmarajan et al. 1988; Ozgurel et al. 2005; Ahossin Guezo et al. 2014; Vipulanandan et al. 1985, 1986, 1987, 1994, 1995a, b, c, 1997a, b, c, 2000a, b, c, 2001a, b, 2012b, 2013b). Also, both cement grouts and polymer grouts are used in soil stabilisation and in repairing rocks, stabilising slopes and tunnels, and also solidifying contaminated soils (Bowen 1981; Nelson 1990; Vipulanandan et al. 1992, 1995a, b, and 1997a; Chun et al. 2008; Anagnostopoulos 2014; Vipulanandan et al. 2014b, c, d, e, f, 2018a, g; Mahmood et al. 2020). Polymer grouts are broadly characterised as organic and inorganic (Vipulanandan et al. 1985, 1997a, b, c, 2000a, b, c and 2005a, b, e). Sodium silicate–based grouts are popular inorganic grouts used in soil stabilisation (Krizek et al. 1985; Vipulanandan et al. 1985, 1986, 2000a, b, c). With a wide spectrum of applications, there is a need to make cement grouts and polymer grouts sensing so that they can be monitored from the time of mixing to the entire service life, and also make repaired sections highly sensing for real-time monitoring (Vipulanandan et al. 2016e and 2018a).

Among the different grouts used as sealers, sodium silicate–based concrete sealers have become popular because sodium silicate reacts with the $Ca(OH)_2$ in hydrating cement and forms inorganic calcium-silicate hydrates, and hence increases the durability of the grout. Also, a study has shown that sodium meta-silicate (SS)-based cement grout significantly reduced the permeability of the grouted region (Chun et al. 2008). In that study, sodium silicate cement grout with ordinary Portland cement (OPC) showed compressive strength increase with an increase in sodium meta-silicate (SMS) concentration. Other studies have reported that the addition of 1% SMS to Class G cement slurry cured at 100°F showed a reduction in compressive strength from 16.1 MPa to 13.8 MPa (72 hours), whereas with the same amount of SMS and cured at 200°F, the slurry showed an increase in compressive strength from 16.1 MPa to 18.6 MPa (Heinold et al. 2002). One of the major variables in cement grouts is the water-to-cement (w/c) ratio, and based on the application it can vary from 0.6 to about 3. Also, many organic and inorganic additives are used in cement grouts (Ata et al. 1997; Vipulanandan 1997a; Heinold et al. 2002; Anagnostopoulos 2014; Vipulanandan 2018a).

Studies have shown that smart polymer–grouted sand columns can be integrated with earth dams and embankments to be used for real-time monitoring (Vipulanandan et al. 2018g). In recent years, smart cement grouts and smart polymer grouts have been modified with nanoparticles to enhance performance (Vipulanandan et al. 2016e).

In this study, smart cement grouts were developed using various types of cements including ultrafine cements, oil-well cements, and Portland cements at various w/c ratios. Also, smart cement grouts with and without sodium silicate have been tested and characterised using Vipulanandan models. In this chapter, a few selected results are presented to characterise and quantify the performance of smart cement grouts.

10.2 MATERIALS AND METHODS

(a) Materials

> **Smart cement grout:** Class H oil-well cement was mixed with carbon fibers with a diameter in the micrometer range to make it a chemo-thermo-piezoresistive material.
> **Sodium silicate (SS):** Commercially available sodium meta silicate powder was mixed with water and then used to prepare the cement grouts. In this study, 1% and 3% sodium silicate water solutions were used. The pH was 12.5 for the 1% sodium silicate solution at room condition.

(b) Methods

Cement Grout Mixture

Sensing cement grout was prepared using a w/c ratio of 0.8 in this study. A commercially available table top mixture was used at a speed of 1,000 rpm for five minutes, and 0.075% carbon fillers (based on cement weight) were added during the mixing process.

Cement Specimen

The test specimens were prepared in plastic cylindrical molds. All specimens were capped to minimise moisture loss and were cured under room condition (23°C and relative humidity of 50%) up to the day of testing.

Repairing Damaged Sensing Cement

After testing the 1-, 7-, and 28-day cured cylindrical smart cement specimens to failure, the samples with cracks along the length of the specimens were submerged in the cement grout solutions with and without sodium silicate (SS) for three hours. The damaged smart cement specimens had wires in place inside the cement at 50 mm apart. After repair, the specimens were cured under room temperature for one day before testing. The weight and resistance of the specimens were monitored to determine moisture loss and change in electrical resistivity before the compressive piezoresistivity tests.

Results and Discussion

The unit weight of the smart cement grout with a 0.8 w/c ratio was 15.79 kN/m^3. It increased to 15.89 kN/m^3 with 1% SS and 15.93 kN/m^3 with 3% SS.

10.3 CURING

The change in electrical resistivity with curing time for the sensing cement grout with and without SS up to 28 days of curing was monitored. The initial resistivity (ρ_o), minimum resistivity (ρ_{min}), and time to reach minimum resistivity (t_{min}) are summarised in Table 10.1. Also the percentages of maximum change in resistivity at the end of 24 hours (RI_{24hr}), 7 days $\left(RI_{7days}\right)$, and 28 days $\left(RI_{28\,days}\right)$ are defined in Eqn. (10.1), Eqn. (10.2), and Eqn. (10.3) as follows:

$$RI_{24\ hr} = \frac{\rho_{24hr} - \rho_{min}}{\rho_{min}} * 100 \tag{10.1}$$

$$RI_{7\ days} = \frac{\rho_{7days} - \rho_{min}}{\rho_{min}} * 100 \tag{10.2}$$

$$RI_{28\ days} = \frac{\rho_{28days} - \rho_{min}}{\rho_{min}} * 100 \tag{10.3}$$

The initial sensing property (ρ_o) of the smart cement grout with a w/c ratio of 0.8 was 1.08 Ohm.m, and the sensing property reduced to reach the ρ_{min} of 1.04 Ohm.m after 180 minutes (t_{min}), as summarised in Table 10.1. With 1% and 3% SS, the initial resistivity and the minimum resistivity decreased, and the time to reach the minimum (t_{min}) increased to 300 minutes. With 1% SS, the initial resistivity (ρ_o) decreased by 36% and ρ_{min} decreased by 44%, and with 3% SS, the ρ_o decreased by 51% and ρ_{min} decreased by 60%. The 24-hour resistivity (ρ_{24hr}) of the smart cement grout only was 2.16 Ω.m.

Table 10.1 Summary of bulk resistivity parameters for the smart cement grout with and without SS cured at room temperature up to 28 days

Mix Type	Density (kN/ m³)	Initial resistivity, ρ_o ($\Omega.m$)	ρ_{min} ($\Omega.m$)	t_{min} (min)	ρ_{24hr} ($\Omega.m$)	$\rho_{7\,days}$ ($\Omega.m$)	$\rho_{28\,days}$ ($\Omega.m$)	$RI_{24\,hr}$ (%)	$RI_{7\,days}$ (%)	$RI_{28\,days}$ (%)
Grout (H, w/c = 0.8 only)	15.79	1.08	1.04	180	2.16	6.16	9.37	108	492	801
Grout (H, w/c = 0.8, SS = 1%)	15.89	0.69	0.54	300	1.01	2.20	4.85	87	307	798
Grout (H, w/c = 0.8, SS = 3%)	15.93	0.52	0.41	300	0.56	1.28	3.29	37	212	702

Hence, the maximum change in resistivity after 24 hours (RI_{24hr}) was 108%, as summarised in Table 10.1. The 7- and 28-day resistivity (ρ_{7days} and ρ_{28days}) of the smart cement grout was 6.16 $\Omega.m$ and 9.37 $\Omega.m$; hence, the maximum changes in resistivity after seven days and 28 days (RI_{7days} and $RI_{28\,days}$) were 492% and 801% respectively. The addition of SS decreased the resistivity compared to that of the smart cement grout only. The addition of 1% SS decreased the 24-hour, 7-day, and 28-day resistivity of the smart cement grout by about 52%, 60%, and 48% respectively, and hence the resistivity indices RI_{24hr}, RI_{7days}, and $RI_{28\,days}$ also decreased compared to that of smart cement grout only. The addition of 3% SS decreased the 24-hour, 7-day, and 28-day resistivity of the smart cement grout by about 74%, 79%, and 65% respectively, and hence the resistivity indices RI_{24hr}, RI_{7days}, and $RI_{28\,days}$ also decreased.

The curing model parameter p_1 for the smart cement grout was 0.286 after 24 hours of curing, and it increased to 0.610 after 28 days, as summarised in Table 10.2. With 1% SS, the parameter p_1 decreased to 0.244 after 24 hours of curing, and it further reduced to 0.236 after 28 days of curing. The addition of 3% SS further decreased the parameter p_1 to 0.174 after 24 hours of curing, and then it decreased to 0.113 after 28 days of curing. The curing model parameter q_1 for smart cement grout only was 0.188 after 24 hours of curing, and it increased to 0.295 after 28 days, as summarised in Table 10.2. The addition of 1% SS increased the parameter q_1 value to 0.217 after 24 hours of curing, but it decreased to 0.161 after 28 days. The addition of 3% SS decreased the parameter q_1 value to 0.131 after 24 hours of curing, and then it decreased to 0.092 after 28 days of curing.

Table 10.2 Vipulanandan curing model parameters for the smart cement grout with and without SS cured at room temperature up to 28 days

Mix Type	Curing Time (day)	ρ_{min} ($\Omega.m$)	t_{min} (min)	p_I	q_I	t_o (min)	RMSE ($\Omega.m$)	R^2
Grout (H, w/c = 0.8 only)	1 day	1.04	180	0.286	0.188	63	0.04	0.98
Grout (H, w/ c = 0.8, SS = 1%)		0.54	300	0.244	0.217	70	0.03	0.95
Grout (H, w/c = 0.8, SS = 3%)		0.41	300	0.174	0.131	61	0.03	0.97
Grout (H, w/c = 0.8 only)	28 days	1.04	180	0.710	0.295	110	0.45	0.97
Grout (H, w/c = 0.8, SS = 1%)		0.54	300	0.236	0.161	70	0.09	0.99
Grout (H, w/c = 0.8, SS = 3%)		0.41	300	0.113	0.092	51	0.03	0.99

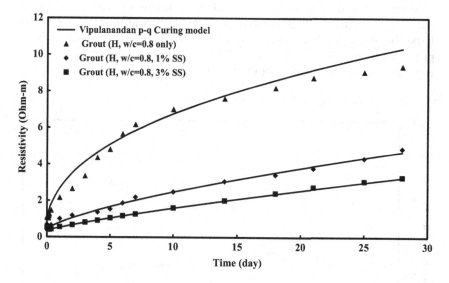

Figure 10.1 Variation of the resistivity with the curing time for the smart cement grouts up to 28 days of curing and modeled with the Vipulanandan p-q curing model.

The Vipulanandan curing model (Eqn. (4.12)) predicted the measured curing resistivity very well, as shown in Figure 10.1. The root-mean-square error (RMSE) varied from 0.03 $\Omega.m$ to 0.45 $\Omega.m$ for 24 hours and 28 days of curing respectively.

10.3.1 Minimum Resistivity

The addition of SS decreased the minimum resistivity of the cement grout. The minimum resistivity (ρ_{min}) of the sensing cement grout decreased from 1.04 Ohm-m to 0.54 Ohm-m with 1% SS, a 44% decrease. By adding 3% SS, the minimum resistivity of the grout was further decreased to 0.41 Ohm-m, a 60% decrease. The relationship between minimum resistivity and SS concentration was modeled using the Vipulanandan correlation model (Vipulanandan et al. 1993, 2018a–k) (three parameters) as follows:

$$\rho_{min} = (\rho_{min})_o - SS/[C_1 + D_1(SS)] \tag{10.4}$$

where

ρ_{min} = minimum resistivity of the grout ($\Omega.m$),
$(\rho_{min})_o$ = minimum resistivity of the grout without SS ($\Omega.m$), and
SS = sodium silicate amount (% by weight).

Material parameters are C_1 and D_1. The parameter C_1 represents the initial rate of change, and parameter D_1 determines the ultimate resistivity. Experimental results matched very well, as shown in Figure 10.2 with the proposed model (Eqn. (10.4)) with R^2 of 0.99, and parameters C_1 and D_1 were 0.62 $Ohm^{-1}\text{-}m^{-1}$ and 1.38 $Ohm^{-1}\text{-}m^{-1}$ respectively.

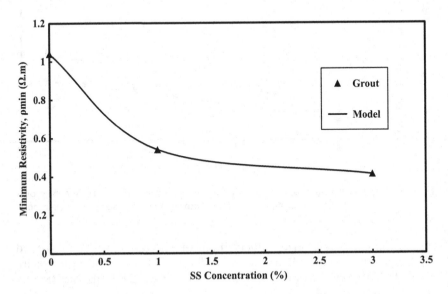

Figure 10.2 Variation of the minimum resistivity with the sodium silicate concentrations in the grouts.

10.3.2 Moisture Loss and Resistivity

During the curing period, the weight of the specimens was monitored to observe any change in moisture loss. The smart cement grout samples lost about 6% moisture during 28 days of curing, whereas the grout made with smart cement and 1% SS lost about 4.2% moisture during 28 days, a 30% less moisture loss with the addition of 1% SS, as shown in Figure 10.3. The grout made with 3% SS lost about 3.7% moisture during 28 days, about a 38% less moisture loss, as shown in Figure 10.3.

Modeling
A relationship has been proposed for the resistivity of the cured specimen with the moisture loss (%) as

$$\rho = \rho_o + A^*(\Delta w/w_o)^n \qquad (10.5)$$

where

ρ = Resistivity of the grout ($\Omega.m$),
ρ_o = Initial resistivity of the grout without moisture loss ($\Omega.m$), and
$\Delta w/w_o$ = moisture loss of the specimen (%).

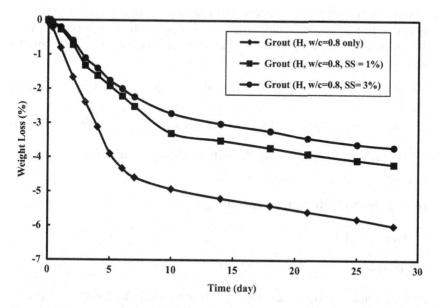

Figure 10.3 Percentage weight loss in the smart cement grout specimens cured at room temperature with and without SS up to 28 days.

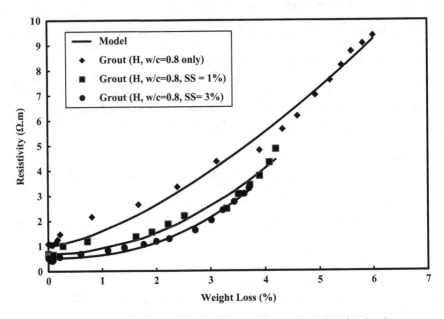

Figure 10.4 Relationship between the resistivity and the moisture loss for the three types of smart cement grouts.

A and n are model parameters that can be determined from the experimental results. For the grouts, the experimental values fit very well with the proposed model, as shown in Figure 10.4. For the grouts, the model parameters are as follows:

(i) For grout without SS, A = 0.587, n = 1.47, $R^2 = 0.99$ (10.6.a)

(ii) For grout with 1% SS, A = 0.198, n = 2.05, $R^2 = 0.97$ (10.6.b)

(iii) For grout with 3% SS, A =0.11, n = 2.4, $R^2 = 0.98$ (10.6.c)

10.4 COMPRESSIVE BEHAVIOR

(a) Piezoresistivity

The piezoresistive axial strain at failure $\left(\dfrac{\Delta \rho}{\rho_o}\right)_f$ for the sensing cement grout only after one, seven, and 28 days of curing were 155%, 156%, and 179%, which decreased to 117%, 116% and 125% respectively for sensing cement grout with 1% SS. Thus, the changes in resistivity with applied stress of the grouts were reduced by 24%, 25%, and 30% after one, seven, and 28 days of curing respectively with the addition of 1% SS. By the addition of 3% SS, the changes in resistivity with applied stress after one day, seven days, and 28 days of curing were 106%, 94%, and 103%, which reduced by 31%, 40%, and 42% respectively.

Based on the experimental results, the Vipulanandan p-q piezoresistive model was used to predict the behavior, and the equation is as follows:

$$\frac{\sigma_c}{\sigma_{cf}} = \left[\frac{\dfrac{x}{x_f}}{q_2 + (1 - p_2 - q_2)\dfrac{x}{x_f} + p_2\left(\dfrac{x}{xf}\right)^{\left(\frac{p_2 + q_2}{p_2}\right)}}\right] \quad (10.7)$$

where the stress is σ_c (MPa); stress at failure is σ_{cf} (MPa); $x = \left(\dfrac{\Delta\rho}{\rho_o}\right) * 100 =$ Percentage of change in piezoresistive axial strain due to the stress; $x_f = \left(\dfrac{\Delta\rho}{\rho_o}\right)_f * 100 =$ Percentage of piezoresistive axial strain at failure; $\Delta\rho$ is the change in the resistivity; ρ_o is the initial resistivity (at $\sigma=0$ MPa); and p_2 and q_2 are model parameters.

Using the p-q piezoresistive model (Eqn. (10.7)), the relationships between compressive stress and the piezoresistive axial strain $\left(\dfrac{\Delta\rho}{\rho_o}\right)$ of the smart cement grout with and without SS for one, seven, and 28 days of curing were modeled. The piezoresistive model (Eqn. (10.7)) predicted the measured stress-change in the resistivity relationship very well, as shown in Figure 10.5 for the smart cement grout, in Figure 10.6 for the smart cement grout with 1% SS, and in Figure 10.7 for the smart cement with 3% SS. The model parameters q_2 and p_2 are summarised in Table 10.3. The R^2 were 0.95 to 0.99. The root-mean-square error (RMSE) varied between 0.04 MPa and 0.64 MPa, as summarised in Table 10.3.

(b) Piezoresistive Secant Modulus

The piezoresistive secant modulus is defined as the ratio of compressive stress at failure to the piezoresistive axial strain at failure $\left[\dfrac{\sigma_{cf}}{\left(\dfrac{\Delta\rho}{\rho_o}\right)_{cf}}\right]$ of the grouts, and was determined up to 28 days. With curing time, the secant

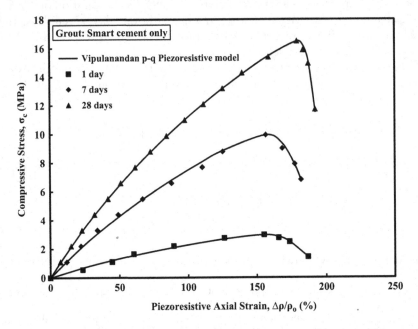

Figure 10.5 Piezoresistive response of the smart cement grout after 1 day, 7 days and 28 days of curing and modeled using Vipulanandan p-q piezoresistivity model.

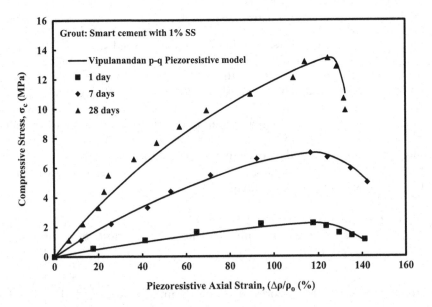

Figure 10.6 Piezoresistive response of the smart cement with 1% sodium silicate after 1 day, 7 days and 28 days of curing and modeled using the Vipulanandan p-q piezoresistivity model.

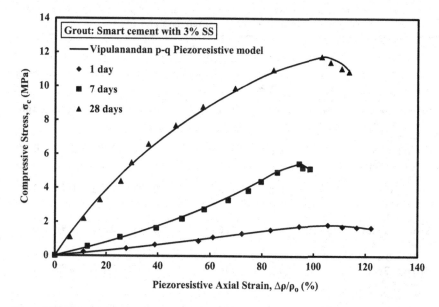

Figure 10.7 Piezoresistive response of the smart cement with 3% sodium silicate after 1 day, 7 days and 28 days of curing modeled using the Vipulanandan p-q piezoresistivity model.

changes in resistivity with applied stress of the cement grout changed. The relationship between the secant modulus and the curing time was modeled using the Vipulanandan correlation model (Vipulanandan et al. 1993, 2017) as follows (t ≥ 1):

$$\frac{\sigma_{cf}}{\left(\dfrac{\Delta\rho}{\rho_o}\right)_{cf}} = t/(G+Ht) \tag{10.8}$$

where $\dfrac{\sigma_{cf}}{\left(\dfrac{\Delta\rho}{\rho_o}\right)_{cf}}$ = Piezoresistive secant modulus is the ratio of strength

to changes in piezoresistive axial strain at peak stress and t = time of curing (day).

Material parameters are G and H in Eqn. (10.8). The parameter G represents the initial rate of change, and parameter H determines the ultimate secant changes in sensing property with applied stress. For the

Table 10.3 Strength, piezoresistive axial strain at failure, model parameters for the smart cement grouts cured after one day, seven days, and 28 days

Mix Type	Curing Time (day)	Strength σ_{cf} (MPa)	Piezoresistivity at peak stress, $(\Delta\rho/\rho_o)_f$ (%)	p_2	q_2	R^2	RMSE (MPa)
Smart Grout w/c = 0.8 SS = 0%	1 day	2.96	155	0.031	0.607	0.99	0.08
Smart Grout w/c = 0.8, SS = 1%		2.23	117	0.037	0.706	0.95	0.17
Smart Grout w/c = 0.8, SS = 3%		1.82	106	0.183	1.193	0.99	0.04
Smart Grout w/c = 0.8 SS = 0%	7 days	9.94	156	0.035	0.642	0.99	0.18
Smart Grout w/c = 0.8, SS = 1%		6.98	116	0.052	0.596	0.99	0.15
Smart Grout w/c = 0.8, SS = 3%		5.45	94	0.07	1.582	0.99	0.16
Smart Grout w/c = 0.8 SS = 0%	28 days	16.47	179	0.012	0.613	0.99	0.10
Smart Grout w/c = 0.8 SS = 1%		13.42	125	0.01	0.561	0.97	0.64
Smart Grout w/c = 0.8 SS = 3%		11.75	103	0.03	0.492	0.99	0.20

cement grouts, the experimental results matched very well, as shown in Figure 10.8, with the proposed model with R^2 of 0.98 to 0.99. For the smart cement grout only, parameters G and H were determined to be 0.44 and 0.09. For the smart cement grout with 1% SS, parameters G and H were determined as 0.62 and 0.07. For the smart cement grout with 3% SS, parameters G and H were determined as 0.74 and 0.06. Parameters G and H were linearly related (Eqn. 10.9a, 10.9b) to SS concentration as follows:

$$G = 0.09 * (\%SS) + 0.47, \quad R^2 = 0.95 \tag{10.9a}$$

$$H = -0.01 * (\%SS) + 0.09, \quad R^2 = 0.95 \tag{10.9b}$$

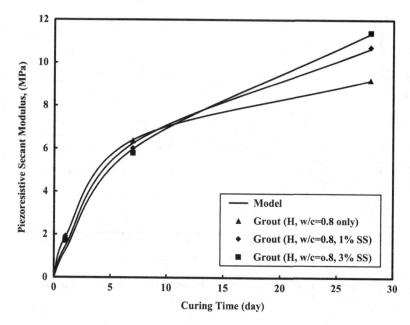

Figure 10.8 Variation piezoresistive secant modulus with the curing time for the grouts modeled using the Vipulanandan correlation model.

(c) Strength

The σ_{cf} of the smart cement grout after one, seven, and 28 days of curing were 2.96 MPa, 9.94 MPa, and 16.47 MPa respectively. With 1% SS, the strength decreased to 2.23 MPa, 6.98 MPa, and 13.42 MPa respectively for the smart cement grout, as summarised in Table 10.3. Thus, the compressive strengths of the grouts were reduced by 24%, 29%, and 19% after one, seven, and 28 days of curing with the addition of 1% SS. With the addition of 3% SS, the compressive strengths after one day, seven days, and 28 days of curing were 1.82 MPa, 5.45 MPa, and 11.75 MPa, a reduction of 38%, 45%, and 28% respectively.

(d) Strength with Curing Time

The strength of the smart cement grouts increased with time. The relationship between compressive strength and time was modeled using the Vipulanandan correlation model (Vipulanandan et al. 1993, 2017) as follows:

$$\sigma_{cf} = t/(E + Ft) \tag{10.10}$$

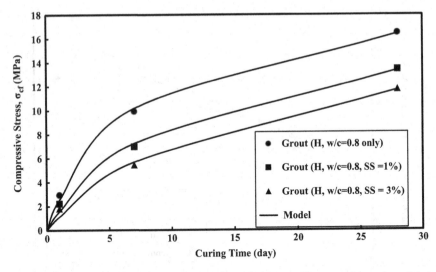

Figure 10.9 Variation of compressive strength with curing time for the grout modeled using the Vipulanandan correlation model.

where

σ_{cf} = Compressive strength of the grout (MPa) and
t = time of curing (day).

In the relationship E and F are material parameters. Parameter E represents the initial rate of change and parameter F determines the ultimate strength. For the cement grouts, the experimental results matched very well, as shown in Figure 10.9, with the proposed model with R^2 of 0.95 to 0.99. For the sensing cement grout only, parameters E and F were 0.347 MPa⁻¹ and 0.048 MPa⁻¹. For the sensing cement grout with 1% SS, parameters E and F were 0.586 MPa⁻¹ and 0.054 MPa⁻¹. For the sensing cement grout with 3% SS, parameters E and F were 0.841 MPa⁻¹ and 0.055 MPa⁻¹. Parameter E and F were linearly related to SS concentration as follows:

$$E = 0.16 * (\%SS) + 0.34, \quad R^2 = 0.97 \tag{10.11a}$$

$$F = 0.002 * (\%SS) + 0.05, \quad R^2 = 0.95 \tag{10.11b}$$

(e) RI_{24} Correlation

Compressive strength was correlated to the resistivity index of the cement grout after 24 hours (RI_{24}), as shown in Figure 10.10. The relationships were linear and are represented as follows:

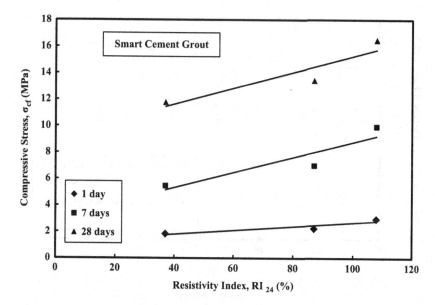

Figure 10.10 Relationship between compressive strength of the cement grouts and resistivity index RI_{24} for the cement grouts.

(i) For 1 day strength, $\sigma_{cf} = 0.015*(RI_{24}) + 1.2$; $R^2 = 0.95$ (10.12a)

(ii) For 7 days strength, $\sigma_{cf} = 0.057*(RI_{24}) + 3.02$; $R^2 = 0.95$ (10.12b)

(iii) For 28 days strength, $\sigma_{cf} = 0.06*(RI_{24}) + 9.2$; $R^2 = 0.85$ (10.12c)

(f) Piezoresistive Axial Strain with Curing Time

The piezoresistivity at failure of the grout specimen made with and without SS was tested up to 28 days. With curing time increase, the piezoresistive axial strain at failure of the cement grout changed based on the SS content. The relationship between the piezoresistivity at failure of the cement grout and curing time was modeled with the Vipulanandan correlation model, and the relationship is as follows:

$$(\Delta\rho/\rho_o)_f = (\Delta\rho/\rho_o)_1 + t/(L + Mt)$$ (10.13)

where

$(\Delta\rho/\rho_o)_f$ = Piezoresistive axial strain at failure (%),
$(\Delta\rho/\rho_o)_1$ = Piezoresistivity at failure of the grout after 1 day (%), and
t = Curing time (day).

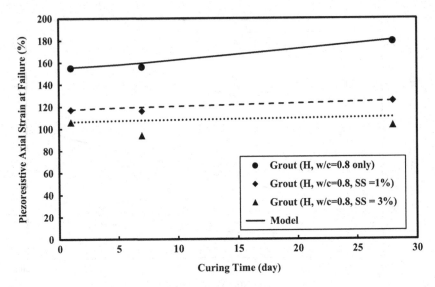

Figure 10.11 Relationship between piezoresistive axial strain at failure and the curing time for the grout modeled using the Vipulanandan correlation model.

Parameters L and M are model parameters, and parameter L represents the initial rate of change and parameters M determines the ultimate piezoresistivity. For the cement grouts, the experimental results matched well, as shown in Figure 10.11 with the proposed model with coefficient of determination (R^2) of 0.95–0.99. For the smart cement grouts, parameters L and M were determined to be 1.64 and 0.019. For the smart cement grout with 1% SS, parameters L and M were found as 2.5 and 0.036. For the smart cement grout with 3% SS, parameters L and M were found as -3.5 and -0.1.

(g) Relationship between RI_{24} and Piezoresistive Axial Strain

A linear relationship between the piezoresistivity axial strain at failure ($\Delta\rho/\rho_o)_f$ and the resistivity index after 24 hours (RI_{24}) was observed, as shown in Figure 10.12 and the relations are as follows:

(i) For 1 day, $(\Delta\rho/\rho_o)_f = 0.8*(RI_{24}) + 60.5,$ $R^2 = 0.95$ (10.14a)

(ii) For 7 days, $(\Delta\rho/\rho_o)_f = 0.6*(RI_{24}) + 79.2,$ and $R^2 = 0.95$ (10.14b)

(iii) For 28 days, $(\Delta\rho/\rho_o)_f = 0.95*(RI_{24}) + 61.7.$ $R^2 = 0.85$ (10.14c)

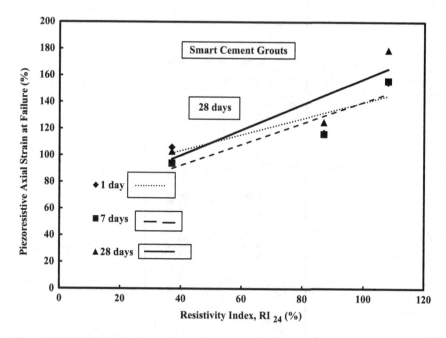

Figure 10.12 Relationship between piezoresistive axial strain at failure and the resistivity index (RI_{24}).

REPAIRED SMART CEMENT

Smart cement specimens were cured for one, seven, and 28 days and tested to failure, and then repaired with the smart cement grouts. Also, the one-day cured and tested specimens were repaired and tested again after one day of curing. The seven-day cured and tested specimens were repaired and tested again after seven more days. The 28-day cured and tested specimens were repaired and tested again after 28 more days of curing.

One-Day Cured

Several one-day cured sensing cement specimens (0.38 w/c ratio) were used for the repair study after failing the specimen under compression loading (Vipulanandan et al. 2014a, b; 2015a, b). The compressive strength of the one-day cured sensing cement varied from 10.81 MPa to 12.00 MPa, as summarised in Table 10.4. The resistivity change of the one-day cured smart cement at peak stress varied from 268% to 300%, and the data on the repaired specimens is summarised in Table 10.4.

Table 10.4 Piezoresistive model parameters for the one-day cured smart cement repaired with grouts

Mix Type	Curing Time (day)	Strength σ_{cf} (MPa)	Piezoresistive strain, $(\Delta\rho/\rho_o)_f$ (%)	p_2	q_2	R^2	RMSE (MPa)	Strength Regain (%)	Piezoresistivity Regain* (%)
Initial smart cement	1 day	12.00	300	0.01	0.693	0.99	0.14	N/A	N/A
Repaired cement with grout (H, w/c = 0.8 only)		10.09	48	0.039	0.601	0.99	0.19	84	21
Initial smart cement	1 day	11.37	294	0.01	0.639	0.99	0.21	N/A	N/A
Repaired cement with grout (H, w/c = 0.8, SS = 1%)		5.80	56	0.046	0.793	0.99	0.17	51	39
Initial smart cement	1 day	10.82	268	0.016	0.684	0.99	0.19	N/A	N/A
Repaired cement with grout (H, w/c = 0.8, SS = 3%)		4.63	44	0.072	0.628	0.99	0.078	43	42

* At the failure stress of the repaired specimen

(a) Piezoresistive Behavior

No Additive

The changes in piezoresistive axial strain $(\Delta\rho/\rho)\%$ at an applied stress of 4 MPa for the repaired sample was 15% compared to the smart cement, which was 86%, a recovery of 18%. The piezoresistive axial strain at failure $(\Delta\rho/\rho)_f\%$ for the repaired specimen was 48% at peak stress of 10.09 MPa, as summarised in Table 10.4. The sensing property change in the smart cement at 10.09 MPa was 234%. Hence, smart grout repair resulted in a 21% recovery of piezoresistive axial strain the change in sensing property with the recovered applied stress, as shown in Figure 10.13.

1% SS

The change in piezoresistive axial strain $(\Delta\rho/\rho)\%$ at an applied stress of 4 MPa for the repaired sample was 42% as compared to the smart cement, which was 88%, a recovery of 47%. The change in piezoresistive axial strain at failure $(\Delta\rho/\rho\%)_f$ for the repaired specimen was 56% at peak stress of 5.80 MPa. The sensing property change in the sensing cement at 5.80 MPa was 142%. Hence, smart grout repair resulted in a 39% recovery of piezoresistive axial strain the change in sensing property with the recovered applied stress, as shown in Figure 10.14.

3% SS

The changes in piezoresistive axial strain $(\Delta\rho/\rho)\%$ at an applied stress of 4 MPa for the repaired sample was 29% compared to the sensing cement, which was 88%, a recovery of 33%. The change in piezoresistive axial strain at failure $(\Delta\rho/\rho)_f\%$ for the repaired specimen was 44% at peak stress of 4.63 MPa. The sensing property change in smart cement at 4.63 MPa was 104%. Hence, smart grout repair resulted in 42% recovery of the piezoresistive axial strain with the applied stress, as shown in Figure 10.15, the highest percentage of change in sensing property with the recovered applied stress.

(b) Strength

No Additive

Smart cement grout was used to repair the sensing cement specimen with a strength of 12.00 MPa with 300% piezoresistive axial strain at failure. After submerging the damaged specimen for three hours in the grout, it was cured for one day and tested under compression load. The percentage weight change in the repaired specimen was 0.24%. The compressive strength of the repaired specimen was 10.09 MPa, a 84% strength recovery, the highest strength recovery.

1% SS

Smart cement grout with 1% SS was used to repair the sensing cement specimen with a strength of 11.37 MPa with 294% piezoresistive axial strain

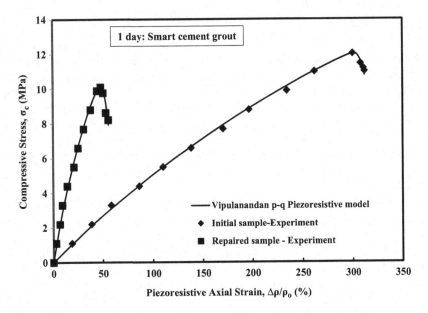

Figure 10.13 Relationship between piezoresistive axial strain at failure and the resistivity index (RI_{24}).

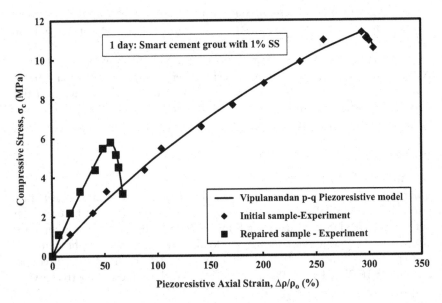

Figure 10.14 Comparing the predicted and measured compressive stress–piezoresistive axial strain relationships for the 1-day cured smart cement before and after repairing using the smart cement grout with 1% silicate after one day of curing.

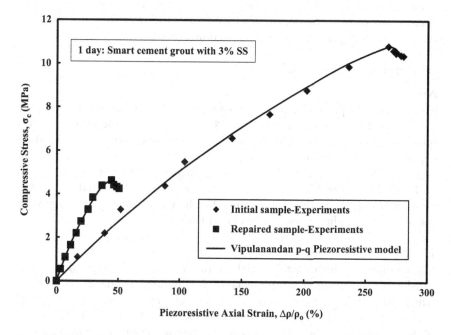

Figure 10.15 Comparing the predicted and measured compressive stress–piezoresistive axial strain relationships for 1-day cured smart cement before and after repairing using the smart cement grout with 3% sodium silicate.

at failure. After submerging the damaged specimen for three hours in the grout, it was cured for one day and tested under compression load. The percentage weight change in the repaired specimen was 0.07%. The strength of the repaired specimen was 5.80 MPa, a 51% strength recovery.

3% SS

Smart cement grout with 3% SS was used to repair the sensing cement specimen with a strength of 10.82 MPa with 268% piezoresistive axial strain at failure. After submerging the damaged specimen for three hours in the grout, it was cured for one day and tested under compression load. The percentage weight change in the repaired specimen was 0.05%. The compressive strength of the repaired specimen was 4.63 MPa, a 43% strength recovery.

7 Days Cured

Smart cement specimens (0.38 w/c ratio) cured for seven days were used for the repairing study after failing the specimen under compression loading (Vipulanandan et al. 2014a, b; 2015a, b). The compressive strength of the seven-day cured smart cement varied from 18.48 MPa to 19.75 MPa, as summarised in Table 10.5. The piezoresistivity axial strain for the seven-day cured smart cement at peak stress varied from 218% to 276%, and the data on the repaired specimens is summarised in Table 10.5.

(a) Piezoresistive Behavior

No Additive

The changes in piezoresistive axial strain $(\Delta\rho/\rho)\%$ at an applied stress of 5 MPa for the repaired sample was 14% compared to the smart cement, which was 64%, a recovery of 22%. The piezoresistive axial strain at failure $(\Delta\rho/\rho)_f\%$ for the repaired specimen was 58% at peak stress of 13.82 MPa, as summarised in Table 10.5. The sensing property change in smart cement at 13.82 MPa was 176%. Hence, smart grout repair resulted in a 33% recovery of piezoresistive axial strain the change in sensing property with the recovered applied stress, as shown in Figure 10.16.

1% SS

The changes in piezoresistive axial strain $(\Delta\rho/\rho)\%$ at an applied stress of 5 MPa for the repaired sample was 18% compared to the smart cement, which was 59%, a recovery of 35%. The change in piezoresistive axial strain at failure $(\Delta\rho/\rho\%)_f$ for the repaired specimen was 62% at peak stress of 14.57 MPa. The sensing property change in sensing cement at 14.57 MPa was 167%. Hence, smart grout repair resulted in a 37% recovery of piezoresistive axial strain the change in sensing property with the recovered applied stress, as shown in Figure 10.17.

3% SS

The change in piezoresistive axial strain $(\Delta\rho/\rho)\%$ at an applied stress of 5 MPa for the repaired sample was 17% compared to the sensing cement 50%, a recovery of 34%. The change in piezoresistive axial strain at failure $(\Delta\rho/\rho)_f\%$ for the repaired specimen was 56% at peak stress of 12.74 MPa. The sensing property change in smart cement at 12.74 MPa was 135%. Hence, smart grout repair resulted in a 41% recovery of the piezoresistive axial strain with the applied stress, as shown in Figure 10.18, the highest percentage of change in sensing property with the recovered applied stress.

(b) Strength

No Additive

Sensing cement grout was used to repair the sensing cement specimen with a strength of 19.75 MPa with 276% piezoresistive axial strain at failure. After submerging the damaged specimen for three hours in the grout, it was cured for one day and tested under compression load. The compressive strength of the repaired specimen was 13.82 MPa, as shown in Figure 10.19, a 70% strength recovery.

Table 10.5 Piezoresistive model parameters for the seven-day cured smart cement repaired with grouts

Mix Type	Curing Time (day)	Strength σ_{cf} (MPa)	Piezoresistivity at peak stress, $(\Delta\rho/\rho_o)_f$ (%)	p_2	q_2	R^2	RMSE (MPa)	Strength Regain (%)	Piezoresistivity Regain* (%)
Initial smart cement	7 days	19.75	276	0.017	0.825	0.99	0.30	N/A	N/A
Repaired cement with grout (H, w/c = 0.8 only)		13.82	58	0.028	0.629	0.99	0.07	70	21
Initial smart cement	7 days	18.92	238	0.09	0.777	0.99	0.31	N/A	N/A
Repaired cement with grout (H, w/c = 0.8, SS = 1%)		14.57	62	0.03	0.73	0.98	0.31	77	26
Initial smart cement	7 days	18.48	218	0.048	0.838	0.99	0.21	N/A	N/A
Repaired cement with grout (H, w/c = 0.8, SS = 3%)		12.74	56	0.05	0.704	0.99	0.19	69	26

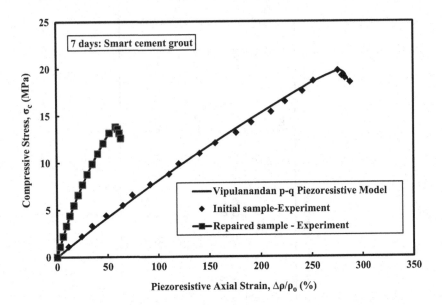

Figure 10.16 Comparing the predicted and measured compressive stress–piezoresistive axial strain relationships for 7-day cured smart cement before and after repairing using the smart cement grout.

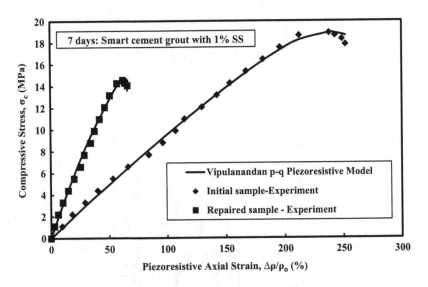

Figure 10.17 Comparing the predicted and measured compressive stress–piezoresistive axial strain relationships for the 7-day cured smart cement before and after repairing using the smart cement grout with 1% sodium silicate.

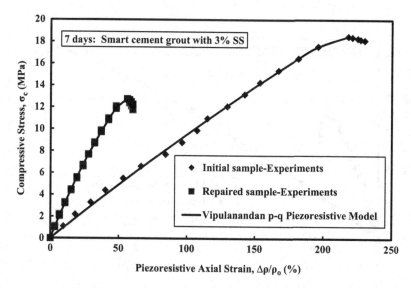

Figure 10.18 Comparing the predicted and measured compressive stress–piezoresistive axial strain relationships for 7-day cured smart cement before and after repairing using the smart cement grout with 3% sodium silicate.

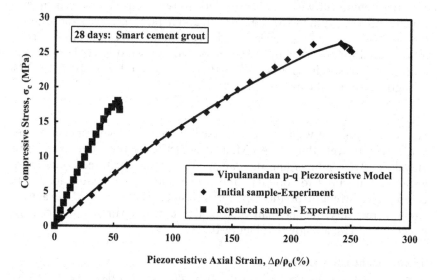

Figure 10.19 Comparing the predicted and measured compressive stress–piezoresistive axial strain relationships for 28-day cured smart cement before and after repairing using the smart cement grout.

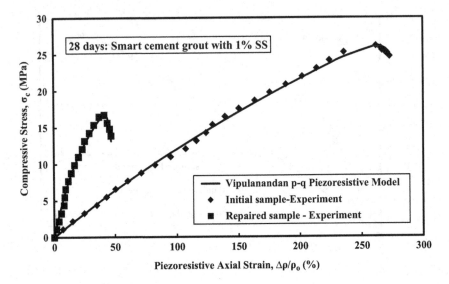

Figure 10.20 Comparing the predicted and measured compressive stress–piezoresistive axial strain relationships for the 28-day cured smart cement before and after repairing using the smart cement grout with 1% sodium silicate.

1% SS

Smart cement grout with 1% SS was used to repair the sensing cement specimen with a strength of 18.92 MPa with 238% piezoresistive axial strain at failure. After submerging the damaged specimen for three hours in the grout, it was cured for one day and tested under compression load. The strength of the repaired specimen was 14.57 MPa, as shown in Figure 10.20, a 77% strength recovery, the highest strength recovery.

3% SS

Smart cement grout with 3% SS was used to repair the sensing cement specimen with a strength of 18.48 MPa with 218% piezoresistive axial strain at failure. After submerging the damaged specimen for three hours in the grout, it was cured for one day and tested under compression load. The compressive strength of the repaired specimen was 12.74 MPa, a 69% strength recovery, higher than the one-day cured repaired smart cement, which was 43%, as shown in Figure 10.21.

Twenty-eight Days Cured

Smart cement specimens (0.38 w/c ratio) cured for 28 days were used for the repairing study after failing the specimen under compression loading (Vipulanandan et al. 2014a, b; 2015a, b). The compressive strength of the 28-day cured smart cement varied from 26.11 MPa to 26.74 MPa, as

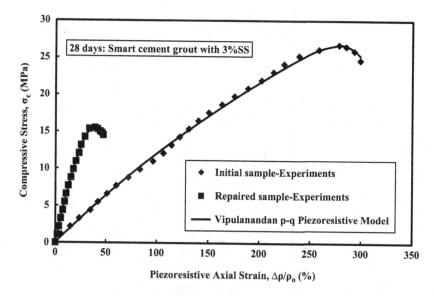

Figure 10.21 Comparing the predicted and measured compressive stress–piezoresistive axial strain relationships for 28-day cured smart cement before and after repairing using the smart cement grout with 3% sodium silicate.

summarised in Table 10.6. The piezoresistivity axial strain for the 28-day cured smart cement at peak stress varied from 241% to 278%, and the data on the repaired specimens is summarised in Table 10.6.

(a) Piezoresistive Behavior

No Additive

The change in piezoresistive axial strain $(\Delta\rho/\rho)\%$ at an applied stress of 5 MPa for the repaired sample was 12% compared to the smart cement, which was 33%, a recovery of 36%. The piezoresistive axial strain at failure $(\Delta\rho/\rho)_f\%$ for the repaired specimen was 53% at peak stress of 18.05 MPa, as summarised in Table 10.4. The sensing property change in smart cement at 18.05 MPa was 137%. Hence, smart grout repairing resulted in a 39% recovery of piezoresistive axial strain, the change in sensing property with the recovered applied stress, as shown in Figure 10.19. This is the highest percentage of recovery of piezoresistive axial strain for the 28-day cured smart cement.

1% SS

The change in piezoresistive axial strain $(\Delta\rho/\rho)\%$ at an applied stress of 5 MPa for the repaired sample was 8% compared to the smart cement, which was 40%, a recovery of 20%. The change in piezoresistive axial strain at failure $(\Delta\rho/\rho\%)_f$ for the repaired specimen was 42% at peak stress

Table 10.6 Piezoresistive model parameters for the 28-day cured smart cement repaired with grouts

Mix Type	Curing Time (day)	Strength σ_{cf} (MPa)	Piezoresistive Axial Strain, $(\Delta\rho/\rho_o)_f$ (%)	p_2	q_2	R^2	RMSE (MPa)	Strength Regain (%)	Piezoresistivity Regain* (%)
Initial smart cement	28 days	26.54	241	0.02	0.672	0.99	0.42	N/A	N/A
Repaired cement with grout (H, w/c = 0.8 only)		18.05	53	0.009	0.747	0.99	0.17	68	22
Initial smart cement	28 days	26.11	262	0.02	0.735	0.99	0.35	N/A	N/A
Repaired cement with grout (H, w/c = 0.8, SS = 1%)		16.71	42	0.02	0.487	0.99	0.49	64	16
Initial smart cement	28 days	26.74	278	0.03	0.707	0.99	0.31	N/A	N/A
Repaired cement with grout (H, w/c = 0.8, SS = 3%)		15.50	39	0.17	0.59	0.99	0.14	58	14

of 16.71 MPa. The sensing property change in sensing cement at 16.71 MPa was 139%. Hence, smart grout repair resulted in a 30% recovery of piezoresistive axial strain the change in sensing property with the recovered applied stress, as shown in Figure 10.20.

3% SS

The change in piezoresistive axial strain $(\Delta\rho/\rho)$% at an applied stress of 5 MPa for the repaired sample was 8% compared to the sensing cement at 40%, a recovery of 20%. The change in piezoresistive axial strain at failure $(\Delta\rho/\rho)_f$% for the repaired specimen was 39% at peak stress of 15.5 MPa. The sensing property change in smart cement at 15.5 MPa was 131%. Hence, smart grout repair resulted in a 30% recovery of the piezoresistive axial strain with the applied stress, as shown in Figure 10.21.

(b) Strength

No Additive

Sensing cement grout was used to repair the sensing cement specimen with a strength of 26.54 MPa with 241% piezoresistive axial strain at failure. After submerging the damaged specimen for three hours in the grout, it was cured for one day and tested under compression load. The compressive strength of the repaired specimen was 18.05 MPa, a 68% strength recovery, the highest strength recovery for the 28-day cured smart cement.

1% SS

Smart cement grout with 1% SS was used to repair the sensing cement specimen with a strength of 26.11 MPa with 262% piezoresistive axial strain at failure. After submerging the damaged specimen for three hours in the grout, it was cured for one day and tested under compression load. The strength of the repaired specimen was 16.71 MPa, a 64% strength recovery.

3% SS

Smart cement grout with 3% SS was used to repair the sensing cement specimen with a strength of 26.74 MPa with 278% piezoresistive axial strain at failure. After submerging the damaged specimen for three hours in the grout, it was cured for one day and tested under compression load. The percentage weight change in the repaired specimen was 0.05%. The compressive strength of the repaired specimen was 15.50 MPa, a 58% strength recovery.

10.6 RELATIONSHIP BETWEEN SS CONCENTRATIONS AND STRENGTH/PIEZORESISTIVITY

The strength of the smart grouts with and without SS were measured at different curing times. The addition of SS decreased the compressive strength of the cement grout. The relationship between the compressive strength of the cement grout and SS concentration has been modeled using the Vipulanandan correlation model and the relationship is as follows:

$$\sigma_{cf} = (\sigma_c)_o - S/(E + FS)$$ (10.15)

where

σ_{cf} = Compressive strength of the grout (MPa),
$(\sigma_c)_o$ = Compressive strength of the grout without SS (MPa), and
S = Concentration of sodium meta-silicate (% by weight).

Parameters E and F are model parameters, and parameter E represents the initial rate of change and parameter F determines the ultimate strength. The experimental results matched very well, as shown in Figure 10.22 with the proposed model with coefficient of determination (R^2) of 0.99. For the one-day strength test, parameters E and F were found as 0.74 MPa^{-1} and 0.63 MPa^{-1}. For the seven-day strength test, parameters E and F were found as 0.17 MPa^{-1} and 0.16 MPa^{-1}. For the 28-day strength test, parameters E and F were found as 0.17 MPa^{-1} and 0.15 MPa^{-1}.

The piezoresistivity at failure of the grout specimen made with and without SS was observed. The addition of SS decreased the piezoresistivity

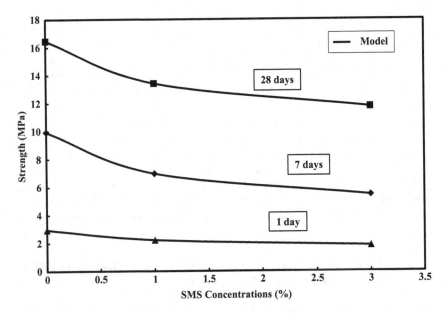

Figure 10.22 Relationship between compressive strength and SS concentrations.

at failure of the cement grout. The relationship between the piezoresistive axial strain at failure of the cement grout and the SS concentration was modeled with the Vipulanandan correlation model and the relationship is as follows:

$$(\Delta\rho/\rho_o)_f = (\Delta\rho/\rho_o)_o - S/(G + HS) \qquad (10.16)$$

where

> $(\Delta\rho/\rho_o)_f$ = Piezoresistive axial strain at failure (%),
> $(\Delta\rho/\rho_o)_o$ = Piezoresistive axial strain at failure for the grout without SS (%), and
> S = Concentration of sodium meta-silicate (% by weight).

Parameters G and H are model parameters, and parameter G represents the initial rate of change and parameter H determines the ultimate piezoresistive axial strain. Experimental results matched very well, as shown in Figure 10.23, with the proposed model with coefficient of determination (R^2) of 0.99. For the one-day test, parameters G and H were found as 0.009 and 0.017. For the seven-day test, parameters G Ƒ and H Ƒ were found as 0.013 and 0.011. For the 28-day strength test, parameters G and H were found as 0.008 and 0.010.

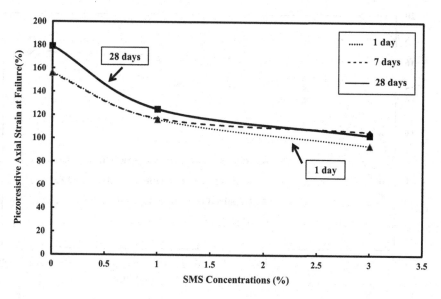

Figure 10.23 Relationship between piezoresistive axial strain at failure and SS concentrations.

10.6.1   Repaired Strength and Cured Age of Smart Cement

The strength of the repaired smart cements varied with the type of grout and also the cured age of the smart cement. The relationship between the repaired smart cement compressive strength of the type of cement grout and the curing time was modeled using the Vipulanandan correlation model, as follows:

$$\sigma_c = t/(J + Kt) \tag{10.17}$$

where

σ_c = Compressive strength of the repaired smart cement (MPa) and
t = Cured time of smart cement (day).

For the cement specimens repaired with grout, experimental results also matched very well, as shown in Figure 10.24, with the proposed model with coefficient of determination (R^2) of 0.95–0.99. For the specimen repaired with smart cement grout only, parameters J and K were found as 0.046 MPa^{-1} and 0.058 MPa^{-1}. For the specimen repaired with smart cement grout with 1% SS, parameters J and K were found as 0.110 MPa^{-1} and 0.055 MPa^{-1}. For the specimen repaired with smart cement grout with 3% SS, parameters J and K were found as 0.149 MPa^{-1} and 0.058 MPa^{-1}.

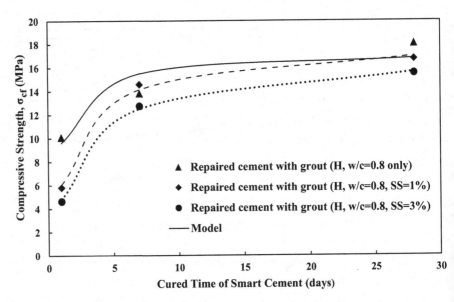

Figure 10.24 Relationship between the recovered compressive strength and the cured time of the repaired smart cement using the Vipulanandan correlation model.

10.6.2 Repaired Piezoresistive Axial Strain at Failure and Cured Age of Smart Cement

The piezoresistivity at failure of the repaired specimen with various grouts was measured up to 28 days. The piezoresistive axial strain at failure varied from 39% to 62%, as shown in Figure 10.25. In all cases, the seven-day cured smart cement had the highest repaired piezoresistive axial failure strain. With curing time increases, the piezoresistivity at failure of the cement grout changed. The relationship between the piezoresistivity at failure of the cement grout and the curing time has been modeled with the Vipulanandan correlation model, as follows:

$$(\Delta\rho/\rho_o)_f = (\Delta\rho/\rho_o)_1 - t/(L + Mt) \qquad (10.18)$$

where

$(\Delta\rho/\rho_o)_f$ = Piezoresistive axial strain at failure (%),
$(\Delta\rho/\rho_o)_1$ = Piezoresistive axial strain at failure of the grout after 1 day (%), and
t = Curing time (day).

For the smart cement specimens repaired with grouts, the experimental results matched well, as shown in Figure 10.25, with the proposed model, with coefficient of determination (R^2) of 0.95–0.99. For the specimen

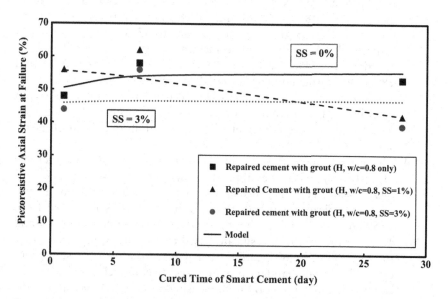

Figure 10.25 Relationship between the recovered piezoresistive axial at failure and the cured time of the repaired smart cement.

repaired with smart cement grout only, parameters L (day) and M were found as 0.270 day and 0.126. For the specimen repaired with smart cement grout with 1% SS, parameters L and M were found as 2.97 day and -0.035. For the specimen repaired with smart cement grout with 3% SS, parameters L and M were found as 0.158 day and 0.367.

10.7 SUMMARY

Based on the testing and modeling of the sensing cement grout with and without SS cured under room condition (23°C, relative humidity 50%), the following conclusions are advanced:

1   The initial sensing resistivity property (ρ_o) of the smart cement grout decreased from 1.08 Ohm-m to 0.52 Ohm-m with 3% sodium silicate (SS), a 51% decrease. The minimum sensing property (ρ_{min}) of the sensing cement grout decreased from 1.04 Ohm-m to 0.54 Ohm-m with 1% SS, a 44% decrease. The changes in the electrical sensing property were higher than the changes in the unit weight of the cement. Hence, the electrical sensing property can also be used for quality control of smart cement grout curing.

2   Electrical resistivity during the curing for the cement grout with the addition of SS followed a similar trend where it first dropped to a minimum value and then gradually increased with time. The electrical property indexes ($RI_{24\ hour}$, $RI_{7\ days}$, $RI_{28\ days}$) for the smart cement grout were 108%, 492%, and 801%, but for the smart cement grout with 3% SS, these indices reduced to 37%, 212%, and 702% respectively. Also, the smart cement grout with 3% SS reduced the resistivity by 74%, 79%, and 65% after one day, seven days, and 28 days of curing compared to that of the smart cement grout.

3   The curing of the grouts was modeled using the Vipulanandan p-q curing model, and the model predicted the experimental results very well based on the coefficient of determination and the root-mean-square error (RMSE).

4   The smart cement grouts with a w/c ratio of 0.8 showed increase in resistivity with the applied compressive stress verifying the piezoresistive cement concept. Without any addition of sodium silicate (SS), piezoresistive axial strain at peak stress increased from 155% to 179% with curing time, which was reduced to 116% and 125% with 1% sodium silicate based on the curing time. The addition of 3% sodium silicate further reduced the piezo sensing property. The Vipulanandan p-q piezoresistive model predicated the compressive stress–change in the sensing property relationship of the smart cement grout with and without SS very well. Also, the changes

in the piezoresistive properties with the curing time were modeled using the Vipulanandan correlation model.

5 The repaired smart cement showed recovered piezoresistive behavior based on the grout type used for repairs and the age of the cured smart cement that was used for repairs. The strength regain varied from 43% to 84%, and the piezoresistive axial strain regain varied from 14% to 42% based on the grout type and the cured smart cement.

6 The variation of strength and secant piezoresistive modulus with time was modeled using the Vipulanandan correlation model, and the model parameters were linearly related to the SS content.

Chapter 11

Smart Foam Cement

11.1 BACKGROUND

Cement applications require materials to be multifunctional. Hence, cement must be modified or treated to enhance different properties such as hydration, thermal conductivity, fluid loss, resistivity, and sensitivity (Vipulanandan et al. 2016b). Lightweight cement slurries with foam additions are used in multiple applications including onshore and offshore deep wells installed in varying geological formations with lower rock strengths (Cobb et al. 2002). Also, foam cement has reduced thermal conductivity, making it a better insulator. It is also important to model the behavior of foam cement for real-time monitoring.

In the past, numerous attempts have been made to use various types of cements in different types of fragile geological formations. The use of ordinary lightweight cements additives such as bentonite, fly ash, and silicates has proved to have limited success when applied to exceptionally fluctuating pressure and temperature environments and lost circulation zones in wells and buried storage facilities (Harms et al. 1985; Harness et al. 1992; Vipulanandan et al. 1996a and 1997b; Labibzadeh 2010; Shadravan et al. 2012). Foam cement is a versatile and economical lightweight cementing slurry that develops useful compressive strength and can be used for primary cementing in fragile formations and lost circulation zones (Rickard et al. 1985). Foam cement is a lightweight and thermal insulation material consisting of cement matrix with a porosity structure created by injecting preformed foam into cement slurry during the mixing process (Wang et al. 2015).

11.2 HISTORY OF FOAM CEMENT

Cellular concrete was first developed in Stockholm, Sweden, in the early 1900s. The original material was known as 'gas concrete', to be used in producing heat-insulated building materials. This led to the development of related lightweight concretes, which are now known as cellular concrete, foamed concrete, aerated concrete, and autoclaved cellular concrete. After World War II, this technology quickly spread to different parts of the world, mostly Europe and the Soviet Union. The applications were for economically large-size structural panel units. These were used in site reconstruction and low-rise structures. It was not until the late 1950s when this was introduced to the United States as foamed or cellular concrete. The applications were for floor, roof, and wall units. Having low compression strengths, this product was limited to fills and insulation only. The insulating property of cement with foam has encouraged its use in the oil and gas well industry.

The practical lower density limit for conventionally extended slurries is between 12.9 and 14.1 kN/m³ (11 and 12 ppg). Below this limit, the compressive strength is too low and the permeability is too high to provide adequate zonal isolation. Foamed cements are less expensive slurries and are easier to design. In addition, foam cement can be mixed at lower densities and yet maintain better properties (Marriot 2005; Vipulanandan 2016b). Foamed cement also has several advantages in addition to its low density. It has relatively high compressive strength with lower density, causing less damage to water-sensitive formations and also reduces the chance of annular gas flow. Since the density is less, and cement losses to potential producing zones are less, increased well productivity may be a benefit.

Advantages of Using Foam Cement
The enhanced mechanical properties of foam cement will allow more flexibility to the cement sheath to respond to the effects of excessive temperatures in the wellbore, therefore maintaining cement sheath integrity and providing zonal isolation and casing protection (Cobb et al. 2002; Marriot 2005). Foamed cements can help support both primary and remedial cementing functions for wells located offshore and onshore, as summarisad as follows:

- Foam cements with multiple enhanced physical and mechanical properties offer a low-density alternative to conventional cements.
- With some cement additives, foaming creates a synergistic effect that enhances the properties of the other additives. This effect is evident with some fluid-loss additives, lost-circulation materials, and latex.
- The density of foamed cement is a variable. Its ductility allows for expansion and pressure maintenance during hydration, thus helping provide long-term zonal isolation of wells. As the cement expands, it can fill washed-out hole sections and megadarcy lost-circulation zones without formation breakdown.
- The improved mud-removal capacity of foamed cement also helps enhance zonal isolation. Because of its ductility, foamed cement can provide casing support for the life of the well.

The following are some of the advantages of foam cement over other lightweight cements for various applications:

- Lower permeability
- More flexibility, lower brittleness and modulus
- Better thermal insulator with lower thermal conductivity
- Low densities
- Can be designed for required compressive strength
- Fluid-loss control
- Higher mud removal and cleaning efficiency
- Cost effective

Importantly, foam cement establishes a tighter bond for a reliable annular seal in wells and around pipelines because the nitrogen bubbles help prevent shrinkage while the cement slurry hydrates. A foamed system, due to its expansion properties, also accommodates challenging wellbore geometries such as washouts.

Carbon dioxide (CO_2) is the primary greenhouse gas emitted through human activities around the world. About 5% to 10% of the world's carbon dioxide emission is attributed to the cement manufacturing industry. Since about the period 1750 to 1950, the CO_2 concentration on the earth has increased from about 275 ppm to 310 ppm. In the last 50 years, the CO_2 concentration has increased from approximately 310 ppm to 375 ppm. The use of foam cement reduces the amount of cement used in the cement slurries. The use of 50% quality foam cement reduces cement usage by about 50%, thus reducing carbon dioxide emission and costs as well. In general applications, foam cement has at least 20% foam by volume.

In this study, commercially available preformed foam was added to smart cement to investigate their effects on the density, thermal conductivity, rheology, curing, fluid loss, piezoresistive behavior, and mechanical properties.

11.3 MATERIALS AND METHODS

11.3.1 Materials

Cement
To study the effect of foam on smart cement, Class H oil-well cement was used.

Foam
Commercially available foam was used. The major constituents of the foam included water, stearic acid, triethanolamine, isobutane, laureth-23, sodium lauryl sulfate, and propane.

Smart Cement
Commercially available oil-well cement (Class H) was modified with carbon fibers to make it a chemo-thermo-piezoresistive material. The cement was

modified by adding about 0.04% of carbon fibers (conductive fibers, CF), by weight of cement, and the water-to-cement (w/c) ratio was 0.38. Smart cement technology can monitor the changes in cement at very high magnification of about 1,000 times after one-day curing (Vipulanandan et al. 2014b, c and 2015b–e). The main property of interest was piezoresistivity, the change in the resistivity of the cement with the application of stress. Also, rheological properties are not affected by the addition of carbon fibers (Vipulanandan et al. 2015d).

11.3.2 Foam Characterisation

The commercially available preformed foam was first tested to evaluate the stability of the foam. The stability of foam is defined by the half-life of the foam, which is the time for the foam to collapse to half its height. The half-life of the foam was greater than 24 hours, showing it to be one of the most stable foams. The foam experienced a drop in height of 0.3 mL in the first six hours. The drop was measured using a tube, as shown in Figure 11.1. The oxidation reduction potential (ORP), representing the tendency of chemical species to acquire electrons and thereby be reduced, was found to be -55 mV, showing that the foam was in a reduced condition. The conductivity of the foam measured using a conductivity probe was in the range of 75 to 120 μS/cm, and the resistivity was in the range of 80 to 135 $\Omega \cdot$m. This showed the presence of air voids since the resistivity of the air was in the range of 1.30×10^{16} to 3.30×10^{16} $\Omega \cdot$m.

Foam Cement

In this study, Class H cement with a w/c ratio of 0.38 was used. The samples were prepared according to API standards. To improve the sensing properties and piezoresistive behavior of the cement modified with 0.04% of carbon fibers (CF) by the weight of cement was mixed with all the samples. Commercially available foam was added to the cement slurry and mixed for at least for five minutes. After mixing, cement slurries with and without foam were used for rheological, fluid loss, curing, and piezoresistivity studies. The foam percentage was varied from 0% to 20% by weight of the sample (cement and water), and the sample cross-sectional images are shown in Figure 11.2.

11.3.3 Methods of Testing

Cement Mixing Procedure

Smart cement with a w/c ratio of 0.38 was used in this study. The cement slurries were prepared using hand mixing, followed by mixing using a high-shear blender-type mixer with bottom-driven blades. First, a measured amount of mixing water was poured into the container. Next, a small amount

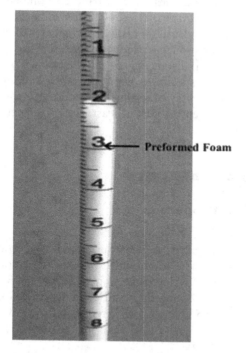

Figure 11.1 Foam stability test.

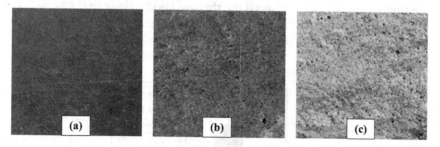

Figure 11.2 Images of the smart foam cement (a) cement only, (b) 5% foam cement, (c) 20% foam cement (20X magnification image).

of conductive filler was added to the water, and then a small amount of cement was added and mixed. Then, little by little, cement and carbon fibers were gradually added to the container and mixed for about one minute so that they could be properly dispersed in the mixing water. Final mixing was done with blender at a low speed for about four to five minutes to ensure homogeneity. The desired weight of foam based on the weight of the slurry was added to the mixture. Then the slurry was mixed for about five minutes until the foam bubbles were distributed homogeneously. Mixing was done at room temperature of 23±2 °C. After mixing, cement specimens were prepared using cylindrical molds with a diameter of 50 mm (2 inches) and a height of 100 mm (4 inches). Four conductive wires were placed in all of the molds. The vertical distances between the two wires of the specimens were kept the same. All specimens were capped to minimise moisture loss and were cured up to the day of testing for quantifying the piezoresistivity. In order to have consistent results, at least three specimens were prepared for each type of mix.

11.4 RESULTS AND ANALYSES

11.4.1 Physical Properties

(a) Density
The unit weight of the smart cement slurry with a w/c ratio of 0.38 was 1.95 g/cc (19.1 kN/m³). The addition of 5% foam (based on total weight of the smart cement slurry) reduced the unit weight to 1.53 g/cc(15 kN/m³), a 21.5% reduction. Increasing the foam content to 20% reduced the unit weight to 1.07 g/cc (10.5 kN/m³), a 45% reduction, as shown in Figure 11.3.

(b) Thermal Conductivity
The thermal conductivity of the smart cement slurry with a w/c ratio of 0.38 was 0.802 W/mK. The addition of 5% foam (based on total weight of the smart cement slurry) reduced the thermal conductivity to 0.482 W/mK, 40% reduction. Increasing the foam content to 20% reduced the thermal conductivity to 0.284 W/mK, a 65% reduction, as shown in Figure 11.4.

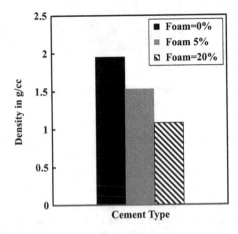

Figure 11.3 Comparing the density of smart cement with foam content.

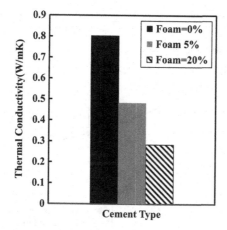

Figure 11.4 Comparing the thermal conductivity of smart cement with foam content.

Table 11.1 Initial bulk resistivity parameters for smart cement with various foam contents

Foam Content (%)	Initial Resistivity, ρ_o (Ω-m)
Foam = 0%	1.05 $\pm$ 0.03
Foam = 5%	1.2 $\pm$ 0.02
Foam = 20%	2.04 $\pm$ 0.04

(c) Electrical Resistivity

The initial resistivity of the smart cement immediately after mixing was 1.05 Ω·m, as summarised in Table 11.1 and also shown in Figure 11.5. With the addition of 5% foam, the initial resistivity increased to 1.20 Ω·m, a 14.2% increase. With the addition of 20% foam, the initial resistivity increased to 2.04 Ω·m, a 94.3% increase, as summarised in Table 11.1 and also shown in Figure 11.5. Hence, electrical resistivity will be good a quality control parameter in the field.

(d) Property Correlation: Density versus Thermal Conductivity

The variation of thermal conductivity with density was found to be nonlinear, as shown in Figure 11.6. Hence, it is represented using the Vipulanandan correlation model, as follows:

$$K = \frac{\gamma}{G_2 - H_2\gamma}$$

(11.1)

where K is the thermal conductivity in W/mK and

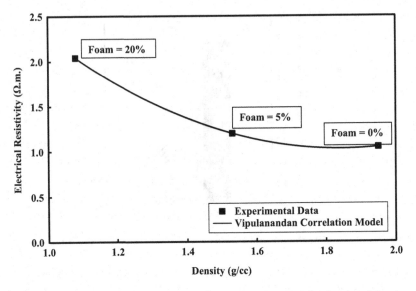

Figure 11.5 Correlation between the initial electrical resistivity to the density of the smart foam cements.

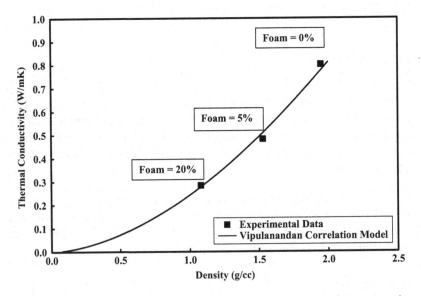

Figure 11.6 Correlation between thermal conductivity and density of the smart foam cements.

γ is the slurry density in g/cc.

The model parameters G_2 and H_2 were 5.68 gmK/Wcm³ and 1.66 mK/W respectively. The root-mean-square error (RMSE) for the thermal conductivity–density correlation model was 0.003 W/mK.

11.4.2 Piezoresistivity of the Slurry

The smart cement slurries with and without foam were subjected to pressure up to 4 MPa in a high-pressure, high-temperature chamber (HPHT) (Figure 8.3) to investigate the piezoresistive behavior.

0% Foam: The resistivity of the smart cement slurry decreased non-linearly with the increase in the pressure, as shown in Figure 11.7. At 4 MPa pressure, the decrease in the resistivity was 8%, indicating the piezoresistivity characteristics of the smart cement slurry. In the slurry, the shear stress effect was very minimal, and that is why the resistivity decreased with the applied pressure.

5% Foam: The resistivity of the smart cement slurry with 5% foam decreased nonlinearly with an increase in the pressure, as shown in Figure 11.7. At 4 MPa pressure, the decrease in resistivity was 12%, indicating the piezoresistivity characteristics of the smart cement slurry. With 5% foam, the piezoresistivity characteristics of the smart foam cement slurry increased by 50%.

20% Foam: The resistivity of the smart cement slurry with 20% foam decreased nonlinearly with an increase in the pressure, as shown in Figure 11.7. At 4 MPa pressure, the decrease in resistivity was 22%, indicating the piezoresistivity characteristics of the smart cement slurry. With 20% foam, the piezoresistivity characteristics of the smart foam cement slurry increased by 175%, making the smart foam cement more sensing.

Vipulanandan Piezoresistivity Slurry Model

The slurry piezoresistivity behavior under confined compressive pressure in the HPHT device was modeled using the Vipulanandan piezoresistive slurry model and the relationship is as follows:

$$p = \frac{\frac{\Delta\rho}{\rho_o}}{A + B\frac{\Delta\rho}{\rho_o}} \tag{11.2}$$

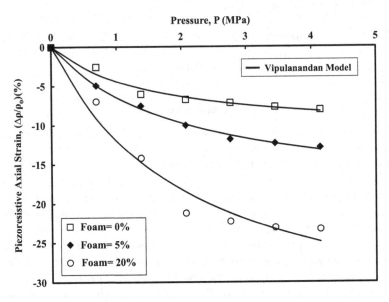

Figure 11.7 Measured and predicted stress–piezoresistive axial strain relationship for the smart cement slurry with and without foam after mixing.

where $(\Delta\rho/\rho_o)$ is the change in bulk resistivity (decrease, negative); p is the pressure applied in MPa; and A, B are model parameters. The model parameters A and B were also sensitive to the foam contents. The parameter A increased with the foam content and the parameter B decreased with the foam content, as summarised in Table 11.2.

11.4.3 Rheological Properties

Shear stress–shear strain rate relationships for the smart foam cement were predicated using the Vipulanandan rheological model and compared with the Herschel–Bulkley model, as shown in Figure 11.8. Also, all the model parameters are summarised in Tables 11.3 and 11.4.

Herschel–Bulkley Model (1926)
The Herschel–Bulkley model (Eqn. 11.3) defines fluid rheology with three parameters, and the model is as follows:

$$\tau = \tau_o + k * (\dot{\gamma})^n \tag{11.3}$$

where τ, τ_o, $\dot{\gamma}$, k and n represent the shear stress, yield stress, shear strain rate, correlation parameter, and flow behavior index respectively.

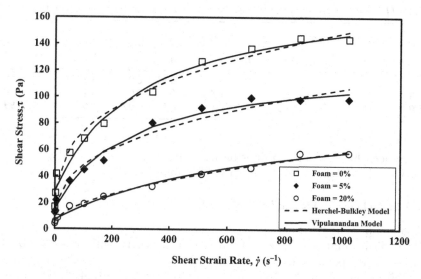

Figure 11.8 Measured and predicted rheological shear stress–shear strain rate relationship for the smart cement with and without foam.

Table 11.2 Vipulanandan piezoresistive slurry model parameters for smart foam cement slurries

Materials	Piezoresistive Slurry Model Parameters			
	A (%.(MPa)$^{-1}$)	B (MPa)$^{-1}$	R^2	RMSE
Foam = 0%	-20.26	0.0875	0.97	0.158
Foam = 5%	-15.13	0.051	0.99	0.145
Foam = 20%	-8.29	0.026	0.97	0.145

Vipulanandan Rheological Model (2014)

The Vipulanandan model for the shear stress–shear strain rate relationship for the oil-well cement slurry was investigated (Vipulanandan et al. 2014a). The relationship is as follows:

$$\tau - \tau_{o2} = \frac{\dot{\gamma}}{C + D * \dot{\gamma}} \tag{11.4}$$

where τ: shear stress (Pa); $\dot{\gamma}$: shear strain rate (s^{-1}); and τ_{o2} is the yield stress (Pa). The model parameters are C (Pa. s)$^{-1}$ and D (Pa)$^{-1}$.

Also when

$$\dot{\gamma} \to \infty \quad then \quad \tau_{max} = \tau_y + \frac{1}{D} \tag{11.5}$$

Table 11.3 Herschel–Bulkley model parameters for smart cement slurries with foam

| Foam Content (%) | Herschel–Bulkley Model (1926) | | | | |
	τ_{ol} (Pa)	K (Pa s^n)	n	RMSE (Pa)	R^2
0	29 ± 1.7	16 ± 0.03	0.31 ± 0.01	5.38	0.98
5	13 ± 1.2	8.7 ± 0.05	0.36 ± 0.01	5.29	0.98
20	5 ± 0.95	1.2 ± 0.06	0.55 ± 0.02	1.58	0.99

Hence, this model has a limit on the maximum shear stress that will be produced at a relatively high shear strain rate.

(a) Herschel–Bulkley Model (1926)

0% Foam

The shear-thinning behavior of smart cement slurry with a w/c ratio of 0.38 at a temperature of 23°C was tested and modeled using the Herschel–Bulkley model (Eqn. (11.3)) up to a shear strain rate of 1,024 s^{-1} (600 rpm), as shown in Figure 11.8. The coefficient of determination (R^2) was 0.98, as summarised in Table 11.3. The root-mean-square error (RMSE) was 5.38 Pa. The average yield stress (τ_{o1}) for the cement slurry at a temperature of 23oC was 29 Pa. The model parameter k for the smart cement slurry with a w/c ratio of 0.38 at 23°C was 16.7 Pa.s^n, as summarised in Table 11.3. The model parameter n for the smart cement slurry was 0.31.

5% Foam

The shear-thinning behavior of smart cement slurry with 5% foam at a temperature of 23oC was tested and modeled using the Herschel–Bulkley model (Eqn. (11.3)) up to a shear strain rate of 1,024 s^{-1} (600 rpm), as shown in Figure 11.8. The coefficient of determination (R^2) was 0.98, as summarised in Table 11.3. The root-mean-square error (RMSE) was 5.29 Pa. The average yield stress (τ_{o1}) for the cement slurry at a temperature of 23°C was 13 Pa, a reduction of 55% compared to smart cement without any foam. The model parameter k for the 5% smart foam cement slurry was 8.7 Pa.s^n, a reduction of 11% compared to smart cement without any foam. The model parameter n for the 5% smart foam cement slurry was 0.36.

20% Foam

The shear-thinning behavior of smart cement slurry with 20% foam at a temperature of 23°C was tested and modeled using the Herschel–Bulkley model (Eqn. (11.3)) up to a shear strain rate of 1,024 s^{-1} (600 rpm), as shown in Figure 11.8. The coefficient of determination (R^2) was 0.99, as summarised in Table 11.3. The root-mean-square error (RMSE) was 1.58 Pa. The average yield stress (τ_{o1}) for the cement slurry at a temperature of

Table 11.4 Vipulanandan rheological model parameters for smart cement slurries with and without foam

Foam Content (%)	Vipulanandan Model (2014)					
	τ_{o2} (Pa)	C (Pa s^{-1})	D (pa)$^{-1}$	τ_{max} (Pa)	RMSE (Pa)	R^2
0	28 ± 2.5	1.97 ± 0.1	0.006	195 ± 3.6	5.68	0.98
5	15 ± 1.3	2.34 ± 0.1	0.009	126 ± 2.1	3.82	0.99
20	7 ± 1.1	8.49 ± 0.15	0.01	107± 1.6	2.3	0.99

23°C was 5 Pa, a reduction of 83% compared to smart cement without any foam. The model parameter k for the 20% foam cement slurry was 1.2 Pa.s^n, a reduction of 93% compared to smart cement without any foam. The model parameter n for the smart foam cement slurry was 0.55, an increase of 77% compared to smart cement without any foam.

(b) Vipulanandan Rheological Model (2014)

0% Foam

The shear-thinning behavior of smart cement slurry with a w/c ratio of 0.38 at a temperature of 23°C was tested and modeled using the Vipulanandan model (Eqn. (11.4)) up to a shear strain rate of 1,024 s^{-1} (600 rpm), as shown in Figure 11.8. The coefficient of determination (R^2) was 0.98, as summarised in Table 11.4. The root-mean-square error (RMSE) was 5.68 Pa. The average yield stress (τ_{o1}) for the cement slurry at a temperature of 23°C was 28 Pa. The model parameter C for the cement slurry with a w/c ratio of 0.38 at 23°C was 1.97 Pa.s^{-1}, as summarised in Table 11.4. The model parameter D for the cement slurry was 0.006 Pa^{-1}.

5% Foam

The shear-thinning behavior of smart cement slurry with 5% foam at a temperature of 23°C was tested and modeled using the Vipulanandan model (Eqn. (11.4)) up to a shear strain rate of 1,024 s^{-1} (600 rpm), as shown in Figure 11.8. The coefficient of determination (R^2) was 0.99, as summarised in Table 11.4. The root-mean-square error (RMSE) was 3.82 Pa. The average yield stress (τ_{o2}) for the cement slurry at a temperature of 23°C was 15 Pa, a reduction of 46% compared to smart cement without any foam. The model parameter C for the 5% foam cement slurry was 2.34 Pa.s^{-1}, an increase of 19% compared to smart cement without any foam. The model parameter D for the 5% foam cement slurry was 0.009.

20% Foam

The shear-thinning behavior of smart cement slurry with 20% foam at a temperature of 23°C was tested and modeled using the Vipulanandan model

(Eqn. (11.4)) up to a shear strain rate of 1,024 s⁻¹ (600 rpm), as shown in Figure 11.8. The coefficient of determination (R^2) was 0.99, as summarised in Table 11.4. The root-mean-square error (RMSE) was 2.3 Pa. The average yield stress (τ_{o1}) for the cement slurry at a temperature of 23°C was 7 Pa, a reduction of 75% compared to smart cement without any foam. The model parameter C for the 20% foam cement slurry was 8.49 Pa.s⁻¹, an increase of 331% compared to smart cement without any foam. The model parameter D for the foam cement slurry was 0.01.

Maximum Shear Stress ($\tau_{max.}$)

Based on Eqn. (11.5), the Vipulanandan model has a limit on the maximum shear stress ($\tau_{max.}$) the slurry will produce at a relatively high rate of shear strains. The τ_{max} for smart cement slurries with 0%, 5%, and 20% foam at a temperature of 23°C were 195 Pa, 126 Pa, and 107 Pa respectively, as summarised in Table 11.4. Hence, with 20% foam, the maximum shear stress was reduced by 45%, another highly sensitive parameter.

11.4.4 Fluid Loss

The total fluid loss from the smart cement slurry at a pressure of 0.7 MPa (100 psi) with a w/c ratio of 0.38 was 134 mL, as shown in Figure 11.9. The addition of 5% foam (based on total weight of the cement slurry) reduced

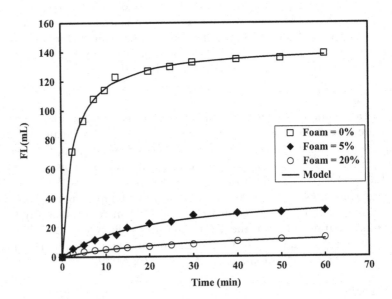

Figure 11.9 Measured and predicted fluid loss–time relationship for the smart cement with and without foam.

Table 11.5 Fluid loss model parameters

Materials	Fluid Loss Model Parameters				
	E (min.(mL)$^{-1}$	F (mL)$^{-1}$	(FL)$_{max}$ (mL)	R^2	RMSE
Foam = 0%	0.017	0.007	142.9	0.99	1.75
Foam = 5%	0.458	0.023	43.5	0.99	1.09
Foam = 20%	1.536	0.056	17.9	0.97	0.61

the fluid loss to 31.4 mL, a 77% reduction. Increasing the foam content to 20% reduced the fluid loss to 13.7 mL, a 90% reduction. The Vipulanandan fluid loss model was used to predict the fluid loss with time (Vipulanandan et al. 2019a, 2020b)

11.4.4.1 Vipulanandan Fluid Loss Model for Cement Slurry

Fluid loss from the slurry was modeled using the Vipulanandan fluid loss model (Vipulanandan et al. 2014d) and the relationship is as follows:

$$FL = t / (E + F * t) \tag{11.6}$$

where FL is the fluid loss in mL; t is the time elapsed in minutes; and E, F are the model parameters. The model parameters are summarised in Table 11.5 and the model predictions are compared to the experimental results in Figure 11.9. Also, the maximum fluid loss (FL_{max}) predicted by the model is summarised in Table 11.5. With the addition of 20%, the FL_{max} reduced from 142.9 mL to 17.9 mL, an 87% reduction.

11.4.5 Curing

The resistivity of the cement slurry with curing time up to 28 days was measured from the samples cured under room condition. The normal trend of the resistivity of the curing cement is that the resistivity decreases up to a certain time (t_{min}) and reaches to a minimum resistivity (ρ_{min}) and then increases with time. The initial resistivities of the cement with 0%, 5%, and 20% foam were 1.05 Ω·m, 1.2 Ω·m, and 2.04 Ω·m, as summarised in Table 11.6. Hence, the addition of 20% foam increased the initial resistivity of the smart foam cement by 94%. Also the RI_{24} for smart foam cement with 0%, 5%, and 20% foam content were 230%, 133%, and 6% respectively, as summarised in Table 11.6.

At least three specimens were tested, and the average results for the variation in the resistivity with time up to one day and 28 days of curing are

Table 11.6 Curing resistivity parameters for smart cement with various foam contents

Foam Content (%)	Initial Resistivity, ρ_o (Ω-m)	ρ_{min} (Ω-m)	t_{min} (min)	ρ_{24h} (Ω-m)	RI_{24} (%)	p_1	q_1	t_0 (min)
0	1.05 ± 0.03	0.96 ± 0.01	98 ± 5.0	3.46 ± 0.02	230	0.65	0.281	80
5	1.2 ± 0.02	1.15 ± 0.03	55 ± 4.5	2.8 ± 0.02	133	0.41	0.151	50
20	2.04 ± 0.04	1.81 ± 0.04	43 ± 3.5	2.16 ± 0.02	6	0.19	0.082	15

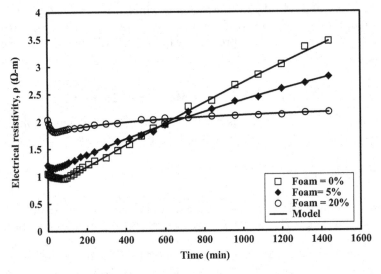

Figure 11.10 Measured and predicted curing resistivity–time relationship for the smart cement with and without foam up to 1 day of curing.

shown in Figure 11.10 and Figure 11.11. The Vipulanandan curing model was used to predict the experimental results and the relationship is as follows:

$$\frac{1}{\rho} = \left(\frac{1}{\rho_{min}}\right)\left[\frac{\left(\dfrac{t+t_o}{t_{min}+t_o}\right)}{q_1+(1-p_1-q_1)*\left(\dfrac{t+t_o}{t_{min}+t_o}\right)+p_1*\left(\dfrac{t+t_o}{t_{min}+t_o}\right)^{\frac{q_1+p_1}{p1}}}\right], \quad (11.7)$$

where, ρ is the electrical resistivity ($\Omega\cdot$m); ρ_{min} is the minimum electrical resistivity (Ω-m); and t_{min} is the time to reach the minimum electrical resistivity (ρ_{min}). The model parameters were t_o, p_1 (t), and q_1 (t), and t was the curing time (min). The parameter q_1 represents the initial rate of change in the resistivity.

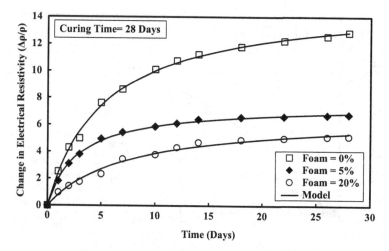

Figure 11.11 Measured and predicted curing resistivity–time relationship for the smart cement with and without foam up to 28 days of curing.

There are three characteristic resistivity parameters that can be used in monitoring the curing (hardening process) of cement. These resistivity parameters are the initial resistivity (ρ_o), minimum electrical resistivity (ρ_{min}), and time to reach the minimum resistivity (t_{min}).

The resistivity showed an increasing trend with curing time, as shown in Figure 11.10 and Figure 11.11, which was modeled using the Vipulanandan curing model. The model parameters are summarised in Table 11.6. The resistivity of smart cement with zero foam after 28 days of curing was 14.5 $\Omega \cdot m$; hence, the percentage change in resistivity was 1,283%. The resistivity of the 20% foam cement after 28 days of curing was 12.5 $\Omega \cdot m$, and the percentage change in resistivity in 28 days was 520%, as shown in Figure 11.10. The two model parameters p_1 and q_1 decreased with the addtion of foam, as summarised in Table 11.6.

11.4.6 Compressive Behavior

(a) Strength

0% Foam

The average compressive strength of the smart cement after one day of curing was 10.3 MPa, as shown in Figure 11.12. The average compressive strength of the smart cement after 28 days of curing was 19.7 MPa, a 91% increase in strength, as summarised in Table 11.7 and also shown in Figure 11.13.

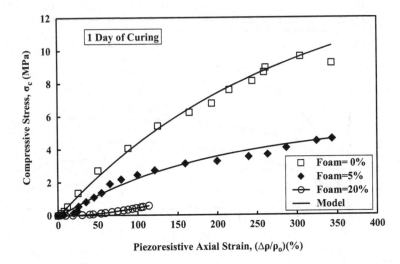

Figure 11.12 Compressive piezoresistive stress–strain relationships for the smart foam cement after 1 day of curing.

Table 11.7 Summary of piezoresistive and compressive properties of smart cement with and without foam

Foam Content	Compressive Strength (MPa)		Piezoresistive axial strain at failure $(\Delta\rho/\rho)$ (%)		Model Parameters			
					p_2		q_2	
	1 Day	28 Days	1 Day	28 Days	1 Day	28 Days	1 Day	28 Days
0	10.29	19.65	343	252.4	0.1	0.001	0.435	0.4133
5	4.6	14.6	304	187.9	0.083	0.001	0.274	0.586
20	0.57	5.27	113	98.2	0.403	0.47	0.605	1.07

5% Foam

The average compressive strength of the smart foam cement after one day of curing was 4.6 MPa, about a 55% reduction in strength compared to smart cement without any foam, as shown in Figure 11.12. The average compressive strength of the smart foam cement after 28 days of curing was 14.6 MPa, a 217% increase in strength. The 28th day compressive strength of the smart foam cement was 26% less than the compressive strength of smart cement without any foam, as summarised in Table 11.7 and also shown in Figure 11.13.

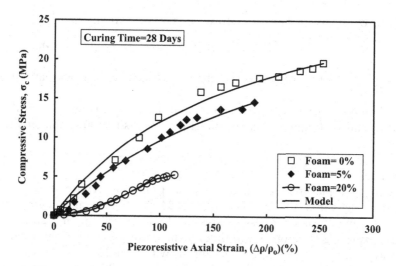

Figure 11.13 Compressive piezoresistive stress–strain relationships for the smart foam cement after 28 days of curing.

20% Foam

The average compressive strength of the smart foam cement after one day of curing was 0.57 MPa, about a 94% reduction in strength compared to smart cement without any foam (Figure 5.10). The average compressive strength of the smart foam cement after 28 days of curing was 5.27 MPa, a 825% increase in strength. The 28th day compressive strength of the smart foam cement was 73% less than the compressive strength of smart cement without any foam, as summarised in Table 11.7 and also shown in Figure 11.13.

(b) Piezoresistivity

The addition of 0.04% CF substantially improved the piezoresistive behavior of the cement. Based on the experimental results, the Vipulanandan piezoresistive p-q model was used to predict the change in the electrical resistivity of cement with applied stress for one day and 28 days of curing. The model is defined as

$$\frac{\sigma_c}{\sigma_{cf}} = \left[\frac{\dfrac{x}{x_f}}{q_2 + (1 - p_2 - q_2)\dfrac{x}{x_f} + p_2 \left(\dfrac{x}{xf} \right)^{\left(\frac{p_2}{p_2 - q_2} \right)}} \right] \qquad (11.8)$$

where σ_c stress (MPa); σ_{cf} stress at failure (MPa); $x = \left(\dfrac{\Delta \rho}{\rho_o}\right) * 100 =$ Percentage of change in piezoresistive axial strain due to the stress; $x_f = \left(\dfrac{\Delta \rho}{\rho_o}\right)_f * 100 =$ Percentage of change in electrical resistivity at failure; $\Delta \rho$: change in electrical resistivity; ρ_o: Initial electrical resistivity ($\sigma = 0$ MPa); and p_2 and q_2 are piezoresistive model parameters that are summarised in Table 11.7.

0% Foam

The average piezoresistive axial strain at peak compressive stress of the smart cement after one day of curing was 343%. The average piezoresistive axial strain at peak compressive stress of the smart cement after 28 days of curing was 252%, about 1,250 times higher than the compressive strain at failure (0.2%). On average, the piezoresistivity after 28 days of curing was 12.8%/MPa. The model parameter p_2 after one and 28 days of curing were 0.1 and 0.001 respectively. The model parameter q_2 after one and 28 days of curing were 0.435 and 0.413 respectively, as summarised in Table 11.7.

5% Foam

The average piezoresistive axial strain at peak compressive stress of the smart cement after one day of curing was 304%, a 11% reduction compared to smart cement without any foam. The average piezoresistive axial strain at peak compressive stress of the smart cement after 28 days of curing was 188%, about 940 times higher than the compressive strain at failure (0.2%) for smart cement without any foam. On average, the piezoresistivity after 28 days of curing was 12.9%/MPa, comparable to smart cement without any foam. The model parameter p_2 after one and 28 days of curing were 0.083 and 0.001 respectively. The model parameter q_2 after one and 28 days of curing were 0.274 and 0.586 respectively, as summarised in Table 11.7.

20% Foam

The average piezoresistive axial strain at peak compressive stress of the smart cement after one day of curing was 113%, a 67% reduction compared to smart cement without any foam. The average piezoresistive axial strain at peak compressive stress of the smart cement after 28 days of curing was 98%, about 490 times higher than the compressive strain at failure (0.2%) for smart cement without any foam. On average, the piezoresistivity after 28 days of curing was 18.6%/MPa, higher than for smart cement without any foam. The model parameter p_2 after one and 28 days of curing were 0.403 and 0.47 respectively. The model parameter q_2 after one and 28 days of curing were 0.605 and 1.07 respectively, as summarised in Table 11.7.

11.5 SUMMARY

In this study, smart foam cement was tested for density, thermal conductivity, rheological, fluid loss, curing, and compressive piezoresistivity behavior. Smart foam cement piezoresistive behavior was studied up to 28 days of curing at room condition (temperature 23±2 °C). Based on the experimental results and analytical modeling of the rheological and piezoresistivity behavior on smart cement foam, the following conclusions are advanced:

1. The addition of foam reduced the density and thermal conductivity of the smart foam cement. The thermal conductivity was correlated to the density using the Vipulanandan correlation model.

2. The addition of 20% foam increased the initial electrical resistivity of the smart cement from 1.05 $\Omega \cdot m$ to 2.04 $\Omega \cdot m$, a 93% increase. The addition of foam changed the curing path of the smart cement based on the resistivity measurements.

3. Smart foam cement slurry was piezoresistive, and with the applied compressive pressure the resistivity decreased due to very low shear effect. The change increased with the foam content and applied pressure. With the addition of 20% foam, the resistivity change at 4 MPa (600 psi) increased from 8% for the smart cement slurry with no foam to 22% with 20% foam, about a 175% increase in the piezoresistivity.

4. A rheological test showed that the smart foam cement had shear-thinning behavior and the Vipulanandan rheological model predicted the test results very well compared to the Herschel–Bulkley model.

5. The yield stress and maximum shear stress limit reduced with the addition of foam. The maximum shear stress limit for smart cement slurry was 195 Pa, and it reduced to 107 Pa with the addition of 20% foam, a 45% reduction.

6. The total fluid loss for the smart cement at 0.7 MPa (100 psi) pressure was reduced from 134 mL to 13 mL with the addition of 20% foam, about a 90% reduction.

7. The average compressive strength of the smart foam cement after one day of curing was 0.57 MPa, about a 94% reduction in strength compared to smart cement without any foam. The average compressive strength of the smart foam cement after 28 days of curing was 5.27 MPa, a 825% increase in strength. The 28th day compressive strength of smart foam cement was 73% less than the compressive strength of smart cement without any foam.

8. The solidified smart cement foam was piezoresistive and verified piezoresistive cement theory, and with applied compressive stress, the resistivity increased. The average percentage change in resistance

at peak compressive stress of the smart foam cement with 20% foam after one day of curing was 113%, a 67% reduction compared to smart cement without any foam. The average percentage change in resistance at peak compressive stress of the smart foam cement after 28 days of curing was 98%, about 490 times higher than the compressive strain at failure (0.2%) for smart cement without any foam. On average, the piezoresistivity per unit applied stress after 28 days of curing was 18.6%/MPa, higher than smart cement without any foam.

Chapter 12

Concrete with Smart Cement Binder

12.1 BACKGROUND

Concrete with varying amounts of aggregate contents are used in the construction of different structures such as highways, roads, houses, bridges, pipes, dams, canals, storage facilities, tunnels, missile silos, and nuclear waste containers (Wei et al. 1995; Gani, 1997; Liu et al. 2001 and 2003; Wei et al. 2012; Vipulanandan et al. 2001a, 2002, 2005 b, c, d, e, 2009a, 2014f). To attain the required levels of safety and durability of such structures, mixing proportions and especially aggregate content must be adjusted according to the application in order to achieve the mechanical property requirements which will significantly affect the performance during its life time (Hou et al. 2017; Vipulanandan 2011). In preparing the concrete and cement slurries, the water-to-cement (w/c) ratios were varied from 0.38 to 0.6 based on the mixing method, constituents of the concrete mix and applications (Vipulanandan et al. 2008, 2015a, 2016a, 2018a). The flow characteristics of concrete mixes are tested using the flow cone method (ASTM D6994). Many different testing techniques such as ultrasound, fiber optics, electronic microscopy and X-ray diffraction have been used to study the aging of concretes and for damage detection (Ramachandran 1984; Sayers et al. 1993; Mantrala et al. 1995; Zhang et al. 2009; Petro et al. 2012; Vipulanandan 2018b). Also the ultrasonic method has been used to characterise the cement slurries (Sayers et al. 1993) The ultrasonic method is difficult to adopt in some field conditions where accessibility becomes an issue in piles, wells, dams, canals and pipelines. (Vipulanandan et al. 2018e, f)

Research studies on the electrical resistivity of the concretes over the past two decades have shown that out of all the different methods of characterising the concretes, electrical resistivity is one of the highly sensitive and economical nondestructive method to monitor the serviceability of the concretes throughout the entire service life (Saleem et al. 1996; McCarter et al. 2000 and 2003; Polder 2001; Sett et al. 2004; Shi

2004; Presuel-Moreno at al. 2013; Vipulanandan et al. 2016b; Hou et al. 2017). Also studies have investigated the surface resistivity as an indicator of concrete chloride penetration resistance for a wide range of concrete compositions (Ramezanianpour et al. (2011). Studies on conductive cement-based materials which are also piezoresistive have proved that the electrical resistivity decreased with the applied compressive stresses and were nonlinear and rate dependent (Banthia et al. 1994; Azhari et al. 2012). Vipulanandan et al. (2008, 2009a, 2011, 2014b, c, d) characterised modified cementitious and polymer composites using electrical resistivity measurements. Also, newly developed piezoresistive smart cement studies have shown that the resistivity increased with applied compressive and tensile stresses (Chapter 5). Electrical resistivity is a sensitive parameter for the characterisation of concrete, which can be affected by the curing conditions, concrete composition, and cement type (Presuel-Moreno et al. 2013; Medeiros-Junior et al. 2016; Chu et al. 2016). As previously mentioned, aggregates play a vital role in concrete strength, and the effects on the electrical resistivity should be considered when using concrete for various monitoring applications. Aggregate content, size, and type of aggregates are parameters that have been documented as affecting the electrical resistivity of concrete using different methods of measurements (Princigallo et al. 2003). Also, the effect of up to 30% coarse aggregates (based on the total weight) on the electrical resistivity of concrete using AC measurements at 1 Hz frequency has been investigated (Hou et al. 2017).

In this study, smart cement was used as the binder in concrete to evaluate the sensitivity of chemo-thermo-piezoresistive smart cement concrete developed for real-time monitoring of the performance of the concrete. Also, property changes with time were measured and quantified, and new models were developed to predict concrete behavior and verified using the experimental data.

12.2 MODELING

(a) Mixture Theory

The mixture theory is used to predict composite properties using the constituents in a series of or in parallel resistance circuits, and in concrete it will be the gravel and cement binder (McLachlan et al. 1990; McCarter et al. 2000 and 2003). The series equation describes the resistances of the two constituents, and the electrical resistivity of the concrete (A) is represented as follows:

$$\rho = \rho_s \left(\frac{V_s}{V_G + V_s} \right) + \rho_G \left(\frac{V_G}{V_G + V_s} \right) \qquad (12.1)$$

where, V_G is the aggregate volume content; V_s is the smart cement slurry volume content; ρ_G is the electrical resistivity of the dry aggregate, which was about 900 Ω.m; and ρ_s is the electrical resistivity of the smart cement slurry, which was 1.02 Ω.m immediately after mixing. The parallel circuit equation describes the two resistive materials in parallel, and the electrical resistivity of the composite (ρ) is represented as follows:

$$\frac{1}{\rho} = \frac{1}{\rho_s}\left(\frac{V_s}{V_G + V_s}\right) + \frac{1}{\rho_G}\left(\frac{V_G}{V_G + V_s}\right) \tag{12.2}$$

The applicability of these two predictions will be verified using the experimental results.

(b) Vipulanandan Composite Resistivity Model
In order to quantify the effect of aggregate content on the initial electrical resistivity of the smart concrete, the following composite resistivity model was used and the relationship is as follows:

$$\rho - \rho_s = (\rho_G - \rho_s)\left[\frac{\dfrac{\left(\dfrac{V_G}{V_G+V_s}\right)}{\left(\dfrac{V_G}{V_G+V_s}\right)_c}}{q_4 + (1-p_4-q_4)\dfrac{\left(\dfrac{V_G}{V_G+V_s}\right)}{\left(\dfrac{V_G}{V_G+V_s}\right)_c} + p_4\left[\dfrac{\left(\dfrac{V_G}{V_G+V_s}\right)}{\left(\dfrac{V_G}{V_G+V_c}\right)_c}\right]^{\left(\frac{p_4}{p_4-q_4}\right)}}\right] \tag{12.3}$$

where σ is the electrical resistivity in Ω.m for the concrete mixture; V_G is the gravel volume content; and V_s is the smart cement slurry volume content. The electrical resistivity of the dry gravel (ρG) (ρ_G) was 900 Ω.m; p_4 and q_4 are model parameters; and the maximum gravel volume content $\left(\dfrac{V_G}{V_G+V_s}\right)_c$ was equal to 1 in this study.

(c) Vipulanandan Pulse Velocity-Curing time Model

In order to represent the development of the pulse velocity of the hardening concrete with curing time, the newly developed Vipulanandan pulse velocity-curing time model was used and the relationship is as follows:

$$V_p - V_{p0} = \left(V_{pc} - V_{p0}\right)\left[\frac{\dfrac{t}{t_c}}{q_3 + \left(1 - p_3 - q_3\right)\dfrac{t}{t_c} + p_3\left(\dfrac{t}{t_c}\right)^{\frac{p_3}{(p_3 - q_3)}}}\right], \qquad (12.4)$$

where V_{p0} is the initial pulse velocity of the concrete slurries; V_{pc} is the ultimate pulse velocity of the concretes; t_c is the curing time of the smart cement concrete, which was 28 days in this study; and the material parameters p_3 and q_3 are influenced by the material constituents, density, and testing conditions.

(d) Vipulanandan Curing Model

In order to represent the electrical resistivity changes during curing of the concrete, the Vipulanandan curing model was used (Vipulanandan et al. 2015e and 2016a), and the relationship is as follows:

$$\frac{1}{\rho} = \frac{1}{\rho_{min}}\left[\frac{\left(\dfrac{t + t_0}{t_{min} + t_0}\right)}{q_1 + \left(1 - p_1 - q_1\right)\left(\dfrac{t + t_0}{t_{min} + t_0}\right) + p_1\left(\dfrac{t + t_0}{t_{min} + t_0}\right)^{\left(\frac{p_1 + q_1}{p_1}\right)}}\right], \qquad (12.5)$$

where ρ is the electrical resistivity of the cement in $\Omega.m$; ρ_{min} is the minimum electrical resistivity in $\Omega.m$; t_{min} is the time to reach the minimum electrical resistivity (ρ_{min}); t represents the curing time; t_0 is a model parameter influenced by the initial resistivity; and p_1 and q_1 are time-dependent model parameters as follows:

$$p_1 = p_{10} + \frac{t}{A + B.t} \qquad (12.6)$$

$$q_1 = q_{10} + \frac{t}{A' + B'.t} \qquad (12.7)$$

where p_{10}, A, B, q_{10}, A', and B' are model material parameters influenced by composition and material properties such as density and initial resistivity and curing temperature.

(e) Vipulanandan Piezoresistivity Model

In order to represent the piezoresistive behavior of the hardened piezoresisitive concrete, the Vipulanandan piezoresistivity model (Vipulanandan et al. 2015, 2016) was used, and the relationship is as follows:

$$\sigma = \cfrac{\sigma_{max} \times \left(\cfrac{\left(\cfrac{\Delta \rho}{\rho} \right)}{\left(\cfrac{\Delta \rho}{\rho} \right)_0} \right)}{q_2 + \left(1 - p_2 - q_2\right) \times \left(\cfrac{\left(\cfrac{\Delta \rho}{\rho} \right)}{\left(\cfrac{\Delta \rho}{\rho} \right)_0} \right) + p_2 \times \left(\cfrac{\left(\cfrac{\Delta \rho}{\rho} \right)}{\left(\cfrac{\Delta \rho}{\rho} \right)_0} \right)^{\left(\frac{p_2 + q_2}{p_2} \right)}}$$

(12.8)

where σ_{max} is the maximum stress; $(\Delta \rho / \rho)_0$ is the piezoresistivity of the hardened cement under the maximum stress; and p_2 and q_2 are model parameters influenced by the material properties and testing conditions.

12.3 MATERIALS AND METHODS

Sample Preparation

In this study, table top concrete blenders were used to prepare the cement and concrete specimens. Smart cement (Portland cement Type 1 (ASTM C 150)) specimens were prepared using a w/c ratio of 0.38 to minimise the bleeding, and aggregates were added and mixed for at least 10 minutes. Concrete specimens were prepared using up to 75% coarse aggregates based on the total volume of the concrete. After mixing, the cement and concrete were placed in 100 mm height and 50 mm diameter cylindrical molds, and two conductive flexible wires 1 mm in diameter (representing the probes) were placed 50 mm apart vertically to measure the electrical resistance. The specimens were cured up to 28 days under relative humidity of 90%. At least three specimens were tested under each condition, and the average values are presented in the figures, tables, and discussion.

12.4 RESULTS AND DISCUSSION

12.4.1 Physical Properties

Aggregate Particle Size Distribution

Sieve analysis test (ASTM C136) was performed to determine the gradation of the aggregates, and the gradation is shown in Figure 12.1 (Vipulanandan et al. 2009c). The median diameter, which is represented as D_{50} (ASTM; represents the size of 50% of the particles), was about 4.2 mm, representing gravel. The D_{10} was 2 mm, and hence only 10% of the particles were less than 2 mm, representing the sand.

Density

The average density of smart cement (w/c ratio of 0.38) without any gravel was 1.97 g/cm³. The density of the silica aggregates was 2.65 g/cm³, and the density of the cement was 3.14 g/cm³. The concrete densities with 10%, 25%, 40%, 50%, 60%, and 75% aggregates and smart cement binder were 2.09 g/cm³, 2.19 g/cm³, 2.35 g/cm³, 2.31 g/cm³, 2.25 g/cm³, and 2.04 g/cm³ respectively, as summarised in Table 12.1. The density of the concrete increased by 6.1%, 11.2%, 19.3%, 17.2%, 14.3%, and 3.6% for the concretes with 10%, 25%, 40%, 50%, 60%, and 75% aggregate content respectively. The density peaked at an aggregate content of 40%. Concrete is a three-phase material (Chapter 4), and based on the mixing method and aggregate contents there will be air voids influencing the density of the concrete.

12.4.2 Quality Control

It is important to develop a monitoring parameter that can be easily adopted in the field.

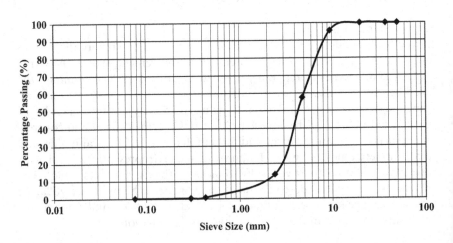

Figure 12.1 Particle size distribution of the coarse aggregates.

Table 12.1 Density of the smart concretes

Smart Concrete (by volume)	Density (g/cm³)	Changes in density (%)
No Gravel	1.97±0.02	0%
10% Gravel	2.09±0.06	6.1%
25% Gravel	2.19±0.13	11.2%
40% Gravel	2.35±0.11	19.3%
50% Gravel	2.31±0.09	17.2%
60% Gravel	2.25±0.10	14.3%
75% Gravel	2.04±0.10	3.6%

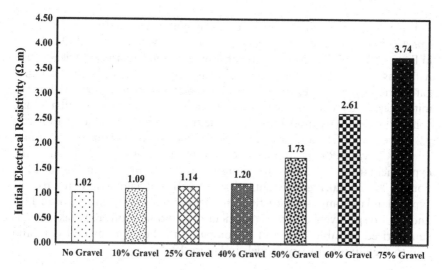

Figure 12.2 Comparing the initial electrical resistivity of the smart concretes.

Initial Resistivity

The initial electrical resistivities of the smart concretes were measured and are compared in Figure 12.2.

(a) **Smart Cement (No Gravel):** The average initial electrical resistivity of the smart cement was 1.02 Ω.m. This will be used to compare the changes in resistivity with the addition of aggregates.

(b) **10% Gravel Smart Concrete:** The average initial electrical resistivity of the 10% gravel smart concrete was 1.09 Ω.m, a 7% increase compared to the smart cement with no gravel.

(c) **25% Gravel Smart Concrete:** The average initial electrical resistivity of the 25% gravel smart concrete was 1.14 Ω.m, a 12% increase compared to the smart cement with no gravel.

(d) **40% Gravel Smart Concrete:** The average initial electrical resistivity of the 40% gravel smart concrete was 1.20 Ω.m, a 18% increase compared to the smart cement with no gravel.

(e) **50% Gravel Smart Concrete:** The average initial electrical resistivity of the 50% gravel smart concrete was 1.73 Ω.m, a 70% increase compared to the smart cement with no gravel.

(f) **60% Gravel Smart Concrete:** The average initial electrical resistivity of the 60% gravel smart concrete was 2.61 Ω.m, a 156% increase compared to the smart cement with no gravel.

(g) **75% Gravel Smart Concrete:** The average initial electrical resistivity of the 75% gravel smart concrete was 3.74 Ω.m, a 267% increase compared to the smart cement with no gravel.

The changes in the initial and the 28-day cured electrical resistivity of the smart concrete with different aggregate content were evaluated. As shown in Figure 12.2, with the low aggregate content concrete, the electrical resistivity was mainly due to the electrical resistivity of cement. With the aggregate content over 40%, the electrical resistivity increased rapidly. For the initial resistivity changes with the aggregate contents, the Vipulanandan composite resistivity model predicted the trend well with model parameters p_4 and q_4 were 3.1 and 3.2 respectively and the coefficient of determination (R^2) was 0.99, as summarised in Table 12.2 and as compared to the experimental data and also to the mixture theory models in Figure 12.3. For the 28-day curing resistivity changes with the aggregate contents, the Vipulanandan composite resistivity model parameters p_4 and q_4 were 11.5 and 19.5 respectively and the coefficient of determination (R^2) was 0.99, as summarised in Table 12.2 and as compared to the experimental data and also to the mixture theory models in Figure 12.4.

12.4.3 Curing

The electrical resistivity of concrete is influenced by the porosity and conductive ion concentrations in the pore solution in the cement binder. From the standpoint of conductivity, concrete can be regarded as a two-component composite material, pore solution and solid phase (aggregate + hydration products + unhydrated binders) (Chapter 4 and Xiao et al. 2008). During the setting of the cement, the capillary porosity and pore solutions result in changes in slurry resistivity (Liu et al. 2014; Vipulanandan et al. 2014b, c and 2015b, c, d). The pore solution resistivity decreased initially and reached a minimum resistivity of ρ_{min} at specific time of t_{min}, which is due to the increment of ionic concentration in the pore solution. With the hydration, the production of a C-S-H gel network results in an increase in the bulk resistivity (Zhang et al. 2009; Vipulanandan et al. 2014b, c).

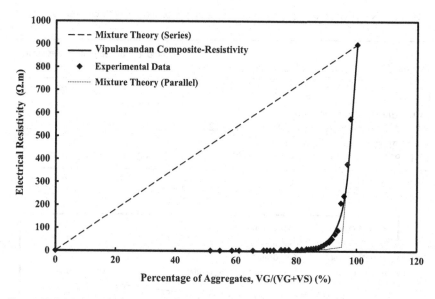

Figure 12.3 Measured and predicted initial electrical resistivity of the smart concretes with different gravel contents (volume).

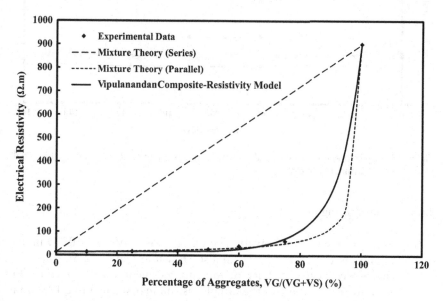

Figure 12.4 Measured and predicted electrical resistivity of the smart concretes with different gravel contents (volume) after 28 days of curing.

Table 12.2 Resistivity model parameters for the smart concretes with different gravel contents

Mixture Theory (Series)				Mixture Theory (Parallel)				Vipulanandan Composite-Resistivity Model					
ρ_S	ρ_G	R^2	RMSE	ρ_S	ρ_G	R^2	RMSE	ρ_S	ρ_G	p_4	q_4	R^2	RMSE
Immediately After Mixing													
1.0	900	NA	664	1.0	900	0.57	125	1.0	900	3.1	3.2	0.99	23
After 28 Days of Curing													
13.3	900	NA	354	13.3	900	0.99	5	13.3	900	11.5	19.5	0.99	2

NA: Not available.

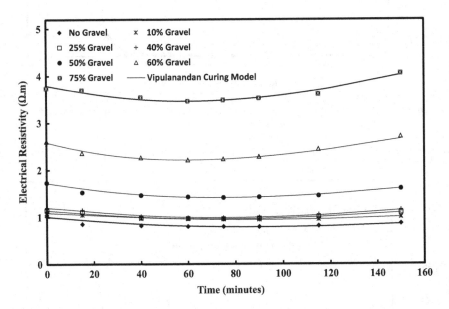

Figure 12.5 Development of electrical resistivity in the smart concretes during the initial 150 minutes of curing.

(a) One Day Curing

(a) **Smart Cement (No Gravel):** The initial resistivity was 1.02 Ω.m, as summarised in Table 12.3 and also shown in Figure 12.5. The minimum electrical resistivity of the smart cement was 0.79 Ω.m, 90 minutes after mixing of the sample, as summarised in Table 12.3 and also shown in Figure 12.5. In 24 hours the resistivity increased to 5.14 Ω.m, as shown in Figure 12.6, about a 404% increase, a clear indication of the sensitivity of resistivity for monitoring the concrete.

Table 12.3 Curing parameters for the smart concretes during the initial 24 hours of curing

Smart Concrete (by volume)	ρ_0 ($\Omega .m$)	ρ_{min} ($\Omega .m$)	t_{min} (minute)	ρ_{24} ($\Omega .m$)	$\frac{\rho_{24} - \rho_{min}}{\rho_{min}}$ %
No Gravel	1.02	0.79	90	5.14	550%
10% Gravel	1.09	0.94	90	6.52	594%
25% Gravel	1.14	0.96	75	6.88	617%
40% Gravel	1.20	0.99	75	7.25	632%
50% Gravel	1.73	1.41	75	8.63	512%
60% Gravel	2.61	2.21	60	12.97	487%
75% Gravel	3.74	3.46	60	20.01	478%

(b) **10% Gravel Smart Concrete:** The initial resistivity was 1.09 Ω.m, as summarised in Table 12.3 and also shown in Figure 12.5. The minimum electrical resistivity of the 10% gravel smart concrete increased by 20% to 0.94 Ω.m, compared to the smart cement, and also happened 90 minutes after the mixing of the sample. In 24 hours the resistivity increased to 6.52 Ω.m, as shown in Figure 12.6, about a 498% increase, a clear indication of the sensitivity of resistivity for monitoring the concrete.

(c) **25% Gravel Smart Concrete:** The initial resistivity was 1.14 Ω.m as summarised in Table 12.3 and also shown in Figure 12.5. The minimum electrical resistivity of the 25% gravel smart concrete increased by 22% to 0.96 Ω.m, compared to the smart cement. The time corresponds to the minimum resistivity of the 25% gravel smart concrete reduced by 15 minutes to 75 minutes, compared to the smart concrete with no gravel. In 24 hours the resistivity increased to 6.88 Ω.m, as shown in Figure 12.6, about a 504% increase, a clear indication of the sensitivity of resistivity for monitoring the concrete.

(d) **40% Gravel Smart Concrete:** The initial resistivity was 1.20 Ω.m, as summarised in Table 12.3 and also shown in Figure 12.5. The minimum electrical resistivity of the 40% gravel smart concrete increased by 24% to 0.99 Ω.m, compared to the smart cement. The time corresponds to the minimum resistivity of 40% gravel smart concrete reduced by 15 minutes to 75 minutes, compared to the smart concrete with no gravel. In 24 hours the resistivity increased to 7.25 Ω.m, as shown in Figure 12.6, about a 504% increase, a clear indication of the sensitivity of resistivity for monitoring the concrete.

(e) **50% Gravel Smart Concrete:** The initial resistivity was 1.73 Ω.m, as summarised in Table 12.3 and also shown in Figures 12.2 and 12.5. The minimum electrical resistivity of the 50% gravel smart concrete increased by 79% to 1.41 Ω.m, compared to the smart cement. The

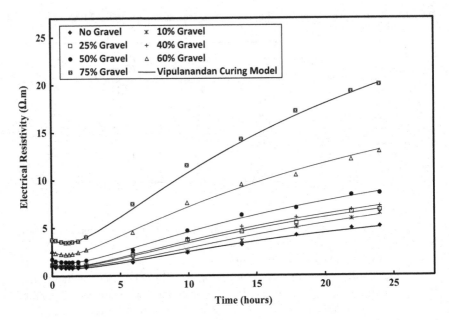

Figure 12.6 Development of electrical resistivity in the smart concretes during the initial 24 hours of curing.

time corresponds to the minimum resistivity of 50% gravel smart concrete reduced by 15 minutes to 75 minutes, compared to the smart cement with no gravel. In 24 hours the resistivity increased to 8.63 Ω.m as shown in Figure 12.6, about a 399% increase, a clear indication of the sensitivity of resistivity for monitoring the concrete.

(f) **60% Gravel Smart Concrete:** The initial resistivity was 2.61 Ω.m, as summarised in Table 12.3 and also shown in Figure 12.5. The minimum electrical resistivity of the 60% gravel smart concrete increased by 180% to 2.21 Ω.m, compared to the smart cement. The time corresponds to the minimum resistivity of 60% gravel smart concrete reduced by 30 minutes to 60 minutes, compared to the smart cement. In 24 hours the resistivity increased to 12.97 Ω.m, as shown in Figure 12.6, about a 397% increase, a clear indication of the sensitivity of resistivity for monitoring the concrete.

(g) **75% Gravel Smart Concrete:** The initial resistivity was 3.74 Ω.m, as summarised in Table 12.3 and also shown in Figures 12.2 and 12.5. The minimum electrical resistivity of the 75% gravel smart concrete increased by 339% to 3.46 Ω.m, compared to the smart cement. The time corresponds to the minimum resistivity of 75% gravel smart concrete reduced by 30 minutes to 60 minutes, compared to the smart concrete with no gravel. In 24 hours the resistivity increased to 20.01 Ω.m, as shown in Figure 12.6, about a 435% increase, a clear indication of the sensitivity of resistivity for monitoring the concrete.

(b) Electrical Resistivity Index (RI_{24})

Vipulanandan et al. (2014b, c, d, 2015e, 2016a, 2019a) suggested the resistivity index (RI_{24}) as an indicator of hydration up to 24 hours, which is the maximum percentage change in the resistivity after 24 hours of curing. This index represents the volume fraction of C-S-H, which has a significant influence on the strength of the hardening cement and concrete. The resistivity indexes for the smart concretes are compared in Figure 12.7.

(a) **Smart Cement (No Gravel):** The RI_{24} for the smart cement was 550%.

(b) **10% Gravel Smart Concrete:** The RI_{24} for the 10% gravel smart concrete increased by 8% to 594%, compared to the smart cement, as shown in Figure 12.7.

(c) **25% Gravel Smart Concrete:** The RI_{24} of the 25% gravel smart concrete increased by 12% to 617%, compared to the smart cement, as shown in Figure 12.7.

(d) **40% Gravel Smart Concrete:** The RI_{24} of the 40% gravel smart concrete increased by 15% to 632%, compared to the smart cement, as shown in Figure 12.7.

(e) **50% Gravel Smart Concrete:** The RI_{24} of the 50% gravel smart concrete decreased by 7% to 512%, compared to the smart cement, as shown in Figure 12.7.

(f) **60% Gravel Smart Concrete:** The RI_{24} of the 60% gravel smart concrete decreased by 11% to 487%, compared to the smart cement, as shown in Figure 12.7.

(f) **75% Gravel Smart Concrete:** The RI_{24} of the 75% gravel smart concrete decreased by 13% to 478%, compared to the smart cement, as shown in Figure 12.7.

(c) Twenty-eight Day Curing

During the setting of the cement in the concrete, the formation of the C-S-H gel results in higher resistivity (Liu et al. 2014; Vipulanandan et al. 2014b, c, d, 2015c, 2019a). The changes in resistivity during curing up to 28 days are compared in Figure 12.8.

(a) **Smart Cement (No Gravel):** After 28 days of curing, the electrical resistivity of the smart cement was 14.14 Ω.m, as shown in Figure 12.8. From one day to 28 days of curing, the change in resistivity was about 175%.

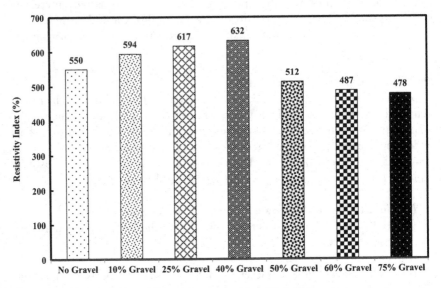

Figure 12.7 Comparing the electrical resistivity index (RI$_{24}$) of the smart concretes.

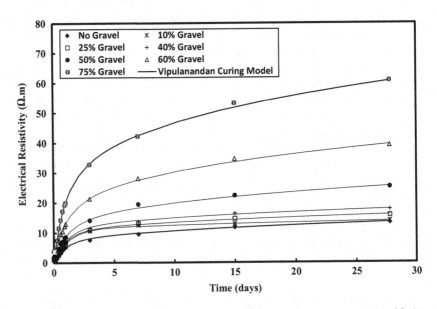

Figure 12.8 Development of electrical resistivity in the smart concretes during 28 days of curing.

(b) **10% Gravel Smart Concrete:** After 28 days of curing, the electrical resistivity of 10% gravel smart concrete increased by 7% to 15.09 Ω.m, as compared to the smart cement. From one day to 28 days of curing, the change in resistivity was about 131%.

(c) **25% Gravel Smart Concrete:** After 28 days of curing, the electrical resistivity of 25% gravel smart concrete increased by 20% to 17.00 Ω.m, as compared to the smart cement. From one day to 28 days of curing, the change in resistivity was about 147%.

(d) **40% Gravel Smart Concrete:** After 28 days of curing, the electrical resistivity of 40% gravel smart concrete increased by 35% to 19.11 Ω.m, as compared to the smart cement. From one day to 28 days of curing, the change in resistivity was about 164%.

(e) **50% Gravel Smart Concrete:** After 28 days of curing, the electrical resistivity of 50% gravel smart concrete increased by 79% to 25.26 Ω.m, as compared to the smart cement. From one day to 28 days of curing, the change in resistivity was about 193%.

(f) **60% Gravel Smart Concrete:** After 28 days of curing, the electrical resistivity of 60% gravel smart concrete increased by 177% to 39.14 Ω.m, as compared to the smart cement. From one day to 28 days of curing, the change in resistivity was about 202%.

(g) **75% Gravel Smart Concrete:** After 28 days of curing, the electrical resistivity of 75% gravel smart concrete increased by 333% to 61.24 Ω.m, as compared to the smart cement. From one day to 28 days of curing, the change in resistivity was about 206%.

12.4.4 Ultrasonic Pulse Velocity

Pulse velocities of the smart concretes with varying aggregate contents were measured during the 28 days of curing. The compressive pulse velocities were measured using 150 kHz transducers. The test results are compared to the Vipulanandan pulse velocity model predictions in Figure 12.9.

(a) **Initial Pulse Velocity (Slurry)**

(a) **Smart Cement (No Gravel):** The initial pulse velocity of the smart cement was 1,050 m/s, which is less than the pulse velocity of water because of the air voids produced during the mixing of the cement.

(b) **10% Gravel Smart Concrete:** The initial pulse velocity of the 10% gravel smart concrete increased by 3%, compared to smart cement, to 1,080 m/s, which was due to the gravel content in the slurry.

(c) **25% Gravel Smart Concrete:** The initial pulse velocity of the 25% gravel smart concrete increased by 10%, compared to smart cement, to 1,160 m/s, which was due to the gravel content in the slurry.

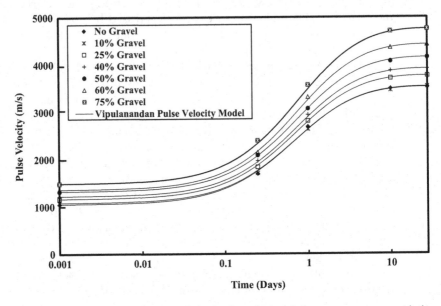

Figure 12.9 Comparing the development of pulse velocity of the smart concretes with the Vipulanandan pulse velocity model during 28 days of curing.

(d) **40% Gravel Smart Concrete:** The initial pulse velocity of the 40% gravel smart concrete increased by 16%, compared to smart cement, to 1,220 m/s, which was due to the gravel content in the slurry.

(e) **50% Gravel Smart Concrete:** The initial pulse velocity of the 50% gravel smart concrete increased by 26%, compared to smart cement, to 1,320 m/s, which was due to the gravel content in the slurry.

(f) **60% Gravel Smart Concrete:** The initial pulse velocity of the 60% gravel smart concrete increased by 30%, compared to smart concrete, to 1,360 m/s, which was due to the gravel content in the slurry.

(g) **75% Gravel Smart Concrete:** The initial pulse velocity of the 75% gravel smart concrete increased by 42%, compared to smart concrete, to 1,490 m/s, which was due to the gravel content in the slurry.

(b) Twenty-eight Day Pulse Velocity

Test results are shown in Figure 12.9, with the model predictions and the model parameters summarised in Table 12.4.

(a) **No Gravel Smart Cement (No Gravel):** The pulse velocity of the smart cement was 3,520 m/s after 28 days of curing, a 235% increase based on the initial pulse velocity. The model parameters p_3 and q_3 were 3.174 and 0.0237 respectively, as summarised in Table 12.4.

Table 12.4 The pulse velocity p-q model parameters for the smart concretes during the 28 days of curing

Smart Concrete	P_3	q_3	t_c (days)	V_{p0} (m/s)	V_{pc} (m/s)	R^2	RMSE (m/s)
No Gravel	3.174	0.0237	28	1,050	3,520	0.99	40
10% Gravel	3.125	0.0242	28	1,080	3,530	0.99	30
25% Gravel	3.021	0.0249	28	1,160	3,760	0.99	20
40% Gravel	2.984	0.0255	28	1,220	3,910	0.99	20
50% Gravel	2.942	0.0259	28	1,320	4,150	0.99	10
60% Gravel	2.881	0.0261	28	1,360	4,420	0.99	30
75% Gravel	2.851	0.0268	28	1,490	4,750	0.99	50

(b) **10% Gravel Smart Concrete:** The pulse velocity of the 10% gravel smart concrete increased by 0.3%, compared to the smart cement, to 3,530 m/s after 28 days of curing. Also the increase was 227% based on the initial pulse velocity. The model parameters p_3 and q_3 were 3.125 and 0.0242 respectively.

(c) **25% Gravel Smart Concrete:** The pulse velocity of the 25% gravel smart concrete increased by 7%, compared to the smart cement, to 3,760 m/s after 28 days of curing. Also the increase was 224% based on the initial pulse velocity. The model parameters p_3 and q_3 were 3.021 and 0.0249 respectively.

(d) **40% Gravel Smart Concrete:** The pulse velocity of the 40% gravel smart concrete increased by 11%, compared to the smart cement, to 3,910 m/s after 28 days of curing. Also, the increase was 220% based on the initial pulse velocity. The model parameters p_3 and q_3 were 2.984 and 0.0255 respectively.

(e) **50% Gravel Smart Concrete:** The pulse velocity of the 50% gravel smart concrete increased by 18%, compared to the smart cement, to 4,150 m/s after 28 days of curing. Also, the increase was 214% based on the initial pulse velocity. The model parameters p_3 and q_3 were 2.942 and 0.0259 respectively.

(f) **60% Gravel Smart Concrete:** The pulse velocity of the 60% gravel smart concrete increased by 26%, compared to the smart cement, to 4,420 m/s after 28 days of curing. Also, the increase was 225% based on the initial pulse velocity. The model parameters p_3 and q_3 were 2.881 and 0.0261 respectively.

(g) **75% Gravel Smart Concrete:** The pulse velocity of the 75% gravel smart concrete increased by 35%, compared to the smart cement, to 4,750 m/s after 28 days of curing. Also, the increase was 219% based on the initial pulse velocity. The model parameters p_3 and q_3 were 2.851 and 0.0268 respectively.

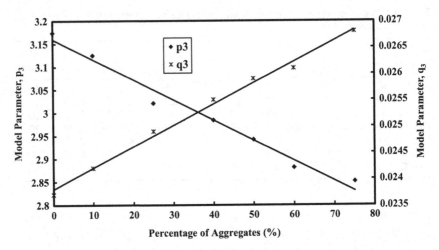

Figure 12.10 Variation of the model parameters of the pulse velocity p-q model with the aggregate contents of the smart concretes.

As shown in Figure 12.10, both model parameters p_3 and q_3 were linearly related to the gravel contents, and the relationships are as follows:

$$p_3 = \left(-0.0044 \times 10^{-5}\right)\left(\frac{V_G}{V_G + V_S}\right) + 3.1597 \qquad \text{with } R^2 = 0.98 \quad (12.9)$$

$$q_3 = \left(4 \times 10^{-5}\right)\left(\frac{V_G}{V_G + V_S}\right) + 0.0024 \qquad \text{with } R^2 = 0.99 \quad (12.10)$$

12.4.5 Resistivity versus Pulse Velocity

Monitoring of the percentage changes in the initial electrical resistivity and the compressive pulse velocity of the smart concrete with different gravel contents is shown in Figure 12.11. For the smart cement with curing, the percentage change in resistivity is compared to the compressive pulse velocity in Figure 12.12. Based on the test results, it can be concluded that electrical resistivity was more sensitive to the changes due to the aggregate content and curing compared to compressive pulse velocity. The addition of up to 75% gravel to smart cement lead to a 267% change in the initial electrical resistivity, while it was 41% for initial compressive pulse velocity. Change in the electrical resistivity of the smart cement in 28 days of curing was 1,286%, while it was only 235% for compressive pulse velocity.

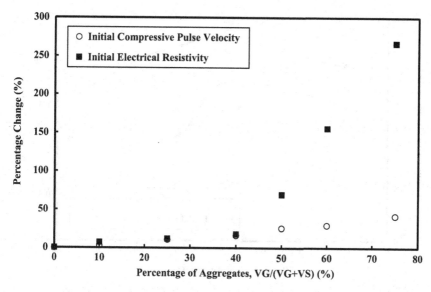

Figure 12.11 Comparing the changes in electrical resistivity and compressive pulse velocity of the smart cement for 28 days of curing.

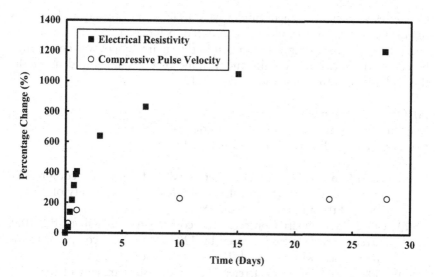

Figure 12.12 Variation of 1-day compressive strength and density of the smart concretes.

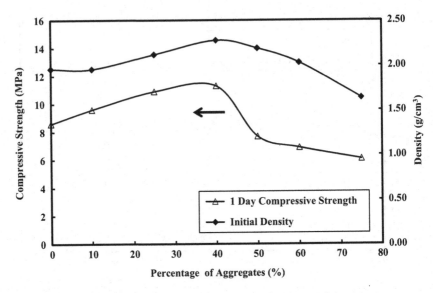

Figure 12.13 Variation of 1-day compressive strength and density of the smart concretes

12.4.6 Compressive Strength

Concretes with smart cement binder were tested after one day and 28 days of curing, and the test results are compared in Figure 12.13, Figure 12.14, and Figure 12.15. Also, the air content in the smart cement was determined based on the measured weight and volume of the sample and using the three-phase diagram calculations (Eqn. 4.12).

(a) One Day Curing
The variation of the compressive strengths with the aggregate (gravel) content are shown in Figure 12.13 and Figure 12.14.

- (a) **Smart Cement (No Gravel):** After one day of curing, the compressive strength of the smart cement was 8.6 MPa with a density of 1.97 g/cm^3, and the air content was 0.38%.
- (b) **10% Gravel Smart Concrete:** The compressive strength of the 10% gravel smart concrete increased by 12% to 9.6 MPa compared to the smart cement with no gravel. The density of the concrete was 2.09 g/cm^3.
- (c) **25% Gravel Smart Concrete:** The compressive strength of the 25% gravel smart concrete increased by 28% to 10.9 MPa, compared to the smart cement with no gravel. The density of the concrete was 2.19 g/cm^3.
- (d) **40% Gravel Smart Concrete:** The compressive strength of the 40% gravel smart concrete increased by 32% to 11.3 MPa, compared to

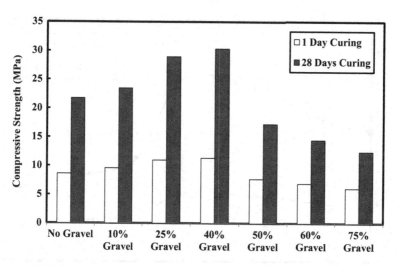

Figure 12.14 Comparing the compressive strengths of the smart concretes with varying gravel contents after 1 day and 28 days of curing.

the smart cement with no gravel. The density of the concrete was 2.35 g/cm³. This had the highest density and compressive strength, as shown in Figure 12.13.

(e) **50% Gravel Smart Concrete:** The compressive strength of the 50% gravel smart concrete decreased by 11% to 7.7 MPa, compared to the smart cement with no gravel. The density of the concrete was 2.31 gr/cm³.

(f) **60% Gravel Smart Concrete:** The compressive strength of the 60% gravel smart concrete decreased by 20% to 6.9 MPa, compared to the smart cement with no gravel. The density of the concrete was 2.25 g/cm³.

(g) **75% Gravel Smart Concrete:** The compressive strength of the 75% gravel smart concrete decreased by 29% to 6.1 MPa, compared to the smart cement with no gravel. The density of the concrete was 2.04 g/cm³.

(b) Twenty-eight Day Curing
The variations of the compressive strength with the aggregate contents are compared and shown in Figure 12.14 and Figure 12.15.

(a) **No Gravel Smart Cement:** After one day of curing, the compressive strength of the smart cement was 21.7 MPa.

(b) **10% Gravel Smart Concrete:** The compressive strength of the 10% gravel smart concrete increased by 8% to 23.4 MPa, compared to the smart cement with no gravel.

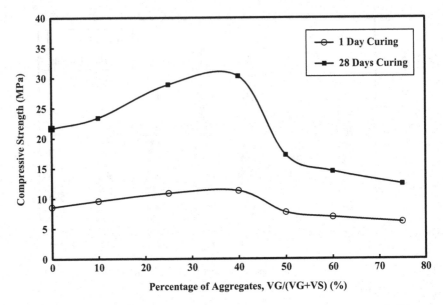

Figure 12.15 Variation of the compressive strengths of the smart concretes with different gravel contents.

(c) **25% Gravel Smart Concrete:** The compressive strength of the 25% gravel smart concrete increased by 33% to 28.9 MPa, compared to the smart cement with no gravel.

(d) **40% Gravel Smart Concrete:** The compressive strength of the 40% gravel smart concrete increased by 40% to 30.3 MPa, compared to the smart cement with no gravel.

(e) **50% Gravel Smart Concrete:** The compressive strength of the 50% gravel smart concrete decreased by 21% to 17.2 MPa, compared to the smart cement with no gravel.

(f) **60% Gravel Smart Concrete:** The compressive strength of the 60% gravel smart concrete decreased by 33% to 14.5 MPa, compared to the smart cement with no gravel.

(g) **75% Gravel Smart Concrete:** The compressive strength of the 75% gravel smart concrete decreased by 43% to 12.4 MPa, compared to the smart cement with no gravel.

12.4.7 Compressive Strength versus RI$_{24}$

The relationships between RI$_{24}$ and the compressive strength of smart concretes were investigated. The variation of RI$_{24}$ with the one-day and 28-day compressive strengths varied linearly, as shown in Figure 12.16. This

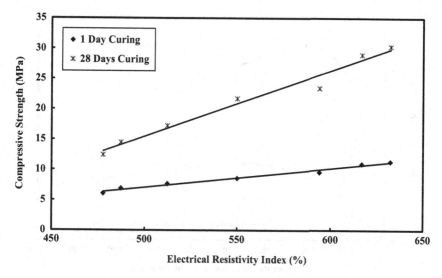

Figure 12.16 Relationship between the electrical resistivity index (RI_{24}) and the compressive strength of the smart concretes after 1 day and 28 days of curing.

is an indicator of the importance of the resistivity index RI_{24} for the smart concrete. Also, by getting the RI_{24} in one day, the strength after 28 days can be predicted. This is a very good quality control for the smart concrete.

12.4.8 Piezoresistivity

The piezoresistive behavior of the smart concretes was evaluated after one day and 28 days of curing.

One Day Curing

(a) **Smart Cement (No Gravel):** After one day of curing, the piezoresistivity of the smart cement was 375%, as shown in Figure 12.17 and summarised in Table 12.5. The Vipulanandan p-q piezoresistive model parameters p_2 and q_2 were 0.61 and 0.57 respectively.

(b) **10% Gravel Smart Concrete:** The piezoresistivity of the 10% gravel smart concrete reduced by 6% to 354%, compared to the smart cement, as shown in Figure 12.18 and summarised in Table 12.5. The Vipulanandan p-q piezoresistive model parameters p_2 and q_2 were 0.55 and 0.72 respectively.

(c) **25% Gravel Smart Concrete:** The piezoresistivity of the 25% gravel smart concrete reduced by 23% to 288%, compared to the smart cement, as shown in Figure 12.19 and summarised in Table 12.5.

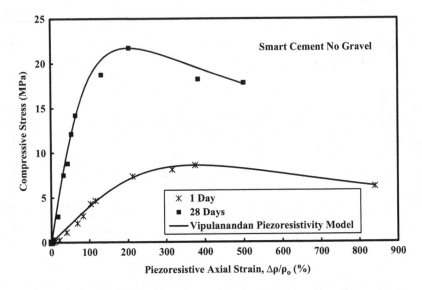

Figure 12.17 Piezoresistive stress–strain of the smart cement after 1 day and 28 days of curing.

The Vipulanandan p-q piezoresistive model parameters p_2 and q_2 were 0.51 and 0.79 respectively.

(d) **40% Gravel Smart Concrete:** The piezoresistivity of the 40% gravel smart concrete reduced by 42% to 217%, compared to the smart cement, as shown in Figure 12.20 and summarised in Table 12.5. The Vipulanandan p-q piezoresistive model parameters p_2 and q_2 were 0.44 and 0.85 respectively.

(e) **50% Gravel Smart Concrete:** The piezoresistivity of the 50% gravel smart concrete reduced by 46% to 200%, compared to the smart cement, as shown in Figure 12.21 and summarised in Table 12.5. The Vipulanandan p-q piezoresistive model parameters p_2 and q_2 were 0.42 and 0.91 respectively.

(f) **60% Gravel Smart Concrete:** The piezoresistivity of the 60% gravel smart concrete reduced by 53% to 175%, compared to the smart cement, as shown in Figure 12.22 and summarised in Table 12.5. The Vipulanandan p-q piezoresistive model parameters p_2 and q_2 were 0.42 and 0.51 respectively.

(g) **75% Gravel Smart Concrete:** The piezoresistivity of the 75% gravel smart concrete reduced by 57% to 163%, compared to smart cement, as shown in Figure 12.23 and summarised in Table 12.5. The Vipulanandan p-q piezoresistive model parameters p_2 and q_2 were 0.40 and 0.80 respectively.

Table 12.5 Piezoresistive p-q model parameters and compressive strengths for the smart concretes

Smart Concrete	P_2	q_2	R^2	Compressive Strength (MPa)	Ultimate Piezoresistivity (%)	RMSE (MPa)
I Day Curing						
No Gravel	0.61	0.57	0.99	8.6	375	0.3
10% Gravel	0.55	0.72	0.98	9.6	354	0.3
25% Gravel	0.51	0.79	0.98	10.9	288	0.4
40% Gravel	0.44	0.85	0.99	11.3	217	0.4
50% Gravel	0.42	0.91	0.99	7.7	200	0.2
60% Gravel	0.42	0.51	0.99	6.9	175	0.2
75% Gravel	0.40	0.80	0.99	6.1	163	0.3
28 Days Curing						
No Gravel	0.83	0.42	0.98	21.7	204	1.0
10% Gravel	0.60	0.45	0.99	23.4	192	1.0
25% Gravel	0.20	0.60	0.98	28.9	147	1.1
40% Gravel	0.11	0.65	0.98	30.3	121	0.9
50% Gravel	0.29	0.73	0.99	17.2	115	0.7
60% Gravel	0.18	0.51	0.99	14.5	103	0.6
75% Gravel	0.81	0.40	0.99	12.4	101	0.4

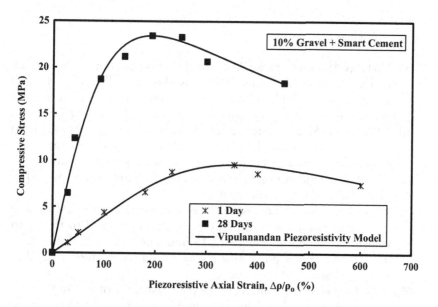

Figure 12.18 Piezoresistive stress–strain of the smart concrete with 10% gravel after 1 and 28 days of curing.

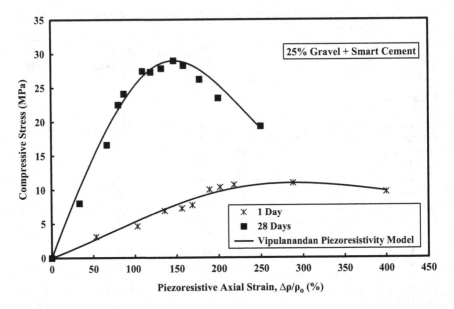

Figure 12.19 Piezoresistive stress–strain of the smart concrete with 25% gravel after 1 and 28 days of curing.

(b) Twenty-eight Day curing

(a) **No Gravel Smart Cement:** After one day of curing, the piezoresistivity of the smart cement was 204%, as shown in Figure 12.17 and summarised in Table 12.5. The Vipulanandan p-q piezoresistive model parameters p_2 and q_2 were 0.83 and 0.42 respectively.

(b) **10% Gravel Smart Concrete:** The piezoresistivity of the 10% gravel smart concrete reduced by 6% to 192%, compared to the smart cement, as shown in Figure 12.18 and summarised in Table 12.5. The Vipulanandan p-q piezoresistive model parameters p_2 and q_2 were 0.60 and 0.45 respectively.

(c) **25% Gravel Smart Concrete:** The piezoresistivity of the 25% gravel smart concrete reduced by 28% to 147%, compared to the smart cement, as shown in Figure 12.19 and summarised in Table 12.5. The Vipulanandan p-q piezoresistive model parameters p_2 and q_2 were 0.20 and 0.60 respectively.

(d) **40% Gravel Smart Concrete:** The piezoresistivity of the 40% gravel smart concrete reduced by 41% to 121%, compared to the smart cement, as shown in Figure 12.20 and summarised in Table 12.5. The Vipulanandan p-q piezoresistive model parameters p_2 and q_2 were 0.11 and 0.65 respectively.

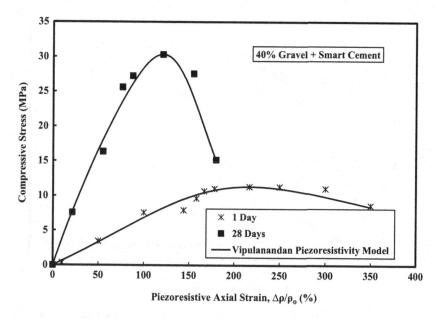

Figure 12.20 Piezoresistive stress–strain of the smart concrete with 40% gravel after 1 and 28 days of curing.

(e) **50% Gravel Smart Concrete:** The piezoresistivity of the 50% gravel smart concrete reduced by 44% to 115%, compared to the smart cement, as shown in Figure 12.21 and summarised in Table 12.5. The Vipulanandan p-q piezoresistive model parameters p_2 and q_2 were 0.29 and 0.73 respectively.

(f) **60% Gravel Smart Concrete:** The piezoresistivity of the 60% gravel smart concrete reduced by 50% to 103%, compared to the smart cement, as shown in Figure 12.22 and summarised in Table 12.5. The Vipulanandan p-q piezoresistive model parameters p_2 and q_2 were 0.18 and 0.51 respectively.

(g) **75% Gravel Smart Concrete:** The piezoresistivity of the 75% gravel smart concrete reduced by 51% to 101%, compared to the smart cement, as shown in Figure 12.23 and summarised in Table 12.5. The Vipulanandan p-q piezoresistive model parameters p_2 and q_2 were 0.81 and 0.40 respectively.

12.5 LOAD AND STRESS DISTRIBUTIONS BETWEEN THE SMART CEMENT BINDER AND THE AGGREGATES

One advantage in using the smart cement as the binder is that it was possible to estimate the percentage of loads and the percentage of stresses carried by

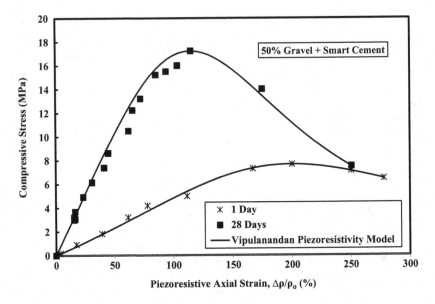

Figure 12.21 Piezoresistive stress–strain of the smart concrete with 50% gravel after 1 and 28 days of curing.

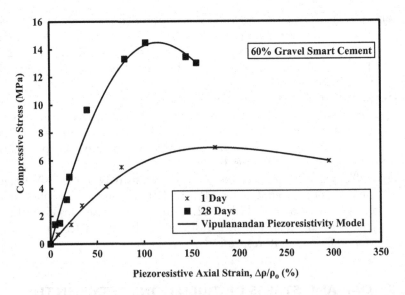

Figure 12.22 Piezoresistive stress–strain of the smart concrete with 60% gravel after 1 and 28 days of curing.

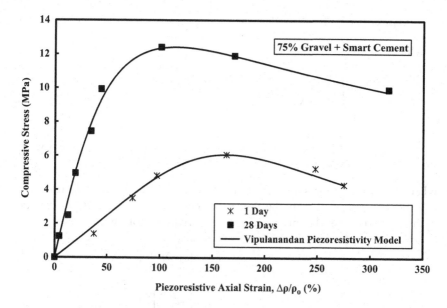

Figure 12.23 Piezoresistive stress–strain of the smart concrete with 75% gravel after 1 and 28 days of curing.

the aggregates and the smart cement binder at the macrostructural level. There is no other method in the literature to estimate the load carried by the aggregates and cement binder when the concrete is loaded. This will also help us better understand the role of the aggregates and the cement binder based on the concrete compositions, related to percentage of aggregates. The predicted stresses in the aggregates will also help in selecting the strength quality of the aggregates to use in the concrete.

The load distribution between the smart cement binder and the aggregates was calculated based on the fact that the piezoresistivity of the concrete was due to the smart cement binder only. The load (P) distribution and the equivalent area (A) were calculated are as follows:

$$P_{total} = P_{Aggregates} + P_{Smart\,Cement} \tag{12.11}$$

$$\sigma_{total} = \frac{P_{total}}{A_{total}} \tag{12.12}$$

The stresses (σ) in the smart cement binder and aggregates are:

$$\sigma_{Aggregates} = \frac{P_{Aggregates}}{A_{Aggregates}} \tag{12.13}$$

$$\sigma_{Smart\,Cement} = \frac{P_{Smart\,Cement}}{A_{Smart\,Cement}} \tag{12.14}$$

in which:

$$A_{Aggregates} = \left(\frac{V_G}{V_G + V_S}\right) \cdot A_{total} \tag{12.15}$$

$$A_{Smart\,Cement} = \left(\frac{V_S}{V_G + V_S}\right) \cdot A_{total} \tag{12.16}$$

where V_S and V_G represent the volume fraction of the smart cement binder (S) and the gravel (G) in the concrete. The smart cement stress can be calculated based on the measured piezoresistivity as follows:

$$\sigma_{Smart\,Cement} = \frac{\sigma_{max} \times \left(\dfrac{\left(\dfrac{\Delta\rho}{\rho}\right)}{\left(\dfrac{\Delta\rho}{\rho}\right)_0}\right)}{q_2 + (1 - p_2 - q_2) \times \left(\dfrac{\left(\dfrac{\Delta\rho}{\rho}\right)}{\left(\dfrac{\Delta\rho}{\rho}\right)_0}\right) + p_2 \times \left(\dfrac{\left(\dfrac{\Delta\rho}{\rho}\right)}{\left(\dfrac{\Delta\rho}{\rho}\right)_0}\right)^{\left(\frac{p_2+q_2}{p_2}\right)}} \tag{12.17}$$

Based on the piezoresistivity measurement, the load and stress distribution between the aggregates and the smart cement can be quantified using the following equations (Equilibrium Equation):

$$P_{Aggregates}\left(P_g\right) = P_{total}\left(P_t\right) - P_{Smart\,Cement}\left(P_s\right) \tag{12.18}$$

Loads can be substituted by stresses and volume factions of smart cement and aggregates as:

$$\sigma_{Aggregates} \cdot A_{Aggregates} = \sigma_{total} \cdot A_{total} - \sigma_{Smart\,Cement} \cdot A_{Smart\,Cement} \tag{12.19}$$

Considering the volume fraction of the smart cement and the aggregates, Eqn. (12.18) can be written as:

$$\sigma_{Aggregates} \cdot \left(\frac{V_G}{V_G + V_S}\right) = \sigma_{total} \cdot A_{total} - \sigma_{Smart\,Cement} \cdot \left(\frac{V_S}{V_G + V_S}\right) \tag{12.20}$$

$$\sigma_{Aggregates} = \frac{\sigma_{total} \cdot A_{total} - \sigma_{Smart\,Cement} \cdot \left(\dfrac{V_S}{V_G + V_S}\right)}{\left(\dfrac{V_G}{V_G + V_S}\right)} \qquad (12.21)$$

The stress in the aggregates can be calculated as:

$$\sigma_{Aggregates} =$$

$$\left[\sigma_{total} - \frac{\sigma_{max} \times \left(\dfrac{\left(\dfrac{\Delta\rho}{\rho}\right)}{\left(\dfrac{\Delta\rho}{\rho}\right)_0}\right)}{q_2 + (1 - p_2 - q_2) \times \left(\dfrac{\left(\dfrac{\Delta\rho}{\rho}\right)}{\left(\dfrac{\Delta\rho}{\rho}\right)_0}\right) + p_2 \times \left(\dfrac{\left(\dfrac{\Delta\rho}{\rho}\right)}{\left(\dfrac{\Delta\rho}{\rho}\right)_0}\right)^{\left(\frac{p_2+q_2}{p_2}\right)}} \cdot \left(\dfrac{V_S}{V_G + V_S}\right)\right] \Bigg/ \left(\dfrac{V_G}{V_G + V_S}\right)$$

$$(12.22)$$

Eqn. (12.16) and Eqn. (12.21) were used to estimate the stress distributions in the smart cement binder and the aggregates respectively.

The change in the resistivity in the concrete during loading is due to stress changes in the smart cement. Based on the experimental results, it varied with the amount of the aggregates in the smart concrete and also the curing time.

12.5.1 Loads and Stress After One Day of Curing

(a) Aggregates Loads

The percentage compressive loads carried by the aggregates in the concretes cured for one day is shown in Figure 12.24. The load in the aggregates increased and decreased with the increase in the piezoresistive axial strain. The maximum load carried by the aggregates was about 79% in the 75% gravel concrete. In the concrete with 50% gravel, the maximum load carried by the aggregates was about 66%, as shown in Figure 12.24. In the concrete with 40% gravel, the maximum load carried by the aggregates was about 58%, as shown in Figure 12.24. In the concrete with 25% gravel,

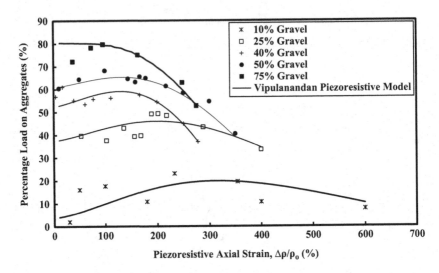

Figure 12.24 Percentage load carried by the aggregates in the smart concrete after 1 day of curing.

the maximum load carried by the aggregates was about 49%, as shown in Figure 12.24. In the concrete with 10% gravel, the maximum load carried by the aggregates was about 23%, as shown in Figure 12.24.

(b) Smart Cement Loads

The percentage of compressive loads carried by the smart cement binder in the concretes cured for one day is shown in Figure 12.25. The load in the smart cement binder decreased, and increased with the increase in the piezoresistive axial strain. The percentage of load carried by the smart cement binder at failure (peak load) was about 25% in the concrete with 75% aggregate, as shown in Figure 12.25. The percentage of load carried by the smart cement binder at failure (peak load) was about 37% in the concrete with 50% aggregate, as shown in Figure 12.25. The percentage of load carried by the smart cement binder at failure (peak load) was about 51% in the concrete with 40% aggregate, as shown in Figure 12.25. The percentage of load carried by the smart cement binder at failure (peak load) was about 57% in the concrete with 25% aggregate, as shown in Figure 12.25. The percentage of load carried by the smart cement binder at failure (peak load) was about 81% in the concrete with 10% aggregate, as shown in Figure 12.25.

(c) Aggregates Stresses

The compressive stresses carried by the aggregates in the concretes cured for one day are shown in Figure 12.26. The stress in the aggregates increased,

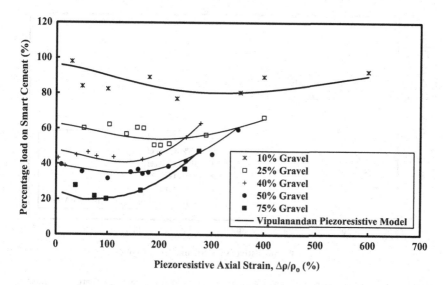

Figure 12.25 Percentage load carried by the smart cement in the smart concrete with various amounts of aggregate contents after 1 day of curing.

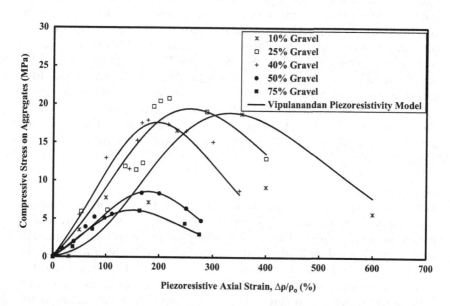

Figure 12.26 Compressive stresses on the aggregates in the smart concrete after 1 day of curing.

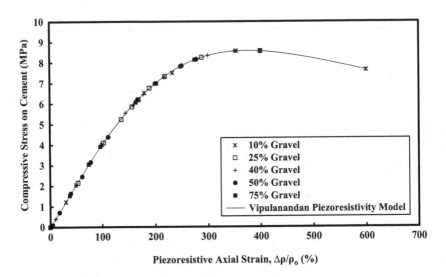

Figure 12.27 Compressive stresses on smart cement in the smart concrete with various aggregate contents after 1 day of curing.

and decreased with the increase in the piezoresistive axial strain. The maximum compressive stress carried by the aggregates was about 6 MPa in the concrete with 75% gravel, as shown in Figure 12.26. The maximum compressive stress carried by the aggregates was about 8.5 MPa in the concrete with 50% gravel, as shown in Figure 12.26. The maximum compressive stress carried by the aggregates was about 18 MPa in the concrete with 40% gravel, as shown in Figure 12.26. The maximum compressive stress carried by the aggregates was about 21 MPa in the concrete with 25% gravel, as shown in Figure 12.26. The maximum compressive stress carried by the aggregates was about 19 MPa in the concrete with 10% gravel, as shown in Figure 12.26.

(d) Smart Cement Stresses

The compressive stresses carried by the smart cement binder in the concretes cured for one day are shown in Figure 12.27. The stress in the smart cement binder increased with the increase in the piezoresistive axial strain. The stress carried at failure by the smart cement binder was about 6 MPa in the concrete with 75% gravel, as shown in Figure 12.27. The stress carried at failure by the smart cement binder was about 7 MPa in the concrete with 50% gravel, as shown in Figure 12.27. The stress carried at failure by the smart cement binder was about 7.3 MPa in the concrete with 40% gravel, as shown in Figure 12.27. The stress carried at failure by the smart cement binder was about 8.3 MPa in the concrete with 25% gravel, as shown in

Figure 12.27. The stress carried at failure by the smart cement binder was about 8.6 MPa in the concrete with 10% gravel, as shown in Figure 12.27.

12.5.2 Comparing the Load and Stress Distribution in Selected Concretes

The compressive strength was the highest with the 40% gravel concrete. The compressive load and stress distribution in the aggregates and the smart cement binder for one-day cured concrete with 40% and 75% aggregates in the concrete are compared in Figures 12.28, 12.29, 12.30, and 12.31. The compressive failure strength of the one-day cured smart cement was 8.6 MPa, as summarised in Table 12.5

(a) 40% Aggregates

With the piezoresistive axial strain measurements in the concrete it was possible to estimate the load distributions in the aggregates and the smart cement binder using Eqns. (12.21) and (12.16) respectively. With the piezoresistive axial strain, the stress in the cement was first calculated and then the stress in the aggregate was determined using Eqn. (12.21). The load distributions in the concrete (P_{total}) with the increased piezoresistivity axial strains are shown in Figure 12.28. The loading on the aggregates peaked and decreased

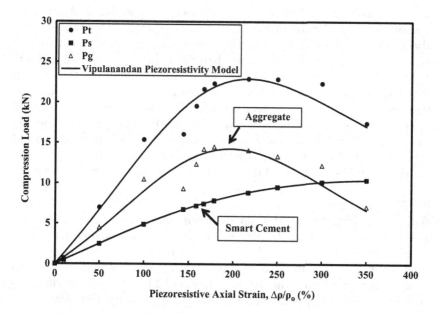

Figure 12.28 Load distributions in the smart cement binder and aggregate in the 40% gravel smart concrete after 1 day of curing.

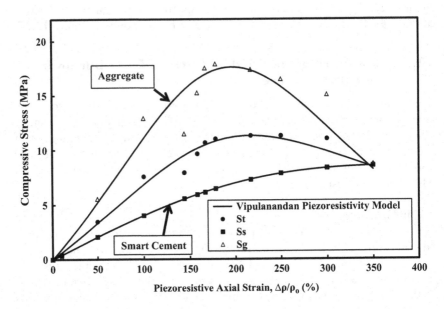

Figure 12.29 Compressive stress distribution in the smart cement and aggregate in the 40% gravel smart concrete after I day of curing.

similar to the total load on concrete, as shown in Figure 12.28. At the peak load of 23 kN, the maximum load carried by the aggregates was about 60%, also shown in Figure 12.28, and the stress was about 17.6 MPa, as shown in Figures 12.29, which was higher than the failure stress of the concrete of 11.3 MPa, as summarised in Table 12.5. This also indicates the strength of the aggregate that is needed to be used in the concrete for one-day performance. The load in the smart cement binder continuously increased, and at the peak load of the concrete it was about 40% and the corresponding stress was about 7.3 MPa, as shown in Figure 12.29. All these estimates of load and stress distributions were possible only because of the piezoresistive smart cement binder. None of this information is available in the literature.

(b) 75% Aggregates

With the piezoresistive axial strain measurements in the concrete it was possible to estimate the load distributions in the aggregates and smart cement binder using Eqns. (12.21) and (12.16) respectively. With the piezoresistive axial strain, the stress in the cement was first calculated and then the stress in the aggregate was determined using Eqn. (12.21). The load distributions in the concrete (P_{total}) with the increased piezoresistivity axial strains are shown in Figure 12.30. The loading on the aggregates peaked and decreased similar to the total load on concrete, as shown in Figure 12.30. At the peak

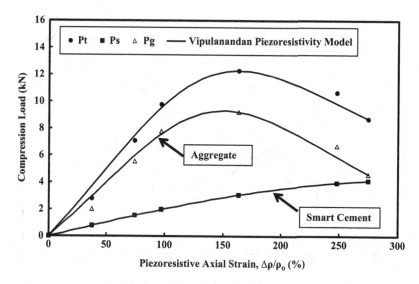

Figure 12.30 Load distribution on the smart cement binder and aggregate in the 75% gravel smart concrete after 1 day of curing.

load of 12.3 kN, the maximum load carried by the aggregates was about 75%, also shown in Figure 12.30, and the stress was about 6.1 MPa, as shown in Figure 12.31, which was equal to the failure stress of the concrete of 6.1 MPa, as summarised in Table 12.5. This also indicates the strength of the aggregate that needs to be used in the concrete for one-day performance. The load in the smart cement binder continuously increased, and at the peak load of the concrete it was about 25% and the corresponding stress was about 6.1 MPa, as shown in Figure 12.31. All these estimates of load and stress distributions were possible only because of the piezoresistive smart cement binder. None of this information is available in the literature.

12.5.3 Loads and Stresses after 28 Days of Curing

The compressive strength of the smart cement was 21.7 MPa, as summarised in Table 12.5.

(a) Aggregates Loads

The percentage compressive loads carried by the aggregates in the concretes cured for 28-day cured concrete is shown in Figure 12.32. The load in the aggregates increased and decreased with the increase in the piezoresistive axial strain. The maximum load carried by the aggregates was about 77% in the 75% gravel concrete. In the concrete with 50% gravel, the maximum

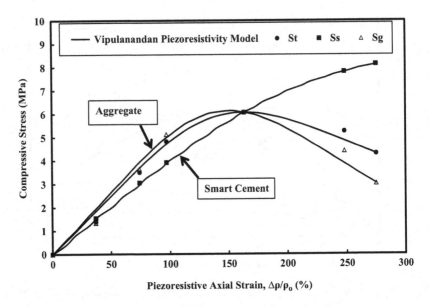

Figure 12.31 Compressive stress distribution in the smart cement and aggregate in the 75% gravel smart concrete after 1 day of curing.

load carried by the aggregates was about 44%, as shown in Figure 12.32. In the concrete with 40% gravel, the maximum load carried by the aggregates was about 61%, as shown in Figure 12.32. In the concrete with 25% gravel, the maximum load carried by the aggregates was about 45%, as shown in Figure 12.32. In the concrete with 10% gravel, the maximum load carried by the aggregates was about 17%, as shown in Figure 12.24. The load distribution will also be influenced by the distribution of aggregates within the concrete.

(b) Smart Cement Loads

The percentage compressive loads carried by the smart cement binder in the concretes cured for 28 days is shown in Figure 12.33. The load in the smart cement binder decreased, and increased with the increase in the piezoresistive axial strain. The percentage of load carried by the smart cement binder at failure (peak load) was about 37% in the concrete with 75% aggregate, as shown in Figure 12.33. The percentage of load carried by the smart cement binder at failure (peak load) was about 56% in the concrete with 50% aggregate, as shown in Figure 12.33. The percentage of load carried by the smart cement binder at failure (peak load) was about 40% in the concrete with 40% aggregate, as shown in Figure 12.33. The percentage of load

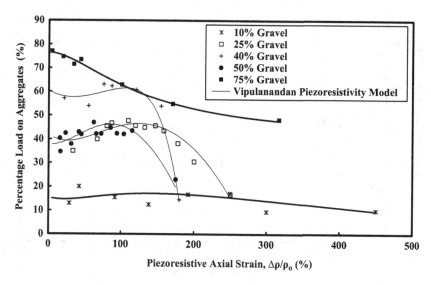

Figure 12.32 Percentage load carried by the aggregates in smart concrete after 28 days of curing.

carried by the smart cement binder at failure (peak load) was about 54% in the concrete with 25% aggregate, as shown in Figure 12.33. The percentage of load carried by the smart cement binder at failure (peak load) was about 83% in the concrete with 10% aggregate, as shown in Figure 12.33.

(c) Aggregates Stresses

The compressive stresses carried by the aggregates in the concretes cured for 28 days are shown in Figure 12.34. The stress in the aggregates increased, and decreased with the increase in the piezoresistive axial strain. The maximum compressive stress carried by the aggregates was about 10 MPa in the concrete with 75% gravel, as shown in Figure 12.34. The maximum compressive stress carried by the aggregates was about 16 MPa in the concrete with 50% gravel, as shown in Figure 12.34. The maximum compressive stress carried by the aggregates was about 46 MPa in the concrete with 40% gravel, as shown in Figure 12.34. The maximum compressive stress carried by the aggregates was about 53 MPa in the concrete with 25% gravel, as shown in Figure 12.34. The maximum compressive stress carried by the aggregates was about 39 MPa in the concrete with 10% gravel, as shown in Figure 12.34. The compressive stresses in the aggregates showed large variation.

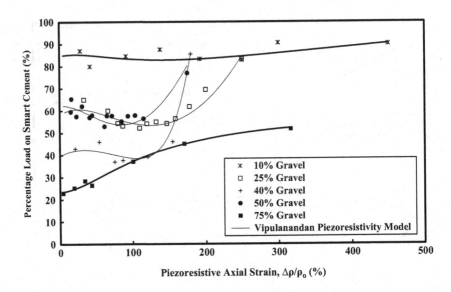

Figure 12.33 Percentage load carried by the smart cement in the smart concrete after 28 days of curing.

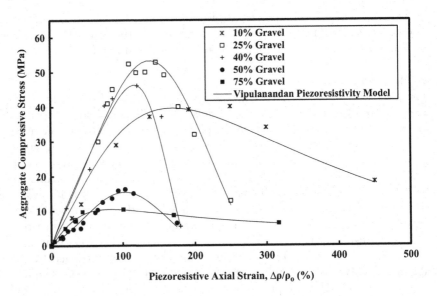

Figure 12.34 Compressive stresses on the aggregates in the smart concrete after 28 days of curing.

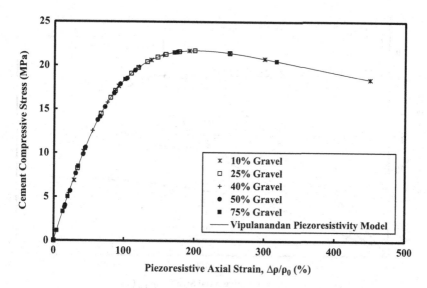

Figure 12.35 Compressive stresses on smart cement in the smart concrete with various aggregate contents after 28 days of curing.

(d) Smart Cement Stresses

The compressive stresses carried by the smart cement binder in the concretes cured for 28 days are shown in Figure 12.35. The stress in the smart cement binder increased with the increase in the piezoresistive axial strain. The stress carried at failure by the smart cement binder was about 17 MPa in the concrete with 75% gravel, as shown in Figure 12.35. The stress carried at failure by the smart cement binder was about 19.5 MPa in the concrete with 50% gravel, as shown in Figure 12.35. The stress carried at failure by the smart cement binder was about 20 MPa in the concrete with 40% gravel, as shown in Figure 12.35. The stress carried at failure by the smart cement binder was about 21 MPa in the concrete with 25% gravel, as shown in Figure 12.35. The stress carried at failure by the smart cement binder was about 21.7 MPa in the concrete with 10% gravel, as shown in Figure 12.35.

12.5.4 Comparing the Load and Stress Distribution in Selected Concretes

The compressive strength was the highest with the 40% gravel concrete. The compressive load and stress distribution in the aggregates and smart cement binder for one day cured concrete with 40% and 75% aggregates in the concrete are compared in Figure 12.36, Figure 12.37, Figure 12.38, and

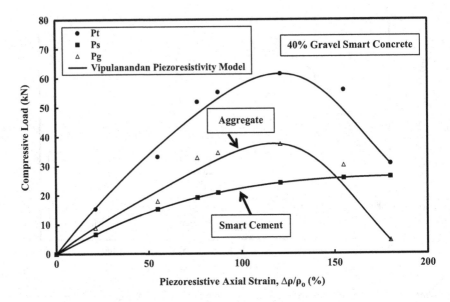

Figure 12.36 Load distribution in the smart cement binder and aggregates in the 40% gravel smart concrete after 28 days of curing.

Figure 12.39. The compressive failure strength of the 28-day cured smart cement was 21.7 MPa, as summarised in Table 12.5

(a) 40% Aggregates

With the piezoresistive axial strain measurements in the concrete it was possible to estimate the load distributions in the aggregates and smart cement binder using Eqns. (12.21) and (12.16) respectively. With the piezoresistive axial strain, the stress in the cement was first calculated and then the stress in the aggregate was determined using Eqn. (12.21). The load distributions in the concrete (P_{total}) with the increasing piezoresistivity axial strains are shown in Figure 12.36. The loading on the aggregates peaked and decreased similar to the total load on concrete, as shown in Figure 12.36. At the peak load of 61 kN, the maximum load carried by the aggregates was about 61%, also shown in Figure 12.36, and the stress was about 46 MPa, as shown in Figure 12.37, which was higher than the failure stress of the concrete of 30.3 MPa, as summarised in Table 12.5. This also indicates the strength of the aggregate that is needed to be used in the concrete for 28-day cured performance. The load in the smart cement binder continuously increased, and at the peak load of the concrete it was about 39% and the corresponding stress was about 19.8 MPa, as shown in Figure 12.37. All these estimates of

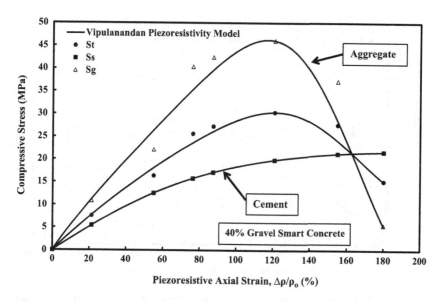

Figure 12.37 Compressive stress distribution in the smart cement binder and aggregates in the 40% gravel smart concrete after 28 days of curing.

load and stress distributions were possible only because of the piezoresistive smart cement binder. None of this information is available in the literature.

(b) 75% Aggregates

With the piezoresistive axial strain measurements in the concrete it was possible to estimate the load distributions in the aggregates and the smart cement binder using Eqns. (12.21) and (12.16) respectively. With the piezoresistive axial strain, the stress in the cement was first calculated, and then the stress in the aggregate was determined using Eqn. (12.21). The load distributions in the concrete (P_{total}) with the increasing piezoresistivity axial strains are shown in Figure 12.38. The loading on the aggregates peaked and decreased similar to the total load on concrete, as shown in Figure 12.38. At the peak load of 25 kN, the maximum load carried by the aggregates was about 63%, also shown in Figure 12.38, and the stress was about 10.4 MPa, as shown in Figure 12.39, which was lower than the failure strength of the concrete of 12.4 MPa, as summarised in Table 12.5. This also indicates the strength of the aggregate that needs to be used in the concrete for 28 days performance. The load in the smart cement binder continuously increased, and at the peak load of the concrete it was about 37% and the corresponding stress was about 18.4 MPa, as shown in Figure 12.39. All these estimates of the

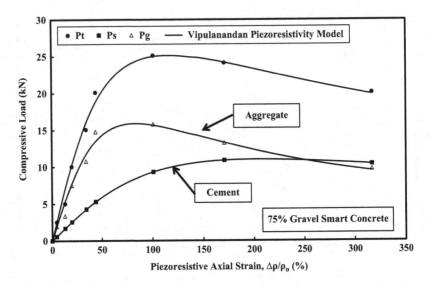

Figure 12.38 Load distribution in the smart cement binder and aggregate in the 75% gravel smart concrete after 28 days of curing.

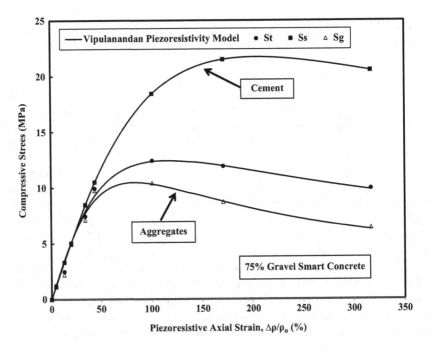

Figure 12.39 Compressive stress distributions in the smart cement binder and aggregates in the 75% gravel smart concrete after 28 days of curing.

load and stress distributions were possible only because of the piezoresistive smart cement binder. None of this information is available in the literature.

12.6 CONCRETE FAILURE MODELS

(a) Drucker–Prager Failure Model (1950)

The Drucker–Prager criterion has been widely adopted for the modeling of confined concrete due to its simplicity (involving only two parameters) and its capability to capture shear strength increases as per result of hydrostatic pressure increases, which is a unique property of concrete under confinement.

The Mohr-Coulomb failure surface has corners on the hexogen, which is not mathematically convenient. Drucker and Prager have smoothened the Mohr-Coulomb by the simple modification of the von Mises criterion. According to the Drucker and Prager model, $\sqrt{J_2}$ increases linearly with the increase of principal stresses. The model equation is

$$\sqrt{J_2} - \alpha I_1 - K = 0 \tag{12.22}$$

where I_1 is the sum of normal stresses (first stress invariant), J_2 is second deviator stress invariant, and α and K are material constants.

$$\text{When} \qquad I_1 \to \infty, \text{then } \sqrt{J_2} \to \infty \tag{12.23}$$

There is a limit to shear tolerance for cement materials and hence, the Drucker–Prager model does not satisfy this condition.

(b) Vipulanandan Failure Model (2018)

The Vipulanandan failure relationship (Eqn. (5.20)) satisfying all the basic conditions is as follows (Vipulanandan et al. 2018h; Mayooran et al. 2018)

$$\sqrt{J_2} = \tau_0 + \frac{I_1}{L + NI_1} \tag{12.24}$$

Hence, this model has a limit on the maximum shear stress the concrete will tolerate at relatively high mean stress.

$$\sqrt{J2}_{max} = \tau_0 + \frac{1}{L} \quad \text{When} \quad I_1 \to \infty \tag{12.25}$$

The Vipulanandan failure model will represent the Drucker and Prager model when B = 0 and the von Mises Criterion when A = 0, a generalised failure model.

12.6.1 Model Verifications

Drucker–Prager Model (1950)

(a) Unconfined Compressive Strength of 23 MPa

Based on the concrete test data, the maximum I_1 was about 220 MPa and the maximum $\sqrt{J_2}$ was about 60 MPa, as shown in Figure 12.40. Based on the analyses of 114 test data, the model parameters α and K were 0.234 and 6.24 MPa respectively, as summarised in Table 12.6. The $(\sqrt{J_2})_o$ (parameter K) representing the pure shear strength of concrete ($I_1 = 0$) was 6.24 MPa, 27% of the compressive strength of the concrete. Additionally, the concrete pure shear strength was higher than the pure shear strength of cement (w/c ratio of 0.38) of 1.82 MPa (Chapter 5), and also the pure shear strength of shale rock and limestone rock of 3.0 MPa and 2.6 MPa respectively (Vipulanandan et al. 2018h). Also the pure tensile strength ($\sqrt{J_2} = 0$, equal tension in three perpendicular directions) for the concrete was 8.88 MPa, 39% of the unconfined compressive strength of concrete. The cement pure tension was 1.25 MPa using the Drucker–Prager model (Chapter 5), only 14% of the pure tensile strength of concrete. The root-mean-square error (RMSE) for the data prediction was 4.20 MPa, and the coefficient of determination (R^2) was 0.92.

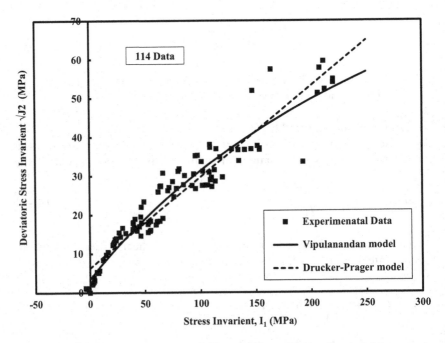

Figure 12.40 Comparing the concrete failure models predictions for the plain concrete with unconfined compressive strength of 23 MPa.

Table 12.6 Failure model parameters for the plain concrete with varying strength data (28 days of curing)

Concrete Strength (MPa) (Data)	Drucker-Prager Linear Model				Model Parameters				
					Vipulanandan Model				
	Slope (α)	K (MPa)	R^2	RMSE (MPa)	L	N $(MPa)^{-1}$	$(\sqrt{J_2})$ (MPa)	$(\sqrt{J_2})_{max}$ (MPa)	RMSE (MPa)	R^2
23 (114)	0.234	6.24	0.92	4.20	2.46	0.009	3.0	114	3.90	0.95
46 (14)	0.249	18.26	0.94	4.10	2.31	0.00425	8.27	14	3.20	0.97

(b) Unconfined Compressive Strength of 46 MPa

Based on the concrete test data, the maximum I_1 was about 290 MPa and the maximum $\sqrt{J_2}$ was about 90 MPa, as shown in Figure 12.41. Based on the analyses of 14 test data, representing unconfined compressive strength of 46 MPa, the model parameters α and K were 0.249 and 18,26 MPa respectively, as summarised in Table 12.6. So doubling the unconfined compressive strength of concrete increased the two parameters. The $(\sqrt{J_2})_o$ (parameter K) representing the pure shear strength of concrete ($I_1 = 0$) was 18.26 MPa, about 40% of the unconfined compressive strength of the concrete. In addition, the concrete pure shear strength was higher than the pure shear strength of cement (w/c ratio of 0.38) of 1.82 MPa (Chapter 5) and also the pure shear strength of shale rock and limestone rock of 3.0 MPa and 2.6 MPa respectively (Vipulanandan et al. 2018). Moreover, the pure tension strength ($\sqrt{J_2} = 0$, equal tension in three perpendicular directions) for the concrete was 24.44 MPa, 53% of the unconfined compressive strength of concrete. The cement pure tension was 1.25 MPa using the Drucker–Prager model (Chapter 5), only 5% of the pure tensile strength of concrete. The maximum deviatoric stress tolerance for the concrete was 114 MPa. The root-mean-square error (RMSE) for the data prediction was 4.10 MPa, and the coefficient of determination (R^2) was 0.94.

Vipulanandan Model (2018)

(a) Unconfined Compressive Strength of 23 MPa

Based on the concrete test data the maximum I_1 was about 220 MPa and the maximum $\sqrt{J_2}$ was about 60 MPa, as shown in Figure 12.40. Based on the analyses of 114 test data, the model parameter τ_o, L, and N were 3.0 MPa, 2.46, and 0.009 respectively, as summarised in Table 12.6. The parameter τ_o represents the pure shear strength of concrete ($I_1 = 0$) and was 3.0 MPa, 13% of the compressive strength of the concrete. Also, the pure shear strength predicted by the Vipulanandan model was 48% of the pure shear strength predicted by the Drucker–Prager model. The concrete pure shear strength was higher than the pure shear strength of cement (w/c ratio of 0.38) of 1.82 MPa (Chapter 5), and also the pure shear strength of shale rock and limestone rock of 3.0 MPa and 2.6 MPa respectively (Vipulanandan et al. 2018). Additionally, the pure tensile strength ($\sqrt{J_2} = 0$, equal tension in 3 perpendicular directions) for the concrete was 2.40 MPa, 10% of the unconfined compressive strength of concrete. The cement pure tension was 1.14 MPa using the Vipulanandan model (Chapter 5), only 48% of the pure tensile strength of concrete. Based on the Vipulanandan model, the $(\sqrt{J_2})_{max}$ for concrete was 114 MPa, compared to the smart cement of 144.7 MPa (Chapter 5), and for the Drucker–Prager model it will be infinity. So the maximum $(\sqrt{J_2})_{max}$ for the concrete was five times the unconfined compressive strength of concrete. It has been shown that for shale rock and limestone rock the maximum shear stress limit will be 103 MPa and 102 MPa respectively (Vipulanandan et al. 2018h). The root-mean-square error (RMSE) for the data prediction was 3.90 MPa, and the coefficient of determination (R^2) was 0.95.

(b) Unconfined Compressive Strength of 46 MPa

Based on the concrete test data the maximum I_1 was about 290 MPa and the maximum $\sqrt{J_2}$ was about 90 MPa, as shown in Figure 12.41. Based on the analyses of 14 test data, representing unconfined compressive strength of 46 MPa, the model parameter τ_o, L, and N were 8.3 MPa, 2.31, and 0.00425 respectively, as summarised in Table 12.6. The parameter τ_o represents the pure shear strength of concrete ($I_1 = 0$) and was 8.3 MPa, 18% of the compressive strength of the concrete. Also, the pure shear strength predicted by the Vipulanandan model was 45% of the pure shear strength predicted by the Drucker–Prager model. The concrete pure shear strength was higher than the pure shear strength of cement (w/c ratio of 0.38) of 1.82 MPa (Chapter 5), and also the pure shear strength of shale rock and limestone rock of 3.0 MPa and 2.6 MPa respectively (Vipulanandan et al. 2018). Also, the pure tensile strength ($\sqrt{J_2} = 0$, equal tension in 3 perpendicular directions) for the concrete was 6.15 MPa, 13% of the unconfined

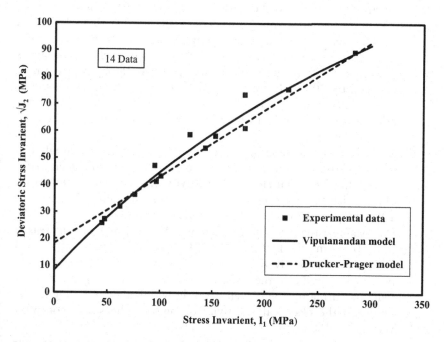

Figure 12.41 Comparing the concrete failure models predictions for the plain concrete with unconfined compressive strength of 46 MPa.

compressive strength of concrete. The cement pure tension was 1.14 MPa using the Vipulanandan model (Chapter 5), only 19% of the pure tensile strength of concrete. Based on the Vipulanandan model, the $(\sqrt{J_2})_{max}$ for the concrete was 244 MPa, compared to the smart cement of 144.7 MPa (Chapter 5), and for the Drucker–Prager model it will be infinity. So the maximum $(\sqrt{J_2})_{max}$ for concrete is about five times the unconfined compressive strength of concrete. It has been shown that for shale rock and limestone rock the maximum shear stress limit will be 103 MPa and 102 MPa respectively (Vipulanandan et al. 2018h). The root-mean-square error (RMSE) for the data prediction was 3.20 MPa, and the coefficient of determination (R^2) was 0.97 as summarised in Table 12.6.

12.7 SUMMARY

The smart cement was used as the binder in the concrete to make it a highly bulk sensing material. Based on experimental and analytical study on the behavior of concrete with varying amounts of aggregates and smart cement binder, the following conclusions are advanced:

1. In the concrete with smart cement binder, the aggregate content and curing time increased the pulse velocity of smart cement concrete. The pulse velocities after 28 days of curing for 0%,10%, 40%, and 75% gravel content smart cement aggregate composites were 3,520 m/s, 3,530 m/s, 3,910 m/s, and 4,750 m/s respectively. The changes in pulse velocity from initial measurements to 28-day measurement changed by 200% to 235% based on the aggregate contents. The new Vipulanandan pulse velocity-curing model predicted the variation of the pulse velocities with the aggregate content and curing time very well.

2. The addition of coarse aggregate (gravel) increased the initial electrical resistivity of the smart concrete as well as long-term electrical resistivity. The initial electrical resistivity of smart cement was 1.02 Ω.m, which increased to 1.09 Ω.m., 1.20 Ω.m., and 3.74 Ω.m. for 10%, 40%, and 75% gravel respectively. After 28 days of curing, the electrical resistivity of smart cement was 14.14 Ω.m, which increased to 15.09 Ω.m., 19.11 Ω.m., and 61.24 Ω.m. for 10%, 40%, and 75% gravel respectively. The new Vipulanandan composite resistivity model predicted the resistivity changes with the aggregate content and curing time very well. Also the Vipulanandan curing model predicted the electrical resistivity changes of the smart concretes very well.

3. The smart concrete also verified the piezoresistive cement concept, and the resistivity increased with the applied compressive stress.

The piezoresistive axial strain at failure for the smart cement aggregate composites with 10%, 40%, and 75% gravel content after 28 days of curing were 192%, 121%, and 101% respectively. The Vipulanandan piezoresistivity model predicted the piezoresistivity behavior of the smart cement concrete very well.

4. The electrical resistivity index (RI_{24}) correlated linearly with the smart concrete compressive strengths after one day and 28 days of curing.

5. The failure strain of concrete is 0.3%; hence, piezoresisitive smart concrete has magnified the monitoring resistivity parameter by 336 times (33,600%) or more higher based on the aggregate content and making the smart concrete a bulk sensor.

6. Using the smart cement in the concrete, it was possible to predict the load and stress distributions in the aggregates and cement in the concrete with varying gravel contents. This is another advancement in quantifying the stress distribution within the concrete and will help in selecting the quality of the aggregate strength based on the highest stresses in the aggregates during the loading to achieve the concrete strengths.

7. The Vipulanandan failure model predicted the concrete pure tension, pure shear, and the maximum deviatoric stress (new property) very well compared to the Drucker–Prager model. Also, the Vipulanandan model root-mean-square error (RMSE) was lower and the coefficent of determination (R^2) was higher, compared to the Drucker–Prager model for the 23 MPa (114 data) and 46 MPa (14 data) unconfined compressive strengths of concretes.

Chapter 13

Conclusions

Cement is an inorganic binding material that has been used in construction for over 5,000 years. Over the past hundred years, it has been manufactured to meet the specifications related to various applications. Health monitoring related to the maintenance and safety of various infrastructures, both onshore and offshore, has become a major issue. Hence, in cement-based infrastructures, it is not only important to develop a highly sensing material with the critical parameters to monitor but also develop a real-time monitoring system that can be adopted in the field. These are two of the major challenges. In recent years, highly sensing chemo-thermo-piezoresistive smart cement has been developed with integrated real-time monitoring technology to address the major issues related to health monitoring of cement-based materials. This is a paradigm shift in making the cement to be a 3D sensor with critical electrical properties for monitoring that has been identified and verified in various model tests. Also, several analytical constitutive models have been developed to represent smart cement behaviors and verified with experimental studies using Portland cements and oil-well cements. In this book, many chapters are included to introduce smart cement with potential applications including grouts, foam cement, and also concrete. Based on smart cement studies, the following conclusions are advanced:

1 A new method has been developed to characterise materials based on electrical properties since electrical property changes can be easily monitored in the field. Using the two-probe method with the alternative current (AC), the impedance-frequency relationship can be experimentally developed to characterise the material. If the bulk material properties are influenced by the resistivity and permittivity (representing the resistance and capacitance), the response was identified as CASE-1. If the bulk material property is represented by resistivity, it was identified as CASE-2. The Vipulanandan impedance model can be used to quantify the two-probe contacts and the bulk material resistance and capacitance. This is a very general

method of electrical characterisation of materials and can be used with any material. Also, the two-probe characterisation can used to detect and quantify surface and bulk corrosion in non-metallic and metallic materials.

2 The smart cement slurry and solidified cement were identified as CASE-2 based on the impedance-frequency responses. This unified the characterisation of the cement slurry and the solidified cement with the changes in electrical material property resistivity with time and other variables. Hence, for the smart cement, electrical resistance will be the parameter to be monitored at a frequency of 300 kHz or higher based on the type of the two probes used for monitoring.

3 Resistivity is a material property and also a second-order tensor, and can be used to characterise cement behavior in three orthogonal directions (3D).

4 The smart cement with metakaoline, fly ash, foam, and aggregate additions were also represented by CASE-2. Hence, the electrical resistivity was used to characterise the cementitious materials investigated in this study.

5 Using the Vipulanandan contact index, the changes in the two-probe contacts with the cement curing with and without additives were quantified. In the smart cement with and without fly ash and aggregates, the contact index increased with the curing time, an indication of higher corrosion potential for the metal probes placed in the cement. Cement with 10% metakaolin and 20% foam showed that the Vipulanandan contact index decreased with time, and hence there was less potential for corrosion of the probes placed in the cement for monitoring.

6 Characterising the rheological properties of smart cement slurry is important to understand its flowability and also mixing efficiency with various additives used in the smart cement. Also, the rheological models are important to better characterise cement slurries related to the initial and long-term pumping pressure gradients, and also the potential erosion of formations based on the shear stresses developed and the maximum shear stress tolerance.

7 The new Vipulanandan rheological model can be used to predict the behavior of both shear-thinning and shear-thickening behaviors of all types of fluids. It can also be used to predict the behavior of Newtonian fluids. This model is used by researchers around the world to characterise various types of fluids.

8 The rheological tests showed that the cement slurries had shear-thinning behavior, and the new Vipulanandan rheological model was used to predict the shear stress–shear strain rate relationship after comparing the predicted performance with the other rheological models in the literature. For shear-thinning fluids, the Vipulanandan

rheological model has a limit to the maximum shear stress tolerance of the fluids. The new Vipulanandan rheological model predicted the test results very well compared to the Herschel–Bulkley model based on the statistical parameters such as the root-mean-square error (RMSE) and the coefficient of determination (R^2).

9 The yield stress and the maximum shear stress limit of the cement slurry reduced with the increase in the water-to-cement (w/c) ratio, and also increasing the temperature increased these rheological properties. Additives such as metakaolin and fly ash (replacing cement) affected the rheological properties, including the maximum shear stress tolerance of the cement slurries. Adding metakaolin increased the yield stress and maximum shear stress limit of the cement slurry, while fly ash increased the yield stress but reduced the maximum shear stress tolerance.

10 Resistivity was a highly sensing parameter to monitor the mixing of the cement with various additives, and also the curing of the cement and the experimental results clearly demonstrated this. Differences in the initial resistivity clearly indicated the addition of the amount and type of additives and also the w/c ratios. Initial resistivity can be used as a quality control measure in the field.

11 During the initial 24 hours of curing, several parameters such as minimum resistivity (ρ_{min}), time to reach minimum resistivity (t_{min}), resistivity after 24 hours of curing (ρ_{24}), and resistivity index (RI_{24}) can be used to characterise the cement curing .

12 The continuous changes in the resistivity of the curing cement under different conditions were measured and modeled using the new Vipulanandan curing model for up to one year. The model also captured all the important initial 24-hour curing parameters. The resistivity changes were highly sensitive to the curing environment of the cement. Also, the effects of additives including carbon dioxide (CO_2) contamination on the curing characteristics of the cement were measured and quantified using the curing model.

13 The thermal conductivity of cement with and without additives was quantified. The thermal conductivity decreased with the increase in the curing time. Also, increasing the w/c ratio decreased the thermal conductivity of the smart cement cured under room condition.

14 The electrical resistivity increased with the curing time after reaching the minimum resistivity, and the percentage changes were much higher than the changes in the thermal conductivity.

15 The mechanical properties of the standard cement (without any carbon fiber addition) under compression, direct tension, and three-point bending loading were tested and quantified. The uniaxial compression and direct tension failure strains after 28 days of curing were about than 0.2% and 0.023% respectively.

16 The new Vipulanandan piezoresistive cement theory was developed and verified with the experiments. The highly sensing smart cement was developed using less than 0.1% carbon fibers or basaltic fibers (by the weight of cement). Based on the efficiency of mixing and random dispersion and distribution of carbon fibers, some studies used about 0.03% carbon fibers. This modification is very cost effective with all the benefits of making the highly sensing chemo-thermo-piezoresisitive smart cement be a 3D sensor throughout its entire service life.

17 Electrical resistivity was identified as the monitoring parameter for the smart cement. The long-term piezoresistive compressive axial strain at failure was over 500 times (50,000%) higher than the compressive failure strain of 0.2%. The piezoresistive direct tensile axial strain at failure was over 400 times (40,000%) higher than the tensile failure strain of 0.023% for the smart cement.

18 Vipulanandan p-q models were used to characterise the stress-strain and stress-piezoresisitive strain of the standard cement and the smart cement respectively. Also the new Vipulanandan failure model was used to characterise the failure stresses of the smart cement with limited data.

19 Additives (inorganic) and contaminants (organic and inorganic) including CO_2 (organic) used in this study changed the density and initial resistivity of the smart cement. All the changes in the resistivity were quantified. The smart cement with the additives and contaminants was a highly sensing chemo-piezoresistive cement.

20 The effect of temperature on the smart cement without and with additives was investigated, and the experimental results showed that the smart cement was chemo-thermo-piezoresistive cement.

21 The relationship between the compressive strength of the smart cement and curing time was modeled using the Vipulanandan property correlation model, and the experimental values matched very well with the model predictions based on the coefficient of determination (R^2) and the root-mean-square error (RMSE). The relationship between the piezoresistive axial strain at failure and curing time was also modeled using the Vipulanandan property correlation model, and the predictions agreed very well with the experimental results.

22 Based on the XRD analyses, the addition of 1% $NanoSiO_2$ changed the cement mineralogy, and new constituents such as magnesium silicate sulfate ($Mg_5 (SiO_4)_2SO_4$) and quartz (SiO_2) were present. Hence, some of the changes observed in the modified smart cement with $NanoSiO_2$ behavior could have been due to the changes in the cement mineralogy. Thermogravimetric analyses (TGA) also showed a reduction in the total weight loss in the cement at 800°C with the addition of 1% $NanoSiO_2$.

23 The addition of 1% of different types of nanoparticles increased the initial resistivity of the smart cement after mixing, and also increased the compressive strength and modulus of the smart cement. Additionally, the addition of 1% nanoparticles reduced the piezoresistive axial strain of the smart cement at failure. All the changes in the properties based on the type of nanoparticles were quantified.

24 A linear correlation was observed between the resistivity index (RI_{24}) and compressive strength at different curing ages. A nonlinear model was used to correlate the constitutive model parameters to the curing time and NanoSiO$_2$ contents.

25 The generalised Vipulanandan fluid flow model was developed based on theory and verified with gas and liquid flow through porous sand and smart cement. The three parameters in the fluid flow model are related to the properties of the porous medium, flowing gas, and liquid and their interaction. Compared to the other models in the literature, the Vipulanandan fluid flow model predicted the experimental results with the water and gas very well.

26 Smart cement slurry and solid detected the gas leak. For the hardened smart cement with the application of pressure and without any fluid loss, the resistivity increased but with the gas leak it decreased. Changes in the resistivity were correlated to the gas flux (discharge per unit area). For the smart cement slurry (one-hour cured), the resistivity changing trends were opposite. With the application of pressure with no fluid loss, the resistivity decreased, and with the gas leak, the resistivity increased.

27 The addition of 3% styrene betadine rubber (SBR) polymer increased the resistivity of the smart cement. The addition of 1% polymer increased the initial resistivity by 4%, and adding 3% polymer increased the resistivity by 12%. Hence, resistivity could be a good quality control parameter in the field. The Vipulanandan curing model predicted the curing trends very well for the smart cement with the polymer modifications.

28 The addition of styrene betadine rubber (SBR) polymer increased the compressive strength of the smart cement after 24 hours of curing. With the addition of polymer, the smart cement was piezoresisitive. The fluid loss was reduced with the addition of polymer. The addition of polymer reduced the gas leak. Smart cement resistivity changes were highly sensitive to the gas leak rates and were used to detect and quantify gas leaks.

29 Based on the laboratory model data and field model data, electrical resistivity showed the largest variation compared to strain and temperature changes. Hence, the electrical resistivity was selected as the monitoring parameter for the smart cement during the 4.5 years of monitoring in the field.

30 The two-probe method was effective in measuring the bulk resistance of the drilling mud, spacer fluid, and smart cement slurries. Based on the changes in resistance measurements, it was possible to identify the fluid rise in the laboratory and field well boreholes.

31 Using the Vipulanandan p-q model, the change in the vertical piezoresistive axial strain in the smart cement was related to the applied pressure in the casing for the laboratory models and field model. The smart cement 3D sensor was very sensitive to the applied pressure in the casing.

32 Artificial intelligence (AI) models with one, two, three, and four layers of artificial neural networks were evaluated using the laboratory models and field model data with the statistical parameter coefficient of determination (R^2) and root-mean-square error (RMSE). The AI model predicted the long-term smart cement curing with the resistivity parameter well and was comparable to the Vipulanandan curing model.

33 Smart cement grouts with a w/c ratio of 0.8 were developed with the piezoresistive axial strain at failure of over 150%. This also verified the piezoresistive cement theory with positive changes in resistivity with the applied compressive stress.

34 The resistivity changes with the curing time for the smart grouts were modeled using the Vipulanandan p-q curing model, and the model predicted the experimental results very well with a very good coefficient of determination and minimized root-mean-square error (RMSE).

35 The smart cement grouts were tested for repairing failed smart cement with a piezoresistive failure strain of over 250%. The repaired efficiencies were quantified based on the percentage recovery of the strength and the piezoresistivity of the failed smart cement.

36 The addition of 20% foam (by the weight of cement slurry) increased the initial electrical resistivity of the smart cement from 1.05 $\Omega \cdot$m to 2.04 $\Omega \cdot$m, a 93% increase. The addition of foam changed the curing path of the smart cement based on the resistivity measurements.

37 The smart foam cement slurry was piezoresistive, and with the applied compressive pressure and the resistivity decreased due to very low shear effect. The change increased with the foam content and applied pressure.

38 The rheological test showed that the smart foam cement had shear-thinning behavior and the Vipulanandan rheological model predicted the test results very well compared to the Herschel–Bulkley model.

39 The total fluid loss for the smart cement at 0.7 MPa (100 psi) pressure was reduced from 134 mL to 13 mL with the addition of 20% foam, about a 90% reduction.

40 The solidified smart cement foam was piezoresistive and verified piezoresistive cement theory, and the resistivity increased with the

applied compressive stress. The strength and the piezoresistive axial strain at failure were lower than the smart cement without foam, and the changes have been quantified. But at lower stresses the smart foam cement was more sensitive than the smart cement without foam.

41 In the concrete with the smart cement binder, the aggregate content and curing time increased the pulse velocity of the smart concrete. The Vipulanandan pulse velocity-curing model predicted the variation of the pulse velocities with the aggregate contents and curing time very well.

42 The addition of coarse aggregates (gravel) increased the initial electrical resistivity of the smart concrete as well as the long-term electrical resistivity. The new Vipulanandan composite resistivity model predicted the resistivity changes with the aggregate content and curing time very well. Also, the Vipulanandan curing model predicted the electrical resistivity changes of the curing smart concretes very well.

43 The smart concrete also verified the piezoresistive cement theory, and the resistivity increased with the applied compressive stress. The piezoresistive axial strain at failure for the smart concrete with 10%, 40%, and 75% gravel content after 28 days of curing were 192%, 121%, and 101% respectively. The Vipulanandan piezoresistivity model predicted the piezoresistive behavior of the smart concrete very well.

44 The electrical resistivity index (RI_{24}) correlated linearly with the smart concrete compressive strengths after one day and 28 days of curing.

45 The compressive failure strain of concrete is 0.3%; hence, piezoresisitive smart concrete magnified the monitoring resistivity parameter by 336 times (33,600%) or more based on the aggregate content and making the smart concrete a bulk sensor.

46 Using the smart cement in the concrete, it was possible to predict the load and stress distributions in the aggregates and smart cement in the concrete with varying gravel contents. This is another advancement in quantifying the stress distribution within the concrete and will help in selecting the quality of the aggregate strength based on the highest stresses in the aggregates during the loading to achieve the concrete strengths.

47 The Vipulanandan failure model predicted the concrete pure tension, pure shear, and the maximum deviatoric stress (new property) very well compared to the Drucker–Prager model. Also the Vipulanandan model root-mean-square error (RMSE) was lower and the coefficient of determination (R^2) was higher compared to the Drucker–Prager model for the 23 MPa (114 data) and 46 MPa (14 data) unconfined compressive strengths of the concretes.

References

Afolabi, R., Esther O., Yusuf, E., Chude V., Okonji, C., and Nwobodo, S. (2019), 'Predictive Analytics for the Vipulanandan Rheological Model and Its Correlative Effect for Nanoparticle Modification of Drilling Mud', *Journal of Petroleum Science and Engineering*, 183, pp. 1–10.

Afolabi, R., Oradu, O. D., Efeovbakhan, V. E., and Rotimi, O. J. (2017), 'Optimizing the Rheological Properties of Silica Nano-Modified Bentonite Mud Using Overlaid Contour Plot and Estimation of Maximum or Upper Shear Stress Limit', *Cognet Engineering*, 4(1), pp. 1–18, DOI:10.1080/23311916.2017.1287248

Ahossin Guezo, Y. J., and Vipulanandan, C, (2014), 'Fracture Behavior of Multilayer Insulation Coatings Used in Subsea Pipelines'. Pipeline Conference 2014, ASCE, CD Proceeding.

Alexander, I., and Morton, H. (1990), *An Introduction to Neural Computing*, 1st ed., Chapman & Hall, pp. x–xv, Padstow, Cornwall.

Amani, N., Vipulanandan, C., and Plank, J. (2020), 'Monitoring the Hydration and Strength Development in Piezoresistive Cement Admixture with Superplasticizer and Retarder Additives', *Journal of Petroleum Science and Engineering*, 195, Issue 3, pp. 1–9.

Anagnostopoulos, C. A. (2014), 'Effect of Different Superplasticisers on the Physical and Mechanical Properties of Cement Grouts', *Construction and Building Materials*, 50, pp. 162–168

API Recommended Practice 10B (1997), *Recommended Practice for Testing Well Cements*, Exploration and Production Department, 22nd edition, American Petroleum Institute, Washington, DC.

API Specification 10A (2002), *Specification for Cements and Materials for Well Cementing*, 23rd edition, American Petroleum Institute, Washington, DC.

ASTM C136 (2016), *Standard Test Method for Sieve Analysis of Fine and Coarse Aggregates*. ASTM International, West Conshohocken, PA.

ASTM C150/C150M-19a (2019), *Standard Specification for Portland Cement*, ASTM Committee C01, West Conshohocken, PA.

ASTM C191 (2019), *Testing Method for Time of Setting of Hydraulic Cement by Vicat Needle*, ASTM Committee C01, West Conshohocken, PA.

ASTM D6994 (2015), *Standard Test Method for Flow of Fine Aggregate Concrete for Fabric Formed Concrete (Flow Cone Method)*, ASTM Subcommittee: D18.25, Book of Standards, Vol. 4.9, West Conshohocken.

Ata, A., and Vipulanandan, C. (1997), 'Silca Fume in Silicate and Cement Grouts and Grouted Sands', *Proceedings, Grouting: Compaction, Remediation and Testing*, GSP 66, ASCE, Logan, UT, pp. 242–257.

Azhari, F., and Banthia, N. (2012), 'Cement-Based Sensors with Carbon Fibers and Carbon Nanotubes for Piezoresistive Sensing', *Cement and Concrete Composites*, 34(7), pp. 866–873.

Banthia, N., and Dubeau, S. (1994), 'Carbon and Steel Micro-Fiber Reinforced Cement Based Composites for Thin Repairs', *Journal of Materials in Civil Engineering*, 6(1), pp. 88–99.

Bao-guo, H., and Jin-ping, O. (2008), 'Humidity Sensing Property of Cements with added Carbon', *New Carbon Materials*, 23, pp. 382–384.

Barbhuiya, S., Mukherjee, S., and Nikraz, H. (2014), 'Effects of Nano-Al 2 O 3 on Early-Age Microstructural Properties of Cement Paste', *Construction and Building Materials*, 52, pp. 189–193.

Bhattacharyya, P. (2011), *Neural Networks: Evolution, Topologies, Learning Algorithms and Applications*. IGI Global Publisher, Hershey, PA.

Bogue, R. H. (1955), *The Chemistry of Portland Cement*, Reinhold Publishing Corporation, New York.

Bonett, A., and Pafitis, D. (1996), 'Getting to the Root of Gas Migration'. *Oilfield Review*, 8(1), pp. 36–49.

Booshehrian, A., and Hosseini, P. (2011), 'Effect of Nano-SiO$_2$ Particles on Properties of Cement Mortar Applicable for Ferrocement Elements', *Concrete Research Letters*, 2(1), pp. 167–180.

Bowen, R. (1981), *Grouting in Engineering Practice*. 2nd edition. Applied Science Pub, London.

Calvert, D. G., and Smith, D. K. (1990), 'API Oil Well Cementing Practices', *Journal of Petroleum Technology*, 42(11), pp. 1364–1373.

Carter, K. M., and Oort, E. (2014), 'Improved Regulatory Oversight Using Real-Time Data Monitoring Technologies in the Wake of Macondo', Society of Petroleum Engineering, Paper Number 170323, pp. 1–51.

Chithra, S., Kumar, S. S., and Chinnaraju, K. (2016), 'The Effect of Colloidal Nano-silica on Workability, Mechanical and Durability Properties of High Performance Concrete with Copper Slag as Partial Fine Aggregate', *Construction and Building Materials*, 113, pp. 794–804.

Choolaei, M., Rashidi, A. M., Ardjmand, M., Yadegari, A., and Soltanian, H. (2012), 'The Effect of Nanosilica on the Physical Properties of Oil Well Cement', *Materials Science and Engineering: A*, 538, pp. 288–294.

Chu, H. Y., and Chen, J. K. (2016), The Experimental Study on the Correlation of Resistivity and Damage for Conductive Concrete', *Cement and Concrete Composites*, 67, pp. 12–19.

Chun, B. S., Yang, H. C., Park, D. H., and Jung, H. S. (2008), 'Chemical and Physical Factors Influencing Behavior of Sodium Silicate-Cement Grout', The Eighteenth International Offshore and Polar Engineering Conference, Vancouver, Canada, July 6–11, 2008.

Chung, D. D. L. (1995), 'Strain Sensors Based on Electrical Resistance Change', *Smart Materials Structures*, 4, pp. 59–61.

Chung, D. D. L. (2000), 'Cement Reinforced with Short Carbon Fibers: A Multifunctional Material', *Composites*, Part B, 31, pp. 511–526.

Chung, D. D. L. (2001), 'Functional Properties of Cement-Matrix Composites', *Journal of Material Science*, 36, pp. 1315–1324.

CIGMAT CT 1-06 (2006), 'Standard Test Method for Chemical Resistance of Coated or Lined Concrete and Clay Bricks', Center for Innovative Grouting Materials and Technology (CIGMAT), University of Houston, Houston, TX.

CIGMAT CT 3-06 (2006), 'Standard Test Method for Bonding Strength of Coatings and Mortars: Sandwich Mehod', Center for Innovative Grouting Materials and Technology (CIGMAT), University of Houston, Houston, TX.

CIGMAT GR 2-02 (2002), 'Standard Test Methods for Unconfined Compressive Strength of Grouts and Grouted Sands', Center for Innovative Grouting Materials and Technology (CIGMAT), University of Houston, Houston, TX.

CIGMAT PC 1-02 (2002), 'Standard Practice for Making and Curing Polymer Concrete Test Specimens in Laboratory', Center for Innovative Grouting Materials and Technology (CIGMAT), University of Houston, Houston, TX.

Cobb, S., Maki, V, and Sabins, F. (2002), 'Method Predicts Foamed Cement Compressive Strength Under Temperature and Pressure', *Oil and Gas Journal*, 100(15), pp. 48–52.

Compendex, S., and Elsevier, G. (2016), 'Influence of Zinc Oxide Nanoparticle on Strength and Durability of Cement Mortar', *International Journal of Earth Sciences and Engineering*, 9(3), pp. 175–181.

Dharmarajan, N., and Vipulanandan, C. (1988), 'Critical Stress Intensity Factor for Epoxy Mortar', *Polymer Engineering and Science, Society of Plastic Engineers*, 28, pp. 1182–1191.

Dom, P. B. et al. (2007), 'Development, Verification, and Improvement of a Sediment-Toxicity Test for Regulatory Compliance', *SPE Drilling & Completion*, 22(2), pp. 90–97.

Durand, C. et al. (1995), 'Influence of Clays on Borehole Stability—A Literature Survey', *Revue de l'Institut Français du Pétrole*, 50(2), pp. 187–218.

Eoff, L., and Waltman, B. (2009), 'Polymer Treatment Controls Fluid Loss while Maintaining Hydrocarbon Flow', *Journal of Petroleum Technology*, 61(7), pp. 28–34.

Florinel-Gabriel, B. (2012), chapter 11, *Chemical Sensors and Biosensors: Fundamentals and Applications*, John Wiley and Sons, Chichester.

Fuller, G., Souza, P., Ferreira, L., and Rouat, D. (2002), 'High-Strength Lightweight Blend Improves Deepwater Cementing', *Oil & Gas Journal*, 100(8), pp. 86–95.

Gani, M. S. J. (1997), *Cement and Concrete*, Chapman and Hall, London.

Gill, C., Fuller, G. A., and Faul, R. (2005), 'Deepwater Cementing Best Practices for the Riserless Section', AADE-05-NTCE-70, AADE. www.aade.org.

Guillot, D. (1990), 'Rheology of Well Cement Slurries'. In: E. B. Nelson and Guillot, D. (Eds.), *Well Cementing*, Schlumberger, Houston, TX, pp. 93–142.

Hammoudi, A., Moussaceb, K., Belebchouche, C., and Dahmoune, F. (2019), 'Comparison of Artificial Neural Network (ANN) and Response Surface Methodology (RSM) Prediction in Compressive Strength of Recycled Concrete Aggregates', Construction and Building Materials, 209(30), pp. 425–436.

Han, B., Guan, X., and Ou, J. (2007), 'Electrode Design, Measuring Method and Data Acquisition System of Carbon Fiber Cement Paste Piezoresistive Sensors', *Sensors and Actuators A: Physical*, 135(2), pp. 360–369.

Harendra, S., and Vipulanandan, C. (2008), 'Degradation of High Concentrations of PCE Solubilized in SDS and Biosurfactant with Bimetallic Fe/Ni Particles', *Colloids and Surfaces A: Physicochemical and Engineering Aspects*, 322(3), pp. 6–13.

Harms, W. M, and Febus, J. S. (1985), 'Cementing of Fragile Formations Wells with Foamed Cement Slurries', *Society of Petroleum Engineers*, SPE-12755, 37(6), pp. 1049–1057.

Harness, P. E., and Sabins, F. L. (1992), 'New Technique Provides Better Low Density Cement Evaluation', *Society of Petroleum Engineers*, SPE-24050, pp. 249–258.

Heinold, T., Dillenbeck, R. L., and Rogers, M. J. (2002), 'The Effect of Key Cement Additives on the Mechanical Properties of Normal Density Oil and Gas Well Cement Systems', Proceeding-SPE Asia Pacific Oil and Gas Conference and Exhibition, Melbourne, Australia. DOI:10.2118/77867-MS

Hou, P., Qian, J., Cheng, X., and Shah, S. P. (2015), 'Effects of the Pozzolanic Reactivity of $NanoSiO_2$ on Cement-Based Materials', *Cement and Concrete Composites*, 55, pp. 250–258.

Hou, T. C., Su, Y. M., Chen, Y. R., and Chen, P. J. (2017), 'Effects of Coarse Aggregates on the Electrical Resistivity of Portland Cement Concrete', *Construction and Building Materials*, 133, pp. 397–408.

Hubbert, M. K. (1957), 'Darcy's Law and Field Equations of Flow of Underground Fluids', *Hydrological Science Journal*, 2(1), pp. 23–59.

Izon, D., Danenberger, E. P., and Mayes, M. (2007), 'Absence of Fatalities in Blowouts Encouraging in MMS Study of OCS Incidents 1992–2006', *Drilling Contractor*, 63(4), pp. 84–89.

John B., (1992), 'Class G and H Basic Oil Well Cements', *World Cement Magazine*, April, pp. 44–50.

Joseph, D., and Vipulanandan, C. (2010), 'Correlation between California Bearing Ratio (CBR) and Soil Parameters', Proceedings, CIGMAT 2010 Conference and Exhibition, Part 2. www.cigmat.uh.edu.

Kim, J., and Vipulanandan, C. (2003), 'Effect of pH, Sulfate and Sodium on the EDTA titration of Calcium', *Cement and Concrete Research*, 33(5), pp. 621–627.

Kim, J., and Vipulanandan, C. (2006), 'Removal of Lead from Contaminated Water and Clay Soil Using a Biosurfactant', *Journal of Environmental Engineering*, 132(7), pp. 857–865.

Krizek, R. J., and Vipulanandan, C. (1985), 'Evaluation of Adhesion in Chemically Grouted Geomaterials', *Geotechnical Testing Journal*, 8(4), pp. 184–190.

Li, F., Vipulanandan, C., and Mohanty, K. (2003), 'Microemulsion and Solution Approaches to Nanoparticle Iron Production for Degradation of Trichloroethylene', *Colloids and Surfaces A: Physicochemical and Engineering Aspects*, 223(1), pp. 103–112.

Liao, Y., and Wei, X. (2014), 'Relationship between Chemical Shrinkage and Electrical Resistivity for Cement Pastes at Early Age', *Journal of Materials in Civil Engineering*, 26, pp. 384–387.

Liu, J., and Vipulanandan, C. (2001), 'Evaluating a Polymer Concrete Coating for Protecting Non-Metallic Underground Facilities from Sulfuric Acid Attack', *Journal of Tunneling and Underground Space Technology*, 16, pp. 311–321.

Liu, J., and Vipulanandan, C. (2003), 'Modeling Water and Sulfuric Acid Transport Through Coated Cement Concrete', *Journal of Engineering Mechanics*, 129(4), pp. 426–437.

Liu, J., and Vipulanandan, C. (2017), 'Effects of Fe, Ni and Fe/Ni Metallic Nanoparticles on Power Production and Biosurfactant Production from Used Vegetable Oil in the Anode Chamber of a Microbial Fuel Cell', *Waste Management*, 66, pp. 169–177.

Liu, Z., Zhang, Y., and Jiang, Q. (2014), 'Continuous Tracking of the Relationship between Resistivity and Pore Structure of Cement Pastes', *Construction and Building Materials*, 53, pp. 26–31.

Mahmood, W., and Mohammed, A. (2020), 'New Vipulanandan p-q Model for Particle Size Distribution and Groutability Limits for Sandy Soils', *Journal of Testing Materials and Evaluation*, 48(5), pp. 3695–3712.

Mangadlao, J. D., Cao, P., and Advincula, R. C. (2015), 'Smart Cements and Cement Additives for Oil and Gas Operations', *Journal of Petroleum Science and Engineering*, 129, pp. 63–76.

Mantrala, S. K., and Vipulanandan, C. (1995), 'Nondestructive Evaluation of Polyester Polymer Concrete', *ACI Materials Journal*, 92(6), pp. 660–668.

Marriot, T. (2005), 'Foamed Conventional Light Weight Cement Slurry for Ultra-Low Density Solves Lost Circulation Problem', SPE Annual Conference, SPE-96108.

Mayooran, K., and Vipulanandan, C. (2018), 'Vipulanandan Failure Model for Plain Concrete and Property Correlations', Proceedings, CIGMAT 2018 Conference and Exhibition, Part 2. www.cigmat.uh.edu.

Mbaba, P. E., and Caballero, E. P. (1983), 'Field Application of an Additive Containing Sodium Metasilicate during Steam Stimulation'. Proceeding, SPE Annual Technical Conference and Exhibition, San Francisco, October 5–8.

McCarter, W. J. (1994), 'A Parametric Study of the Impedance Characteristics of Cement-Aggregate Systems during Early Hydration'. *Cement and Concrete Research*, 24(6), pp. 1097–1110.

McCarter, W. J. (1996), 'Monitoring the Influence of Water and Ionic Ingress on Cover-Zone Concrete Subjected to Repeated Absorption', *Cement Concrete and Aggregates* 18, pp. 55–63.

McCarter, W. J., Chrisp, T. M., Starrs, G., and Blewett, J. (2003), 'Characterization and Monitoring of Cement-Based Systems Using Intrinsic Electrical Property Measurements', *Cement and Concrete Research*, 33(2), pp. 197–206.

McCarter, W. J., Starrs, G., and Chrisp, T. M. (2000), 'Electrical Conductivity, Diffusion, and Permeability of Portland Cement-Based Mortars', *Cement and Concrete Research*, 30(9), pp. 1395–1400.

McLachlan, D. S., Blaszkiewicz, M., and Newnham, R. E. (1990), 'Electrical Resistivity of Composites', *Journal of the American Ceramic Society*, 73(8), pp. 2187–2203.

Mebarkia, S., and Vipulanandan, C. (1992), 'Compressive Behavior of Glass-Fiber Reinforced Polymer Concrete', *Journal of Materials in Civil Engineering*, 4(1), pp. 91–105.

Medeiros-Junior, R. A., and Lima, M. G. (2016), 'Electrical Resistivity of Unsaturated Concrete Using Different Types of Cement'. *Construction and Building Materials*, 107, pp. 11–16.

Mirza, J. (2002), 'Basic Rheological and Mechanical Properties of High Volume Fly Ash Grouts', *Construction and Building Materials*, 16, pp. 353–363.

Mohammed, A., and Vipulanandan, C. (2013), 'Compressive and Tensile Behavior of Polymer Treated Sulfate Contaminated CL Soil', *Journal of Geotechnical and Geological Engineering*, 32, pp. 71–83.

Mohammed, A., and Vipulanandan, C. (2015), 'Testing and Modeling the Short-Term Behavior of Lime and Fly Ash Treated Sulfate Contaminated CL Soil', *Journal of Geotechnical and Geological Engineering*, 33(4), pp. 1099–1114.

Mohammed, A. S. (2018), 'Vipulanandan Model for the Rheological Properties with Ultimate Shear Stress of Oil Well Cement Modified with Nanoclay', *Egyptian Journal of Petroleum*, 27(3), pp. 335–347.

Mondal, P., Shah, S. P., and Marks, L. D. (2008), 'Nanoscale Characterization of Cementitious Materials', *ACI Materials Journal*, 105(2), pp. 174–179.

Montes, D., Orozco, W., Tabora, E. A., Franco, C. A., and Cortes, F. B. (2019), 'Development of Nanofluids for Perdurability in Viscosity Reduction of Extra-Heavy Oils', *MDPI, Energies*, 12(1068), DOI:10.3390/en12061068, MDPI.

Nelson, E. B. (1990), *Well Cementing*. Elsevier Science BV, Amsterdam, The Netherlands, pp. 3–10.

Opeyemi, B., Catalin, T., and Tanveer, Y. (2016), 'Application of Artificial Intelligence Techniques in Drilling System Design and Operations: A State of the Art Review and Future Research Pathways'. SPE Nigeria Annual International Conference and Exhibition, August 2016.

Ozgurel, G., Gonzalez, H. A., and Vipulanandan, C. (2005), 'Two Dimensional Model Study on Infiltration Control at a Lateral Pipe Joint Using Acrylamide Grout', *Proceedings, Pipelines 2005*, ASCE, Houston, TX, pp. 631–642.

Pakeetharan, S., and Vipulanandan, C. (2012), 'Modeling the Piezoresisitive Behavior of Polymeric Material', Proceedings, CIGMAT 2012 Conference and Exhibition, Part 2. www.cigmat.uh.edu.

Pakeetharan, R., and Vipulanandan, C. (2016), 'Effect of Clay Additives on Curing and Sensing Behavior of Smart Cement', Proceedings, CIGMAT 2016 Conference and Exhibition, Part 2. www.cigmat.uh.edu.

Petro, J. T., and Kim, J. (2012), 'Detection of Delamination in Concrete Using Ultrasonic Pulse Velocity Test', *Construction and Building Materials*, 26(1), pp. 574–582.

Petrosk, H. (2003), St. Francis Dam (2003), *American Scientist*, 91(2), pp. 114–118.

Plank, J., Dugonjić-Bilić, F., Lummer, N. R., and Taye, S. (2010), 'Working Mechanism of Poly (Vinyl Alcohol) Cement Fluid Loss Additive', *Journal of Applied Polymer Science*, 117(4), pp. 2290–2298.

Plank, J., Tiemeyer, C., Bülichen, D., and Recalde Lummer, N. (2013), 'A Review of Synergistic and Antagonistic Effects between Oilwell-Cement Additives', *SPE Drilling & Completion*, 28(4), pp. 398–404.

Polder, R. B. (2001), 'Test Methods for On-Site Measurement of Resistivity of Concrete—a RILEM TC-154 Technical Recommendation', *Construction and Building Materials*, 15(2), pp. 125–131.

Presuel-Moreno, F., Wu, Y. Y., and Liu, Y. (2013), 'Effect of Curing Regime on Concrete Resistivity and Aging Factor over Time', *Construction and Building Materials*, 48, pp. 874–882.

Princigallo, A., van Breugel, K., and Levita, G. (2003), 'Influence of the Aggregate on the Electrical Conductivity of Portland Cement Concretes', *Cement and Concrete Research*, 33(11), pp. 1755–1763.

Quercia, G., Brouwers, H. J. H., Garnier, A., and Luke, K. (2016), 'Influence of Olivine Nano-Silica on Hydration and Performance of Oil-Well Cement Slurries', *Materials & Design*, 96, pp. 162–170.

Radenti, G., and Ghiringhelli, L. (1972), 'Cementing Materials for Geothermal Wells', *Geothermics*, 3, pp. 119–123.

Ramachandran, V. S. (1984), *Concrete Admixture Handbook*, Noyes Publication, Park Ridge, NJ.

Ramezanianpour, A. A., Pilvar, A., Mahdikhani, M., and Moodi, F. (2011), 'Practical Evaluation of Relationship between Concrete Resistivity, Water Penetration, Rapid Chloride Penetration and Compressive Strength', *Construction and Building Materials*, 25(5), pp. 2472–2479.

Ravi, K., McMechan, D. E., Reddy, B. R., and Crook, R. (2004), 'A Comparative Study of Mechanical Properties of Density-Reduced Cement Compositions'. In: *SPE Annual Technical Conference and Exhibition*. Society of Petroleum Engineers. DOI:102118/90068-MS

Reddy, B. R., and Riley, W. D. (2004), 'High Temperature Viscosifying and Fluid Loss Controlling Additives for Well Cements, Well Cement Compositions and Methods'. US Patent 6 770 604, Assigned to Halliburton Energy Services, Inc. (Duncan, OK).

Rickard, W. M. (1985), 'Foam Cement for Geothermal Wells', *Transactions*, 9, Part 1, pp. 147–152.

Sadiq, T., and Nashawi, I. S. (2000), 'Using Neural Networks for Prediction of Formation Fracture Gradient', *SPE/PS-CIM International Conference on Horizontal Well Technology*.

Saleem, M., Shameem, M., Hussain, S. E., and Maslehuddin, M. (1996), 'Effect of Moisture, Chloride and Sulphate Contamination on the Electrical Resistivity of Portland Cement Concrete', *Construction and Building Materials*, 10(3), pp. 209–214.

Salib, S., and Vipulanandan, C. (1990), 'Property-Porosity Relationships for Polymer-Impregnated Superconducting Ceramic Composite', *Journal of American Ceramic Society*, 73(8), pp. 2323–2329.

Saridemir, M. (2009), 'Prediction of Compressive Strength of Concretes Containing Metakolin and Silica Fume by Artificial Neural Networks', *Advances in Engineering Software*, 40, pp. 350–355.

Sayers, C. M., and Dahlin, A. (1993), 'Propagation of Ultrasound through Hydrating Cement Pastes at Early Times', *Advanced Cement Based Materials*, 1(1), pp. 12–21.

Scheidegger, A. E. (1974), *The Physics of Flow through Porous Media*. University of Toronto Press, Toronto.

Sett, K., and Vipulanandan, C. (2004), 'Properties of Polyester Polymer Concrete with Glass and Carbon Fibers', *ACI Materials Journal*, 1(1), pp. 30–39.

Shadravan, A., and Amani, M. (2012), 'HPHT 101—What Petroleum Engineers and Geoscientists Should Know about High Pressure High Temperature Wells Environment', *Energy Science and Technology*, 4(2), pp. 36–60.

Shahab, D. (2000), 'Virtual-Intelligence Applications in Petroleum Engineering: Part 1—Artificial Neural Networks', *Journal of Petroleum Technology*, 52(90), pp. 64–73.

Shi, C. (2004), 'Effect of Mixing Proportions of Concrete on Its Electrical Conductivity and the Rapid Chloride Permeability Test (ASTM C1202 or AASHTO T277) Results', *Cement and Concrete Research*, 34(3), pp. 537–545.

Siemens, W. (1871), 'On the Increase in Electrical Resistance in Conductors with Rise of Temperature and Its Applications to Measure the Ordinary and Furnace Temperatures, Also Simple Method of Measuring Electrical Resistance', The Bakerian lecture, Royal Society, Retrieved May 14, 2014. https://archive.org/details/philtrans09056316

Singh, L. P., Karade, S. R., Bhattacharyya, S. K., Yousuf, M. M., and Ahalawat, S. (2013), 'Beneficial Role of Nanosilica in Cement Based Materials—A Review', *Construction and Building Materials*, 47, pp. 1069–1077.

Sivaruban, N., and Vipulanandan, C. (2007), 'Wet Unit Weight and Moisture Content Relationship for Natural Clay Deposits in Houston-Texas', Proceedings, CIGMAT 2007 Conference and Exhibition, Part 2. www.cigmat.uh.edu

Smith, C. S. (1954), 'Piezoresistive Effect in Germanium and Silicon', *Physical Review*, 94, pp. 42–49. DOI: 10.1102/PhysRev. 94.42

Tchameni, A. P., Zhao, L., Ribeiro, J. X. F., and Ting Li, T. (2019), 'Evaluating the Thermal Effect on the Rheological Properties of Waste Vegetable Oil Biodiesel Modified Bentonite Drilling Muds Using Vipulanandan Model', *High Temperatures High Pressures Journal*, 48, pp. 207–232.

Thaemlitz, J. et al. (1999), 'New Environmentally Safe High-Temperature Water-Based Drilling-Fluid System'. *SPE Drilling & Completion*, 14(3), pp. 185–189.

Ubertini, F., Laflamme, S., and D'Alessandro, A. (2016), 'Smart Cement Paste with Carbon Nanotubes', In Kenneth Loh and Satish Nagarajaiah (Eds.), *Innovative Developments of Advanced Multifunctional Nanocomposites in Civil and Structural Engineering*, Elsevier, London, pp. 97–120, DOI: 10.1016/B978-1-78242-326-3.00006-3

U.S. Patent (2019), 'Chemo-Thermo-Piezoresistive Highly Sensing Smart Cement with Integrated Real-Time Monitoring System', Inventor: C. Vipulanandan, Number 10,481,143. Awarded on November 19, 2019.

U.S. Patent (2020), 'Rapid Detection and Quantification of Surface and Bulk Corrosion and Erosion in Metallic and Non-Metallic Materials with Integrated Monitoring System'. Inventor: C. Vipulanandan, Number 10,690,586. Awarded on June 23, 2020.

Usluogullari, O., and Vipulanandan, C. (2011), 'Stress-Strain Behavior and California Bearing Ratio of Artificially Cemented Sand', *ASTM Journal of Testing and Evaluation* 39, pp. 637–645.

Vipulanandan, C., and Krizek, R. J. (1985), 'Tensile Properties of Chemically Grouted Sand', Transportation Research Record 1008, Transportation Research Board, National Research Council, pp. 80–89.

Vipulanandan, C., and Krizek, R. J., (1986), 'Mechanical Behavior of Chemically Grouted Sand', *Journal of Geotechnical Engineering*, American Society of Civil Engineers, 112(9), pp. 869–887.

Vipulanandan, C., and Krizek, R. J. (1987), 'Modeling Grouted Sand Behavior under Torsional Loading', Transportation Research Record 1104, Transportation Research Board, National Research Council, pp. 33–42.

Vipulanandan, C., and Paul, E. (1990), 'Performance of Epoxy and Polyester Polymer Concrete', ACI Materials Journal, 87(3), pp. 241–251.

Vipulanandan, C., and Shenoy, S. (1992), 'Properties of Cement Grouts and Grouted Sands with Additives', Proceedings, ASCE Specialty Conference on Grouting, Soil Improvement and Geosynthetics.

Vipulanandan, C., and Krishnan, S. (1993), 'XRD Analysis and Leachability of Solidified Phenol-Cement Mixtures', Cement and Concrete Research, 23, pp. 792–802.

Vipulanandan, C., Ata, A., and Mebarkia, S. (1994), 'Fracture Behavior of Cement Grouted Sand', Proceedings, Fracture Mechanics Applied to Geotechnical Engineering, Geotechnical Special Publication No. 43, pp. 147–159.

Vipulanandan, C., and Leung, M. (1995a) 'Treating Contaminated Cracked and Permeable Field Clay with Grout', Proceedings, Geoenvironment 2000, ASCE Special Publication, No. 46, pp. 829–843.

Vipulanandan, C., Mamidi, H., Wang, S., and Krishnan, S. (1995b), 'Solidification/ Stabilization of Phenol Contaminated Soils', Proceedings, Geoenvironment 2000, ASCE Special Publication, No. 46, pp. 1408–1421.

Vipulanandan, C., and Jasti, V. (1996a), 'Development and Characterization of Cellular Grouts', Proceedings, Materials for New Millennium, ASCE, pp. 829–839.

Vipulanandan, C. (1996b), 'Effect of Clays and Cement on the Solidification/ Stabilization of Phenol-Contaminated Soils', Waste Management, 15(5/6), pp. 399–406.

Vipulanandan, C. (1997a), Editor, Proceedings, Grouting: Compaction, Remediation and Testing, ASCE, GSP 66, Logan, UT.

Vipulanandan, C., Jasti, V., and Reddy, G. (1997b), 'Behavior of Lightweight Cementitious Cellular Grouts', Proceedings, Grouting: Compaction, Remediation and Testing, GSP 66, ASCE, Logan, UT, pp. 197–211.

Vipulanandan, C., Jasti, V., Magill, D., and Mack, D. (1997c), 'Control of Shrinkage and Swelling in Polymeric Grouts and Grouted Sands', Proceedings, Grouting: Compaction, Remediation and Testing, GSP 66, ASCE, Logan, UT, pp. 271–288.

Vipulanandan, C., and Elton, D. J. (1998), Editor, Proceedings, Recycle Materials in Geotechnical Applications, ASCE, GSP 79, Boston, MA.

Vipulanandan, C., and Ata, A. (2000a), 'Cyclic and Damping Properties of Silicate Grouted Sands', Journal of Geotechnical and Geoenvironmental Engineering, 126(7), pp. 650–656.

Vipulanandan, C., and Neelam Kumar, M. (2000b), 'Properties of Fly Ash-Cement Cellular Grouts for Sliplining and Backfilling Applications', Proceedings, Advances in Grouting and Ground Modification, ASCE, GSP 104, Denver, CO, pp. 200–214.

Vipulanandan, C., Mattey, Y., Magill, D., and Mack, D. (2000c), 'Characterizing the Behavior of Hydrophilic Polyuretane Grout', Proceedings, Advances in Grouting Technologies ASCE, GSP 104, Denver, CO, pp. 234–245.

Vipulanandan, C., and Gerstle, W. H. (2001a), Editor, *Proceedings, Fracture Mechanics for Concrete Materials: Testing and Applications*, ACI, SP 201, San Diego, CA.

Vipulanandan, C., Addison, M. B., and Hansen, M. (2001b), Editors, *Proceedings, Expansive Clay Soils and Vegetative Influences on Shallow Foundations*, ASCE, GSP 115, Houston, TX.

Vipulanandan, C., and Liu, J. (2002), 'Film Model for Coated Cement Concrete'. *Cement and Concrete Research*, 32(4), pp. 1931–1936.

Vipulanandan, C., and Liu, J. (2005a), 'Polyuretane Based Grouts for Deep Off-Shore Pipe-in-Pipe Application', *Proceedings, Pipelines 2005*, ASCE, Houston, TX, pp. 216–227.

Vipulanandan, C., and Ortega, R. (2005b), Editors, *Proceedings, Pipelines 2005, Design, Optimization and Maintenance in Today's Economy*, ASCE, Houston, TX.

Vipulanandan, C., and Townsend, F. C. (2005c), Editors, *Proceedings, Advances in Designing and Testing Deep Foundations*, ASCE, GSP 129, Austin, TX, .

Vipulanandan, C., and Townsend, F. C (2005d), Editor, *Proceedings, Advances in Deep Foundations*, ASCE, GSP 132, Austin, TX.

Vipulanandan, C., and Liu. J. (2005e), 'Sewer-Pipe Joint Infiltration Test Protocol Developed by CIGMAT', *Proceedings, Pipelines 2005*, ASCE, Houston, TX, pp. 553–563.

Vipulanandan, C. (2007a), Editor, *Proceedings, Advances in Measurement and Modeling of Soil Behavior*, ASCE, GSP 173, Denver, CO.

Vipulanandan, C., and Kulkarni, S. P. (2007b) 'Shear Bonding and Thermal Properties of Particle-Filled Polymer Grout for Pipe-in-Pipe Application', *Journal of Materials in Civil Engineering*, 19(7), pp. 583–590.

Vipulanandan, C., and Garas, V. (2008), 'Electrical Resistivity, Pulse Velocity, and Compressive Properties of Carbon Fiber-Reinforced Cement Mortar', *Journal of Materials in Civil Engineering*, 20(2), pp. 93–101.

Vipulanandan, C., and Demircan, E. (2009a), 'Designing and Characterizing LEED Concrete for Drilled Shaft Applications', *Proceedings, Foundation Congress 2009, Contemporary Topics in Deep Foundations*, ASCE, GSP 185, pp. 55–62.

Vipulanandan, C., and Usluogullari, O. (2009b), 'Field Evaluation of a New Down-Hole Penetrometer', *Proceedings, Contemporary Topics in In Situ Testing, Analysis, and Reliability of Foundations, Foundation Congress 2009*, ASCE, GSP 186, pp. 119–126.

Vipulanandan, C., and Ozgurel, H. G. (2009c), 'Simplified Relationships for Particle-Size Distribution and Permeation Groutability Limits for Soils', *Journal of Geotechnical and Geoenvironmental Engineering*, 135(9), pp. 1190–1197.

Vipulanandan, C., Parihar, C., and Issac, M. (2011), 'Testing and Modeling Composite Coatings with Silanes for Protecting Reinforced Concrete in Saltwater Environment', *Journal of Materials in Civil Engineering*, 23(12), pp. 1602–1608.

Vipulanandan, C., Kazez, M. B., and Henning, S. (2012a), 'Pressure-Temperature-Volume Change Relationship for a Hydrophilic Polyurethane Grout', *Proceedings, Grouting and Deep Mixing*, Geo-Institute, ASCE, GSP 228, pp. 1808–1818.

Vipulanandan, C., Stevens, T., Marinshaw, R., and Fedricks, R. (2012b), 'Environmental Technology Verification Program for Grouts Used in Infrastructure Rehabilitation', *Proceedings, Grouting and Deep Mixing*, Geo-Institute, ASCE, GSP 228, pp. 1829–1840.

Vipulanandan, C., and Sunder, S. (2012c), 'Effects of Meta-Kaolin Clay on the Working and Strength Properties of Cement Grouts', *Proceedings, Grouting and Deep Mixing*, Geo-Institute, ASCE, GSP 228, pp. 1739–1747.

Vipulanandan, C., and Prashanth, P. (2013a), 'Impedance Spectroscopy Characterization of a Piezoresistive Structural Polymer Composite Bulk Sensor', *Journal of Testing and Evaluation*, 41(6), pp. 898–904.

Vipulanandan, C., and Sundar, S. (2013b), 'Leak Control in Wastewater Lateral Joint Using a Polymer Grout', *Proceedings, ASCE Pipeline Conference 2013*, CD, June.

Vipulanandan, C., and Mohammed, A. (2014a), 'Hyperbolic Rheological Model with Shear Stress Limit for Acrylamide Polymer Modified Bentonite Drilling Muds', *Journal of Petroleum Science and Engineering* 122, pp. 38–47.

Vipulanandan, C., Heidari, M., Qu, Q., Farzam, H., and Pappas, J. M. (2014b), 'Behavior of Piezoresistive Smart Cement Contaminated with Oil Based Drilling Mud Offshore', Offshore Technology Conference (OTC), 25200-MS1-14.

Vipulanandan, C., Krishnamoorti, R., Saravanan, R., Qi, Q., and Pappas, J. (2014c), 'Development of Smart Cement for Oil Well Applications', Offshore Technology Conference (OTC), OTC-25099-MS.

Vipulanandan, C., Raheem, A., Basirat, B., Mohammed A. S., and Richardson, D. A. 2014d), 'Kinematic Modeling to Characterize Filter Cake Formation and Fluid Loss in HPHT Process', Offshore Technology Conference (OTC) 2014, OTC-25100-MS.

Vipulanandan, C., and Wei, Q. (2014e), 'National Survey on the Trends in Small Diameter Water Pipeline Failures', Pipeline Conference 2014, ASCE, Portland, OR, CD Proceeding, August.

Vipulanandan, C., and Burak, K. (2014f), 'Test Protocol for Evaluating Grout Materials to Repair Water Leaks in Concrete Pipes', Pipeline Conference 2014, ASCE, Portland, OR, CD Proceeding, August.

Vipulanandan, C., and Mohammed, A. (2015a), 'Effect of Nanoclay on the Electrical Resistivity and Rheological Properties of Smart and Sensing Bentonite Drilling Muds', *Journal of Petroleum Science and Engineering*, 130, pp. 86–95.

Vipulanandan, C., Ramanathan, P., Ali, M., Basirat, B., and Pappas, J. (2015b), 'Real Time Monitoring of Oil Based Mud, Spacer Fluid and Piezoresistive Smart Cement to Verify the Oil Well Drilling and Cementing Operation Using Model Tests', Offshore Technology Conference (OTC), OTC-25851-MS.

Vipulanandan, C., and Amani, N. (2015c), 'Behavior of Nano Calcium Carbonate Modified Smart Cement Contaminated with Oil Based Drilling Mud', Offshore Technology Conference (OTC), OTC-25845-MS.

Vipulanandan, C., and Mohammed, A. (2015d), 'Smart Cement Rheological and Piezoresistive Behavior for Oil Well Applications', *Journal of Petroleum Science and Engineering*, 135, pp. 50–58.

Vipulanandan, C., and Mohammed, A. (2015e), 'Smart Cement Modified with Iron Oxide Nanoparticles to Enhance the Piezoresistive Behavior and Compressive Strength for Oil Well Applications', *Journal of Smart Materials and Structures*, 24(12), pp. 1–11.

Vipulanandan, C., and Ali, K. (2016a), 'Smart Cement Piezoresistive Behavior with and without Sodium Meta-silicate under Temperature and Curing Environments for Oil Well Applications', *Journal of Civil Engineering Materials*, 28(9), pp. 1–8, DOI: 10.1061/(ASCE)MT.1943-05533.0001667

Vipulanandan, C., and Reddy, A. (2016b), 'Smart Foam Cement Characterization for Real Time Monitoring of Ultra-Deepwater Oil Well Cementing Applications', AADE Fluids Technical Conference and Exhibition. AADE-16-FTCE-84.

Vipulanandan, C., Ali, K., Basirat, B., Reddy, A., Amani, N., Mohammed, A., Dighe, S., Farzam, H., and W. J. Head (2016c), 'Field Test for Real Time Monitoring of Piezoresistive Smart Cement to Verify the Cementing Operations', Offshore Technology Conference (OTC), OTC-27060-MS.

Vipulanandan, C., Ali, K., Basirat, B., Reddy, A., and Callahan, D. (2016d), 'Field Study on Piezoresistive Smart Cement and Drilling Mud for Real Time Monitoring the Installation and Performance of the Cemented Well', American Association of Drilling Engineers (AADE), Conference Paper, AADE-16-FTCE-73, www.aade. org.

Vipulanandan, C., Ali, K., and Ariram, P. (2016e), 'Nanoparticle and Surfactant Modified Smart Cement and Smart Polymer Grouts', Proceedings, *ASCE Geotechnical and Structural Engineering Congress*, DOI:10.1061/9780784479742

Vipulanandan, C., and Mohammed, A., (2016f), 'XRD and TGA, Swelling and Compacted Properties of Polymer treated Sulfate Contaminated CL Soil', *Journal of Testing and Evaluation*, 44(6), pp. 1–16.

Vipulanandan, C., Mohammed, A., and Samuel, R. G. (2017a), 'Smart Bentonite Drilling Muds Modified with Iron Oxide Nanoparticles and Characterized Based on the Electrical Resistivity and Rheological Properties with Varying Magnetic Field Strengths and Temperatures', Offshore Technology Conference (OTC), OTC-28974-MS.

Vipulanandan, C., and Guezo, Y. J. (2017b), 'Effects of Temperature and Strain Rated in the Tensile Behavior of Polypropylene Composites Insulator Coatings Used in Offshore Deepwater Pipelines', *Journal of Applied Polymer Science*, 134(36), pp. 18–27.

Vipulanandan, C., and Ali, K. (2018a), 'Smart Cement Grouts for Repairing Damaged Piezoresistive Cement and the Performances Predicted Using Vipulanandan Models', *Journal of Civil Engineering Materials*, 30(10) pp. 1–8. Article number 04018253.

Vipulanandan, C., and Amani, N. (2018b), 'Characterizing the Pulse Velocity and Electrical Resistivity Changes in Concrete with Piezoresistive Smart Cement Binder Using Vipulanandan Models', *Construction and Building Materials*, 175, pp. 519–530.

Vipulanandan, C., Mohammed, A., and Ganpatye, A. S. (2018c), 'Smart Cement Performance Enhancement with Nano Al2O3 for Real-Time Monitoring Applications Using Vipulanandan Models'. Offshore Technology Conference (OTC), OTC-28880-MS.

Vipulanandan, C., and Chockalingam, C. (2018d), 'Corrosion Detection and Quantification Real-Time Using the New Nondestructive Test with Vipulanandan Impedance Corrosion Model', *Proceedings, American Association of Drilling Engineers* (AADE) 2018, AADE-18-FTCE-116.

Vipulanandan, C., Vembu, K., Brettmann, T., and Gattu, V. (2018e), 'Full-Scale Field Test Study of Skin Friction Development in Sand for ACIP Piles under Compressive and Tensile Loading Conditions for Bridge Support', ASCE, GSP 294 (Installation and Testing of Deep Foundations), pp. 363–374.

Vipulanandan, C., Vembu, K., and Gattu, V. (2018f), 'Highway Bridge Supported on ACIP Piles in Clay Soils: Instrumentation, Monitoring and Performance of Service Piles', ASCE, GSP 299 (Testing and Analysis of Deep Foundations), pp. 50–67.

Vipulanandan, C., Kula, I., Magill, D., and Aguilar, F. (2018g), 'Developing Smart Grouted Sand Columns for Real Time Monitoring of Earth Dams', ASCE, GSP 298 (Case Histories and Lessons Learned), pp. 279–290.

Vipulanandan, C., and Mohammed, A. (2018h), 'New Vipulanandan Failure Model and Property Correlations for Sandstone, Shale and Limestone Rocks', ASCE, GSP 295 (Advances in Geomateriasl Modeling and Site Characterization), pp. 365–376.

Vipulanandan, C., Maddi, A. R., and Ganpatye, A. (2018i), 'Smart Spacer Fluid Modified with Iron Oxide Nanoparticles for In-Situ Property Enhancement Was Developed for Cleaning Oil Based Drilling Fluids and Characterized Using the Vipulanandan Rheological Model', Offshore Technology Conference (OTC) 2018, OTC-28886-MS.

Vipulanandan, C., and Ali, K. (2018j), 'Smart Portland Cement Curing and Piezoresisitive Behavior with Montmorillonite Clay Soil Contamination', Cement and Concrete Composite, 91, pp. 42–52.

Vipulanandan, C., Panda, G., Maddi, A. R., Wong, G., and Aldughather, A. (2019a), 'Characterizing Smart Cement Modified with Styrene Butadiene Polymer for Quality Control, Curing and to Control and Detect Fluid Loss and Gas Leaks Using Vipulanandan Models', Offshore Technology Conference (OTC) 2019, OTC-29581-MS.

Vipulanandan, C., Chockalingam, C., Gebreselassie, K. A., Pan, D., Panda, G., and Ganpatye, A. S. (2019b), 'New Rapid Nondestructive Testing Method for Detecting and Quantifying Corrosion with Material Property Changes Using Vipulanandan Impedance Corrosion Model', Offshore Technology Conference (OTC) 2019, OTC-29378-MS.

Vipulanandan, C., and Mohammed, A. (2020a), 'Magnetic Field Strength and Temperature Effects on the Behavior of Oil Well Cement Slurry Modified with Iron Oxide Nanoparticles and Quantified with Vipulanandan Models', Journal of Testing and Evaluation, 48(6), pp. 4516–4537.

Vipulanandan, C., and Mohammed, A. (2020b), 'Effect of Drilling Mud Bentonite Contents on the Fluid Loss and Filter Cake Formation on a Field Clay Soil Formation Compared to the API Fluid Loss Method and Characterized Using Vipulanandan Models', Journal of Petroleum Science and Engineering. 189, pp. 1–19.

Vipulanandan, C., and Mohammed, A. (2020c), 'Characterizing the Index Properties, Free Swelling, Stress-Strain Relationship, Strength and Compacted Properties of Polymer Treated Expansive CH Clay Soil Using Vipulanandan Models', Geotechnical and Geological Engineering, 38(5), pp. 5589–5602.

Wang, F., and Yang, L. (2015), 'Microstructure and Properties of Cement Foams Prepared by Magnesium Oxychloride Cement', Journal of Wuhan University, 30(2), pp. 331–337.

Wang, S. Y., and Vipulanandan, C. (1996), 'Leachability of Lead from Solidified Cement-Fly Ash Binders', Cement and Concrete Research, 26(6), pp. 895–905.

Wei, S., Mau, S. T., Vipulanandan, C., and Mantrala, S. K. (1995), 'Performance of New Sandwich Tube under Axial Loading: Experiment', *Journal of Structural Engineering*, 121(12), pp. 1806–1814.

Wei, X., Xiao, L., and Li, Z. (2012), 'Prediction of Standard Compressive Strength of Cement by the Electrical Resistivity Measurement', *Construction and Building Materials*, 31, pp. 341–346.

Wilson, A. 2017. 'Real-Time Monitoring of Piezoresistive Smart Cement to Verify Operations', *Journal of Petroleum Technology*, 69(5), pp. 79–80. Paper Number: **SPE-0517-0079-JPT**

Wohltjen, H., Barger, W. R., Snow, A. W., and Jarvis, N. L. (1985), 'A Vapor-Sensitive Chemiresistor Fabricated with Planar Microelectrodes and Langmuir-Biodegett Organic Semiconductor Film', *IEE Transaction Electron Devices*, 32(7), pp. 1170–1174.

Wong, D., O'Neill, M. W., and Vipulanandan, C. (1992), 'Modeling of Vibratory Pile Driving in Sand', *Numerical and Analytical Methods in Geomechanics*, 16, pp. 189–210.

Xiao, L., and Li, Z. (2008), 'Early-Age Hydration of Fresh Concrete Monitored by Non-contact Electrical Resistivity Measurement', *Cement and Concrete Research*, 38(3), pp. 312–319.

Zhang, J., Qin, L., and Li, Z. (2009), 'Hydration Monitoring of Cement-Based Materials with Resistivity and Ultrasonic Methods', *Materials and Structures*, 42(1), pp. 15–24.

Zhang, J., Weissinger, E. A, Peethamparan, S., and Scherer G. W. (2010), 'Early Hydration and Setting of Oil Well Cement', *Cement and Concrete Research*, 40, pp. 1023–1033.

Zhang, L., Ding, S., Dong, S., Li, Z., Ouyang, J., Yu, X., and Han, B. (2017), 'Piezoresistivity, Mechanisms and Model of Cement-Based Materials with CNT/NCB Composite Fillers', *Materials Research Express*, 4(12), pp. 1–12, 125704.

Zhang, M., Sisomphond, K., Ng, T. S, and Sun, D. J. (2010), 'Effect of Superplasticizers on Workability Retention and Initial Setting Time of Cement Pastes', *Construction and Building Materials* 24, pp. 1700–1707.

Zuo, Y., Zi, J., and Wei, X. (2014), 'Hydration of Cement with Retarder Characterized via Electrical Resistivity Measurements and Computer Simulation'. *Journal of Construction and Building Materials*, 53, pp. 411–418.

Index

Printed in the United States
by Baker & Taylor Publisher Services